NONLINEAR PARTIAL DIFFERENTIAL EQUATIONS

Techniques and Recent Developments

NONLINEAR PARTIAL DIFFERENTIAL EQUATIONS

Techniques and Recent Developments

Prasanta Chatterjee
Visva-Bharati University, India

Kaushik Roy
Beluti MKM High School, India

Laxmikanta Mandi
Gour Banga University, India

World Scientific

NEW JERSEY • LONDON • SINGAPORE • BEIJING • SHANGHAI • HONG KONG • TAIPEI • CHENNAI • TOKYO

Published by

World Scientific Publishing Co. Pte. Ltd.
5 Toh Tuck Link, Singapore 596224
USA office: 27 Warren Street, Suite 401-402, Hackensack, NJ 07601
UK office: 57 Shelton Street, Covent Garden, London WC2H 9HE

British Library Cataloguing-in-Publication Data
A catalogue record for this book is available from the British Library.

NONLINEAR PARTIAL DIFFERENTIAL EQUATIONS
Techniques and Recent Developments

ISBN 978-981-98-1673-6 (hardcover)
ISBN 978-981-98-1674-3 (ebook for institutions)
ISBN 978-981-98-1675-0 (ebook for individuals)

For any available supplementary material, please visit
https://www.worldscientific.com/worldscibooks/10.1142/14407#t=suppl

Typeset by Stallion Press
Email: enquiries@stallionpress.com

Preface

This book charts a clear and systematic road map for nonlinear partial differential equations (NLPDEs). Beginning from the definition of a partial differential equation to the recent developments of nonlinear partial differential equations, this book will be a valuable resource for advanced postgraduate students and researchers in applied mathematics, physics, nonlinear optics, and other engineering disciplines where knowledge of nonlinear differential equations is a must. The book begins with an introductory chapter that briefly describes the developments of linear as well as nonlinear partial differential equations. Several nonlinear partial differential equations that have emerged in various fields have also been discussed. Chapter 2 introduces several analytical techniques, including the traveling wave solutions and the similarity solutions of the nonlinear partial differential equations. In Chapter 3, approximate analytical solutions and semi-analytical solutions are presented, in which solutions of non-integrable or non-autonomous nonlinear partial differential equations are investigated after suitable approximation. Some recent breakthroughs in semi-analytical approaches, such as the variational iteration method (VIM), the adomian decomposition method (ADM), the homotopy analysis method (HAM), and the homotopy perturbation method (HPM), are also explained with examples. Chapter 4 deals with modern advancements in NLPDE, Painlevé tests, the inverse scattering method, the Lax pair method, the Darboux transformation, the Bäcklund transformation, and the Hirota direct method. The focus of this comprehensive monograph is to check the integrability and find analytical solutions for important NLPDEs according to recent developments.

Prasanta Chatterjee, Kaushik Roy
Laxmikanta Mandi
India
December 2024

Acknowledgments

Our book, Nonlinear Partial Differential Equations: Technique and Recent Developments, is the culmination of great deal of effort and is a book we have always wanted. Numerous friends and Seniors have assisted us in the writing of this book. Prof. Chatterjee would also like to thank basic science department, IIT Mandi and Prof. Rajendra Kumar Ray for their encouragement. We also acknowledge the contributions of Dr. TusherKanti Das, Dr. Uday Narayan Ghosh, Ms. Snehalata Nashipuri, Mr. Nanda Kumar Pal, Mr. Dipan Saha, Mr. Sougata Dutta, Mr. Jayashree Mondal, and final postgraduate students having a special paper "Nonlinear Differential Equation." We are all very appreciative of our families' unwavering support and encouragement. We want to thank World Scientific Publishing Co. Pte. Ltd. from the bottom of our hearts.

Contents

Chapter 1

Introduction

1.1 Introduction

Differential equations frequently appear in pure and applied mathematics, physics, and engineering, and they enlighten the core of these subjects as per their requirements. Applied mathematics stresses the thorough justification of the methods to find a solution or approximate solutions; on the other hand, pure mathematics emphasizes the existence and uniqueness of solutions. The role of differential equations is of utmost importance in modeling practically all physical, technical, or biological processes, from celestial motion to bridge design to interactions between neurons. Both ordinary and partial differential equations have become essential tools for the development of the present civilization.

The 18th century was a landmark century for the study of partial differential equations (PDEs), which started with the works of Euler, d'Alembert, Lagrange, and Laplace. Their work served as the primary tool in explaining the mechanics of continua. In other words, the primary form of analytical study of models in physical science. PDEs arose to develop models of sciences like vibrating strings, electricity, gravitational fields, electrostatics, fluid flows, head conduction, electromagnetic theory, etc. One of the fundamental concerns of the development of PDEs in recent times is the analysis of physical models. Starting in the middle of the 19th century, mainly with the work of Riemann, PDEs also became an important tool in different branches of mathematics.

Throughout the 19th and 20th centuries, two different directions and related opinion becomes the central study of PDEs. The relationship to models in physics, engineering, and other applied disciplines is on one side, while on the other side, there are the possible functions that have often been revolutionary of PDE's as a tool in the growth of other branches of mathematics. In 1890, this twofold viewpoint was acknowledged for the first time by H. Poincare. According to Poincare, a wide range of physically important problems surface in various areas, such as electricity, hydrodynamics, heat, magnetism, optics, elasticity, etc. He emphasized having full proof, even though the models are only an estimate of the physical reality. Mathematicians want to carry through their research in an accurate and persuasive form, and then the resulting theory is used as a tool in the study of major mathematical areas. PDE is considered a chief link between the physical sciences and the central issues of applied mathematics, as well as the central growth of mathematical opinion in pure mathematics.

Let us now discuss some areas in mathematics that are important for PDEs. Riemann's application of a potential theoretic argument, the Dirichlet principle and its uses, in developing the general theory of analytic functions of a complex variable and the related theory of Riemann surfaces. Differential geometry is also used, especially in global terms. The different areas in differential geometry, like minimal surfaces and imbedding problems giving rise to the Monge-Ampere equations, have been used to analyze PDEs, especially nonlinear PDEs. However, it has been feasible to respond to elementary open questions in differential geometry due to the development of powerful analytical tools in PDEs. The system of first-order PDEs has been important since its interaction with Lie theory. With the introduction and exercise by E. Cartan, the exterior theory of differential forms has become significant. In addition to this, the sheaf theory by Leray has resulted in the interplay of various practices and thoughts, including algebraic geometry, algebraic and differential topology, homological algebra, and microlocal analysis.

Since the beginning of the 20th century, a thorough need for investigating the solution of PDE's and their boundary value problems has led to the growth of the primary tools in real analysis and functional analysis. J. Dieudonné held a similar view on the growth of functional analysis. This methodical analysis of PDE's and their BVP's at the beginning of the 1950s and 1960s led to the birth of a marvelous broadening of techniques in Fourier analysis. The theory of linear PDE got tremendous impetus when Fourier analysis was applied to PDE with the help of tools like pseudo-differential operators and Fourier integral operators. Another example is the interplay between PDEs and topology. It arose initially in the 1920s and 30s from such goals as the desire to find global solutions for nonlinear PDEs, especially those arising in fluid mechanics, as in the work of Leray. Examples in the 1920s are the variational theories of M. Morse and Ljusternik-Schnirelman, and in the 1930s, the Leray–Schauder degree in infinite-dimensional spaces as an extension of the classical Brouwer degree. After 1960, the introduction of a variational viewpoint in the study of differential topology gave rise to such important results as Bott's periodicity theorem and Smale's proof of the Poincaré conjecture for dimension ≥ 5. More recently, the analysis of the Yang–Mills PDE has given rise to spectacular progress in low-dimensional topology.

The relationship between algebraic geometry and soliton theory related to the Korteweg–De Vries (KdV) equation is really remarkable. In 1896, the KdV equation was used for the first time as a representation of water waves. Later, this equation was further studied in 1965 by M. Kruskal and his collaborators. An important area for the interaction between PDE and the recent topics in chaos theory is the study of the asymptotic patterns of solutions, especially fluid flows and gas dynamics. This is a potential approach to the study of the main problem of turbulence, which is also an important problem in the physical sciences. PDE also plays a very major role in other mathematical areas of research like constructive quantum field theory, infinite-dimensional group representations, homogeneous spaces, and mathematical physics. Last but not least, PDE is equally important in scientific computing. Poincare discussed this but also emphasized the limitations of the practical usage of this technique. In recent times, high-speed computers have made the work of science easy.

1.2 Differential Equation

An equation connecting the variables and the differential coefficients of the dependent variable is called a differential equation. Differential equations are of two types, namely ordinary differential equations (ODEs) and partial differential equations (PDEs). $\frac{dy}{dx} + \frac{y}{x} = 2x^2$ and $\frac{\partial^2 u}{\partial x^2} = \frac{\partial^2 u}{\partial y^2}$ are the examples of ODE and PDE, respectively. The solution of an ODE depends on one variable, and the solution of a PDE depends on more than one variable.

The *order* of a differential equation is the order of the highest derivative occurring in it. The *degree* of a differential equation is the degree to which the highest-order derivative is raised when the differential coefficients are cleared of fractions.

$$\frac{\partial^2 u}{\partial y^2} + 2\frac{\partial^2 u}{\partial y \partial x} + \frac{\partial^2 u}{\partial x^2} = 0$$

is an example of a PDE of second order and first degree.

1.3 Classification of First-Order PDE

Let $z = f(x, y)$ be a function of two independent variables x and y. Then, the partial derivative of z with respect to x keeping y constant is $\frac{\partial z}{\partial x}$ and partial derivative of z with respect to y keeping x constant is $\frac{\partial z}{\partial y}$. The notations that are generally used are as follows: $p = \frac{\partial z}{\partial x}$, $q = \frac{\partial z}{\partial y}$, $r = \frac{\partial^2 z}{\partial x^2}$, $s = \frac{\partial^2 z}{\partial x \partial y}$, and $t = \frac{\partial^2 z}{\partial y^2}$.

PDEs are equations that contain partial derivatives. An example of a PDE is $xp + yq = q^2$.

A PDE is considered to be linear if the dependent variable and its partial derivative only appear in the first degree and are not multiplied. If a PDE is not linear, it is generally referred to as a nonlinear PDE. Examples of linear and nonlinear PDEs, respectively, are $p+q = z$ and $xr = p$. However, they are classified as follows:

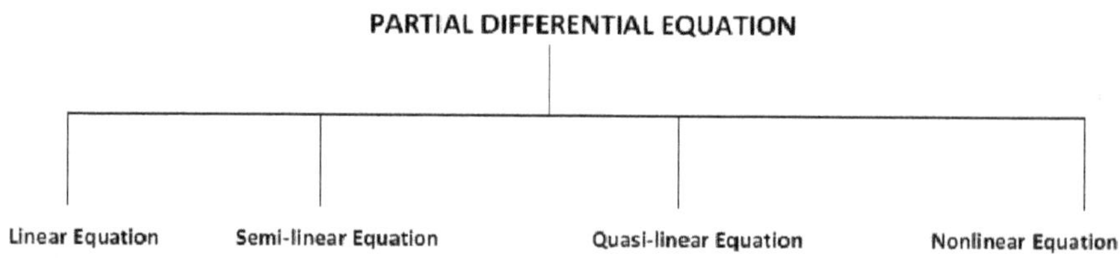

A first-order PDE with two independent variables x, y, and one unknown z is also known as a dependent variable. This equation is in the form of

$$F\left(x, y, z, \frac{\partial z}{\partial y}, \frac{\partial z}{\partial y}\right) = 0 \tag{1.1}$$

Using the notation $p = \frac{\partial z}{\partial x}$, $q = \frac{\partial z}{\partial y}$, Eq. (1.1) which can be written as

$$F(x, y, z, p, q) = 0 \tag{1.2}$$

The classification of the first order of a PDE is determined by the form of the function F.

An equation of the form

$$P(x, y, z)\frac{\partial z}{\partial x} + Q(x, y, z)\frac{\partial z}{\partial y} = R(x, y, z) \tag{1.3}$$

is a *quasi-linear* PDE of the first order, if the derivatives $\frac{\partial z}{\partial x}$ and $\frac{\partial z}{\partial y}$ that appear in the function F are linear. Moreover, the coefficients P, Q and R depend on the independent variables x, y and also on the dependent variables z. $P(z)p + q = 0$ is a quasi-linear PDE of first order.

Similarly, an equation of the form

$$P(x,y)\frac{\partial z}{\partial x}+Q(x,y)\frac{\partial z}{\partial y}=R(x,y,z) \tag{1.4}$$

is called *semi-linear* PDE of first order, if the coefficients P, Q are functions of the independent variables only. $e^x p - yxq = xz^2$ is an example of semi-linear.

An equation of the form

$$a(x,y)\frac{\partial z}{\partial x}+b(x,y)\frac{\partial z}{\partial y}+c(x,y)z=d(x,y) \tag{1.5}$$

is called a *linear* PDE of first order, if the function F is linear in $\frac{\partial z}{\partial x}$, $\frac{\partial z}{\partial y}$, and z, while the coefficients a, b, c, and d depend only on the independent variables x and y. $xp + yq = nz$ is a linear PDE of first order.

An equation which does not fit into any of the above categories is called *nonlinear*. $p^2 + q^2 = 1$ is a nonlinear PDE of first order.

1.4 Types of Solutions

A relationship between the variables is known as a solution to any given PDE if the dependent variable generated from the relationship and its partial derivatives satisfy the given PDE.

Complete integral is a solution of a PDE when it contains an equal number of arbitrary constants as there are independent variables. We get a particular integral of a PDE when particular values of independent constants are present in a complete solution.

It is observed that the envelope of the surfaces represented by the complete integral also satisfies the given PDE, but it is not possible to acquire it by assigning any specific values to the arbitrary constants. This type of solution is known as a *singular integral.*

The complete solution of (1.2) is given by

$$\phi(x,y,z,a,b)=0 \tag{1.6}$$

Since two arbitrary parameters are involved, it belongs to a double infinite system of surfaces or to a single infinite system of a family of surfaces.

The general solution is obtained by eliminating "a" between Eq. (1.6), and

$$b=\psi(a) \tag{1.7}$$

$$\frac{\partial \phi}{\partial a}+\frac{\partial \phi}{\partial b}\psi'(a)=0 \tag{1.8}$$

The *general solution* represents the envelope of a one-parameter family of surfaces.

The singular solution is obtained by eliminating "a" and "b" between Eq. (1.6),

$$\frac{\partial \phi}{\partial a} = 0 \tag{1.9}$$

$$\frac{\partial \phi}{\partial b} = 0 \tag{1.10}$$

The *singular solution* represents the envelope of a two-parameter family of surfaces in the general envelope of all surfaces included in the complete solution.

Numerical solution: It is impossible, and even very difficult, to find an explicit formula for the solution of a PDE. But with the help of the numerical method, the approximate solution of the PDE can be found. The following are the numerical approximation methods:

(i) **Finite Differences**: The domain of the PDE is divided into a grid consisting of small rectangles and the solution $f(x, y, t)$ is approximated at the corner of these rectangles (the "nodes" of the grid). It is easy to implement the method for rectangle domains.

(ii) **Finite Element Method**: It is an advanced method than finite differences. The domain of the PDE is divided into triangles instead of rectangles. For domains of various shapes, the finite method works smoothly.

Example: *Solve*: $z = px + qy + pq$

Solution: The complete integral is

$$z = ax + by + ab \tag{1.11}$$

For singular solution:

Differentiating (1.11) partially with respect to a and b,

$$a = x + b, \; 0 = y + a \tag{1.12}$$

Eliminating a and b between (1.11) and (1.12), we get

$$z = xy \tag{1.13}$$

which is the required singular solution, for it satisfies the given equation.

For a general solution, Take $b = \phi(a)$, where ϕ denotes an arbitrary function. Then (1.11) becomes

$$z = ax + \phi(a)y + a\phi(a) \tag{1.14}$$

Differentiating (1.14) partially with respect to a, we have

$$z = x + \phi'(a)y + a\phi(a)y + \phi(a) - a\phi'(a) \tag{1.15}$$

The general integral is obtained by eliminating a between (1.14) and (1.15).

1.5 Initial Conditions and Boundary Conditions

A PDE subject to certain conditions in the form of initial or boundary conditions is known as an initial value or a boundary value problem. The PDE is termed a boundary value problem (BVP) if the boundary conditions of the dependent variable u and some of its partial derivatives are often prescribed. However, the PDE is called an initial value problem (IVP) if the initial conditions, known as Cauchy conditions, are the values of the unknown function u and an appropriate number of its derivatives at the initial point ($t = t_0$). Moreover, the PDE is termed the Initial-Boundary Value Problem (IBVP) if both initial conditions and boundary conditions are prescribed.

(1) Let us consider the *initial value problem* described by PDE

$$u_{tt} - c^2 u_{xx} = xe^t$$

with $u(x, 0) = \sin x$, $u_t(x, 0) = 0$.

(2) *The boundary conditions* fall into the following three categories:

(i) *Dirichlet conditions* (also known as boundary conditions of the first kind) are the values of the unknown function u prescribed at each point of the boundary ∂D of the domain D under consideration.

Example: Consider the following problem

$$u_t = ku_{xx}, \quad 0 < x < l, t > 0$$

with the boundary condition

$$u(x, 0) = f(x), \quad u_t(x, 0) = g(x), \ \ 0 < x < l,$$

$$u(0, t) = T_1, \quad u(l, t) = T_2, \ \ t > 0$$

(ii) *Neumann conditions* (also known as boundary conditions of the second kind) are the values of the normal derivatives of the unknown function u prescribed at each point of the boundary ∂D.

Example: Consider the following problem

$$u_t = ku_{xx}, \quad 0 < x < l, t > 0$$

with the boundary condition

$$u(x,0) = f(x), \quad u_t(x,0) = g(x), \;\; 0 < x < l,$$

$$u_x(0,t) = T_1, \quad u_x(l,t) = T_3, \;\; t > 0$$

(iii) *Robin conditions* (also known as boundary conditions of the third kind or mixed boundary conditions) are the values of a linear combination of the unknown function u and its normal derivative prescribed at each point of the boundary ∂D.

Example: Consider the following problem

$$u_t = ku_{xx}, \quad 0 < x < l, t > 0$$

with the boundary condition

$$u(x,0) = f(x), \quad u_t(x,0) = g(x), \;\; 0 < x < l,$$

$$u(0,t) + \alpha u_x(0,t) = 0, \quad u(l,t) + \beta u_x(l,t) = 0, \;\; t > 0$$

Note: The PDE admits a well-defined solution if the initial conditions are in agreement with the boundary conditions.

1.6 Formation of PDE

PDE can be formed in one of two ways: (i) by eliminating arbitrary constants, or (ii) by eliminating arbitrary functions.

(i) Formation by eliminating arbitrary constants:
Let

$$\phi(x, y, z, a, b) = 0 \tag{1.16}$$

be any relationship between two independent variables x, y, and a dependent variable z, as well as any arbitrary constants a and b.

Differentiating (1.16) partially with respect to x and y, we get

$$\frac{\partial \phi}{\partial x} + p\frac{\partial \phi}{\partial z} = 0 \tag{1.17}$$

$$\frac{\partial \phi}{\partial y} + q\frac{\partial \phi}{\partial z} = 0 \tag{1.18}$$

Now, eliminating two quantities a and b from these relations (1.16), (1.17) and (1.18), we get a PDE of the first order as $F(x, y, z, p, q) = 0$.

Note: We obtain a PDE with an order greater than two when the function ϕ contains more than two arbitrary constants.

(ii) Formation by the elimination of arbitrary functions:

Let

$$F(u, v) = 0. \tag{1.19}$$

be any relationship between u and v, where u and v are two functions of x, y, and z and x, y being two independent variables and z being a dependent variable.

Now, differentiating (1.19) partially with respect to x and y, we get

$$\frac{\partial F}{\partial u}\left[\frac{\partial u}{\partial x} + \frac{\partial u}{\partial z}p\right] + \frac{\partial F}{\partial v}\left[\frac{\partial v}{\partial x} + \frac{\partial v}{\partial z}p\right] = 0 \tag{1.20}$$

and

$$\frac{\partial F}{\partial u}\left[\frac{\partial u}{\partial y} + \frac{\partial u}{\partial z}q\right] + \frac{\partial F}{\partial v}\left[\frac{\partial v}{\partial y} + \frac{\partial v}{\partial z}q\right] = 0 \tag{1.21}$$

Now eliminating $\frac{\partial F}{\partial u}$ and $\frac{\partial F}{\partial v}$ from Eq. (1.20) and (1.21), we obtain

$$\begin{vmatrix} \frac{\partial u}{\partial x} + \frac{\partial u}{\partial z}p & \frac{\partial v}{\partial x} + \frac{\partial v}{\partial z}p \\ \frac{\partial u}{\partial y} + \frac{\partial u}{\partial z}q & \frac{\partial v}{\partial y} + \frac{\partial v}{\partial z}q \end{vmatrix} = 0$$

which simplifies to

$$p\frac{\partial(u, v)}{\partial(y, z)} + q\frac{\partial(u, v)}{\partial(z, x)} = \frac{\partial(u, v)}{\partial(x, y)} \tag{1.22}$$

or

$$Pp + Qq = R \tag{1.23}$$

where $P = \frac{\partial(u,v)}{\partial(y,z)}$, $Q = \frac{\partial(u,v)}{\partial(z,x)}$, $R = \frac{\partial(u,v)}{\partial(x,y)}$. This is a linear PDE of first order and first degree in p and q. Equation (1.23) is also called Lagrange's PDE of first order.

Note: If the given equation between x, y, z contains two arbitrary functions, then in general, their elimination gives rise to the equation of higher order.

Example: *Obtain a PDE by eliminating* a *and* b *from the equation* $az + b = a^2x + y$.

Solution: The equation is

$$az + b = a^2x + y \tag{1.24}$$

Differentiating (1.24) w.r.t. x and y, we get

$$a\frac{\partial z}{\partial x} = a^2 \tag{1.25}$$

and

$$a\frac{\partial z}{\partial y} = 1 \tag{1.26}$$

From (1.25) and (1.26) eliminating a, we get

$$\frac{\partial z}{\partial x}\frac{\partial z}{\partial y} = 1$$

i.e.,

$$pq = 1$$

which is the required PDE.

Example: *Eliminate the arbitrary function f and ϕ from $y = f(x - at) + g(x + at)$.*

Solution: Here,

$$y = f(x - at) + g(x + at) \tag{1.27}$$

Now, $\frac{\partial y}{\partial x} = f'(x - at) + g'(x + at)$, $\frac{\partial^2 y}{\partial x^2} = f''(x - at) + g''(x + at)$, $\frac{\partial y}{\partial t} = -af'(x - at) + ag'(x + at)$, $\frac{\partial y}{\partial x} = a^2 f'(x - at) + a^2 g'(x + at)$. From (1.27), we get

$$\frac{\partial^2 y}{\partial t^2} = a^2\frac{\partial^2 y}{\partial x^2}.$$

1.7 Solution of Linear PDEs of Order One

Lagrange method:

Consider a linear or semi-linear, or quasi-linear equation

$$Pp + Qq = R \tag{1.28}$$

where P, Q, and R are the functions of x, y, and z. This equation was studied by the French mathematician Joseph Louis Lagrange (1736–1833) and is called Lagrange's equation. The general solution of the corresponding

PDE is a function $\phi(u,v)$=0, where ϕ is an arbitrary function and u and v are two functions of x, y, and z.

To solve the equation $Pp + Qq = R$, we are to find u and v first, and $\phi(u,v) = 0$ will give the general integral of Eq. (1.28). Now, the problem is how to find these u and v.

Let us choose two independent integrals, $u(x,y,z) = C_1$ and $v(x,y,z) = C_2$, where C_1 and C_2 are two arbitrary constants. Therefore, $du = 0$ and $dv = 0$ give

$$\frac{\partial u}{\partial x}dx + \frac{\partial u}{\partial y}dy + \frac{\partial u}{\partial z}dz = 0$$

$$\frac{\partial v}{\partial x}dx + \frac{\partial v}{\partial y}dy + \frac{\partial v}{\partial z}dz = 0$$

From these, we get

$$\frac{dx}{P} = \frac{dy}{Q} = \frac{dz}{R} \tag{1.29}$$

where P, Q, and R are functions of x, y, and z. Thus, $u(x,y,z) = C_1$ and $v(x,y,z) = C_2$ are the solutions of the differential Eq. (1.29). This Eq. (1.29) is known as Lagrange's auxiliary (or subsidiary) equation. Thus, to solve an equation $Pp + Qq = R$, we first solve $\frac{dx}{P} = \frac{dy}{Q} = \frac{dz}{R}$ and get its two independent solutions, u=constant and $v =$ constant. Then the general solution of $Pp + Qq = R$ will be $v = \phi_1(u)$ or $\phi(u,v) = 0$, where ϕ_1 or ϕ is any arbitrary function.

Working Rule:

Step 1: Write the provided PDE in the form $Pp + Qq = R$.

Step 2: Write Lagrange's auxiliary equation $\frac{dx}{P} = \frac{dy}{Q} = \frac{dz}{R}$.

Step 3: Solve Lagrange's auxiliary equation. Let $u(x,y,z) = c_1$ and $v(x,y,z) = c_2$ be the two independent solutions of Lagrange's auxiliary equation.

Step 4: The general solution of $Pp + Qq = R$ can be written as $\phi(u,v) = 0$ or in the form $u = f(u)$ or $v = f(v)$.

Note: The term "special solution" refers to a specific integral that can satisfy the equation $Pp + Qq = R$ but cannot be written as a function of u and v.

Example: *Solve* $x^2p + y^2q = (x+y)z$.

Solution: The given PDE is of the form $Pp + Qq = R$. So, Lagrange's subsidiary equation is

$$\frac{dx}{x^2} = \frac{dy}{y^2} = \frac{dz}{(x+y)z} \tag{1.30}$$

From the first two parts of (1.30), we get

$$\frac{dx}{x^2} = \frac{dy}{y^2}$$

On integration, we get

$$\frac{1}{x} - \frac{1}{y} = c_1 \tag{1.31}$$

c_1 being an integrating constant.

Again, each part of (1.30) is equal to

$$\frac{-\frac{1}{x}dx - \frac{1}{y}dy + \frac{1}{z}dz}{-x - y + x + y} = \frac{-\frac{1}{x}dx - \frac{1}{y}dy + \frac{1}{z}dz}{0}$$

i.e.,

$$-\frac{1}{x}dx - \frac{1}{y}dy + \frac{1}{z}dz = 0$$

Integrating

$$\log\left(\frac{z}{xy}\right) = \log c_2$$

or

$$\frac{z}{xy} = c_2 \tag{1.32}$$

c_2 being an integrating constant. Hence, the general solution of given PDE is

$$\phi\left(\frac{1}{x} - \frac{1}{y}, \frac{z}{xy}\right) = 0$$

ϕ being an arbitrary function.

Geometrical interpretation:

The Lagrange differential equation can be written as $Pp + Qq + R(-1) = 0$.

We know that the direction ratios of the normal at a portion of the surface $f(x, y, z) = 0$ are proportional to $\frac{\partial f}{\partial x} : \frac{\partial f}{\partial y} : \frac{\partial f}{\partial z}$, i.e., $\frac{-\frac{\partial f}{\partial x}}{\frac{\partial f}{\partial z}} : \frac{-\frac{\partial f}{\partial y}}{\frac{\partial f}{\partial z}} : -1$ i.e., $\frac{\partial z}{\partial x} : \frac{\partial z}{\partial y} : -1$ i.e., $p : q : -1$.

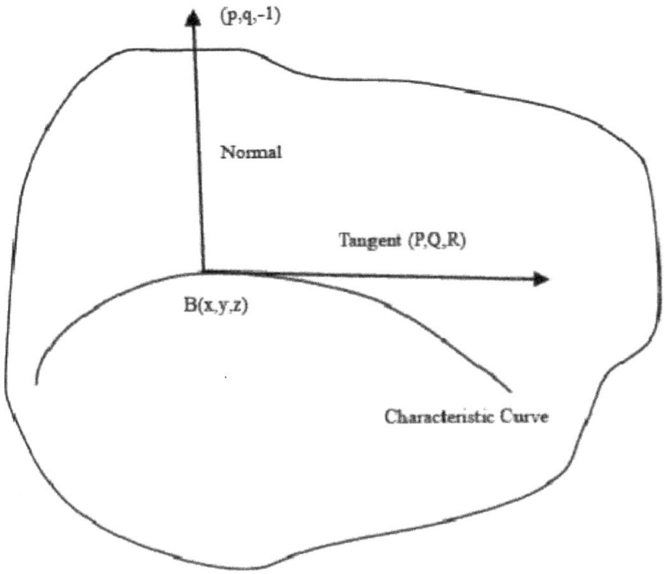

Thus, the geometrical interpretation of $Pp+Qq=R$ is that the normal to a certain surface is perpendicular to a line whose direction ratios are in the ratio $P:Q:R$.

Also, the differential equation (1.29) represents a family of curves such that tangents at any point to the curve have direction cosines in the ratio $P:Q:R$ and that $F(u,v)=0$ represents surfaces through integral curves $u=a$ and $v=b$ of (1.29). Thus, the normal to these surfaces at any point must be perpendicular to the tangent to the curve, i.e., perpendicular to a line whose direction ratios are proportional to $P:Q:R$. Hence, the integral surface of $Pp+Qq=R$ and the surfaces generated by the integral curves of (1.29) are equivalent.

1.8 Integral Surfaces Passing Through a Given Curve

Consider the first-order linear PDE

$$Pp+Qq=R$$

Using Lagrange's method, we obtain two integral curves described by

$$u(x,y,z)=c_1, \quad v(x,y,z)=c_2 \tag{1.33}$$

Now, our goal is to find the integral surface that passes through a curve whose parametric equations are as follows:

$$x=x(t),\ y=y(t),\ z=z(t)$$

where t is a parameter and (1.33) is then written as

$$u(x(t), y(t), z(t)) = c_1, \quad v(x(t), y(t), z(t)) = c_2 \tag{1.34}$$

We eliminate t and obtain a relation involving c_1 and c_2. Finally, by putting $c_1 = u$ and $c_2 = v$, we obtain the integral surface passing through the given curve.

Example: *Find the integral surface of the linear PDE* $x(y^2 + z)p - y(x^2 + z)q = (x^2 - y^2)z$ *containing the straight line* $x + y = 0, z = 1$.

Solution: Lagrange's auxiliary equation is

$$\frac{dx}{x(y^2 + z)} = \frac{dy}{-y(x^2 + z)} = \frac{dz}{(x^2 - y^2)z} \tag{1.35}$$

Using the multiplier xyz, we get

$$yzdx + zxdy + xydz = 0$$

On integration, we get

$$xyz = c_1 \tag{1.36}$$

Again, using the multipliers x, y, and z. Then each fraction in Eq. (1.35) is equal to

$$xdx + ydy + zdz = c_2$$

On integration, we get

$$x^2 + y^2 + z^2 = c_2 \tag{1.37}$$

Here, the integral surface contains the straight line $x + y = 0, \; z = 1$ in the parametric form as

$$x = t, \quad y = -t, \quad z = 1$$

Putting these values in Eqs. (1.36) and (1.37), we obtain

$$-t^2 = c_1, \quad 2t^2 + 1 = c_2$$

Eliminating t between the two, we obtain

$$1 - 2c_1 = c_2, \quad 2c_1 + c_2 - 1 = 0$$

Hence, the required integral surface is

$$x^2 + y^2 + z^2 + 2xyz - 1 = 0.$$

1.9 Cauchy Problem and Characteristics

The existence theorem for the first-order quasi-linear PDE is known as Cauchy's problem. Causchy's problem for a first-order PDE with two independent variables x, y and a dependent variable z can be stated as follows:

If (a) the functions $x_0(t), y_0(t)$, and $z_0(t)$, together with their first derivatives, are continuous in the interval I given by $t_1 < t < t_2$; and if (b) $F(x, y, z, p, q) = P(x, y, z)p + Q(x, y, z)q - R(x, y, z)$ is a continuous function of x, y, z, p, and q in some domain D of the $xyzpq$ space.

Then there exists a function $\phi(x, y)$ having the following properties:

(i) $\phi(x, y)$ and its partial derivatives ϕ_x and ϕ_y are continuous functions of x and y in a region R of the xy space.
(ii) For all values of x and y in R, the point $(x, y, \phi(x, y), \phi_x(x, y), \phi_y(x, y))$ lies in D and $F(x, y, \phi(x, y), \phi_x(x, y), \phi_y(x, y)) = 0$.
(iii) The point $(x_0(t), y_0(t))$ lies in the region R and $\phi(x_0(t), y_0(t)) = z_0$ for all $t \in I$.

Geometrically, the above problem proves that there exists a surface $z = \phi(x, y)$ passing through the curve C, whose parametric equations are as follows:

$$x = x_0(t), y = y_0(t), z = z_0(t) \tag{1.38}$$

and at each point of which the direction $(p, q, -1)$ of the normal satisfies

$$F[x, y, z, p, q] = 0 \implies P(x, y, z)p + Q(x, y, z)q = R(x, y, z) \tag{1.39}$$

The existence of a solution to Eq. (1.39) passing through the curve C given parametrically by Eq. (1.38) can be proved only with extra assumptions about the form of the function F and the nature of the curve C. Thus, several existence theorems are available in the literature based on different assumptions. We state without proof one existence theorem as follows:

Theorem: If (a) $g(y)$ and all its derivatives are continuous for $|y - y_0| < \delta$. (b) x_0 be a given real number and $z_0 = g(y_0)$, $q_0 = g'(y_0)$, and $f(x, y, z, q)$ and all its partial derivatives are continuous in the region S defined by $|x - x_0| < \delta$, $|y - y_0| < \delta$, $|q - q_0| < \delta$ then, there is a unique function $\phi(x, y)$ such that

(i) $\phi(x, y)$ and all its partial derivatives are continuous in the region R defined by $|x - x_0| < \delta_1$, $|y - y_0| < \delta_2$.

(ii) $z = \phi(x, y)$ is a solution of the equation $\frac{\partial z}{\partial x} = f(x, y, z, \frac{\partial z}{\partial y})$ for all $(x, y) \in R$.

(iii) $\phi(x_0, y) = g(y)$ for all values of y in the interval $|y - y_0| < \delta_2$. (proof see page 48 in Ref. [1])

Cauchy's Method of Characteristics:

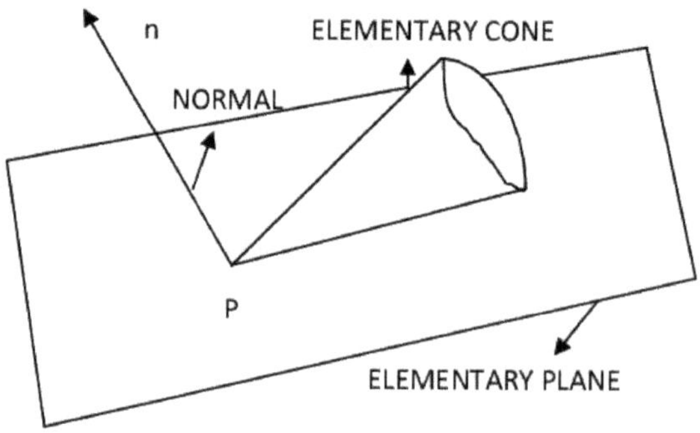

Let us consider a plane Γ passing through any point $P(x_0, y_0, z_0)$ whose equation is

$$A(x - x_0) + B(y - y_0) + C(z - z_0) = 0$$

$(p_0, q_0, -1)$ are the direction ratios of the normal to the plane S. Then $\frac{A}{p_0} = \frac{B}{q_0} = \frac{c}{-1}$ so that the plane is uniquely determined as

$$p_0(x - x_0) + q_0(y - y_0) = z - z_0$$

Conversely, any set of five real numbers defines a plane in three dimensions. We say that a set of five numbers (x, y, z, p, q) defines a plane element of the space. In particular, a plane element $(x_0, y_0, z_0, p_0, q_0)$ whose components satisfy the equation

$$F(x, y, z, p, q) = 0 \tag{1.40}$$

is called an integral element of Eq. (1.40) at the point $P(x_0, y_0, z_0)$.

Equation (1.40) can now be transformed into the equivalent form as

$$q = G(x, y, z, p) \tag{1.41}$$

Fix x_0, y_0, z_0 and allow p to vary. Then we obtain a set of plane elements $(x_0, y_0, z_0, p, G(x_0, y_0, z_0, p))$ which depends only on the parameter p. As p varies, we get a set of plane elements passing through the point P, and these envelopes a cone having its vertex at P. This cone is called the elementary cone of Eq. (1.40) at P.

Let S be a surface given by

$$z = g(x, y) \tag{1.42}$$

If the functions g, g_x, g_y are continuous in a region R of the xy plane then the tangent plane at each point of the surface S defines a plane element of the form

$$(x_0, y_0, g(x_0, y_0), g_x(x_0, y_0), g_y(x_0, y_0)) \tag{1.43}$$

At the position $(x_0, y_0, g(x_0, y_0))$, we refer to this plane element as the tangent element of the surface S. The ensuing theorem should be evident from a geometrical perspective.

Theorem: A surface is an integral surface of a PDE iff at each point, its tangent element touches the elementary cone of the equation.
(proof see p. 62 in Ref. [1])

Integral strip, Characteristic strip:

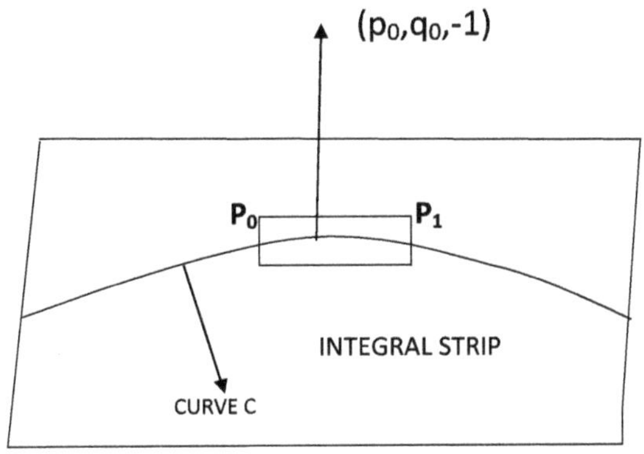

If C is the curve defined by the equation

$$x = x(t), \quad y = y(t), \quad z = z(t) \tag{1.44}$$

then C is said to be on the surface $z = g(x, y)$ if and only if $z(t) \equiv g(x(t), y(t))$ for all t in a suitable interval I. Let P_0 be a point on C in accordance with $t = t_0$. The direction ratios of the tangent line P_0P_1 at P_0 to the curve C are $(x'(t_0), y'(t_0), z'(t_0))$, where $x'(t_0) = [\frac{dx}{dt}]_{t=t_0}$, $y'(t_0) = [\frac{dy}{dt}]_{t=t_0}$, $z'(t_0) = [\frac{dz}{dt}]_{t=t_0}$.

This direction of the tangent line will be perpendicular to the direction $(p_0, q_0, -1)$ if $p_0x'(t_0) + q_0y'(t_0) - z'(t_0) = 0$, i.e., if $z'(t_0) = p_0x'(t_0) + q_0y'(t_0)$. As a result, we state that any set

$$\{x(t), y(t), z(t), p(t), q(t)\} \tag{1.45}$$

of five real functions satisfying the condition

$$z'(t) = p(t)x'(t) + q(t)y'(t) \tag{1.46}$$

define a strip at the point (x, y, z) on C.

An integral strip of Eq. (1.40) is a strip that is also an integral element of Eq. (1.40). In other words, the set of functions (1.45) is an integral strip of Eq. (1.40) if

(a) they satisfy condition (1.46).
(b) $F(x(t), y(t), z(t), p(t), q(t)) = 0$ for all t in I.

A Characteristic strip is a strip that is such that at each point the curve defined by Eq. (1.44) touches a generator of the elementary cone.

To derive the equations determining a characteristic strip:

Let p, q satisfy the equation $F(x, y, z, p, q) = 0$. If

$$dz = pdx + qdy \tag{1.47}$$

then the point $(x + dx,\ y + dy,\ z + dz)$ lies in the tangent plane to the elementary cone at P. Differentiate (1.47) with respect to p, we get

$$0 = dx + \frac{dq}{dp}dy \tag{1.48}$$

Differentiate (1.40) partially with respect to p. Then

$$\frac{\partial F}{\partial p} + \frac{\partial F}{\partial q}\frac{dq}{dp} = 0 \Rightarrow F_p + F_q\frac{dq}{dp} = 0 \tag{1.49}$$

From (1.48) and (1.49),

$$\frac{dq}{dp} = -\frac{dx}{dy} = -\frac{F_p}{F_q}$$

Consequently, using (1.48), we are able to

$$\frac{dx}{F_p} = \frac{dy}{F_q} = \frac{pdx + qdy}{pF_p + qF_q} = \frac{dz}{pF_p + qF_q}$$

Thus,

$$\frac{dx}{F_p} = \frac{dy}{F_q} = \frac{dz}{pF_p + qF_q} \tag{1.50}$$

along a characteristic strip $x'(t), y'(t), z'(t)$ are proportional to $F_p, F_q, pF_p + qF_q$, respectively. Choose the parameter t so that

$$x'(t) = F_p,\ y'(t) = F_q \tag{1.51}$$

then

$$z'(t) = px'(t) + qy'(t) = pF_p + qF_q \tag{1.52}$$

Along a characteristic strip, p is also a function of t.

$$\begin{aligned} p'(t) &= \frac{\partial p}{\partial x}x'(t) + \frac{\partial p}{\partial y}y'(t) \\ &= \frac{\partial p}{\partial x}F_p + \frac{\partial p}{\partial y}F_q \\ &= \frac{\partial p}{\partial x}F_p + \frac{\partial q}{\partial x}F_q, \quad \left(as \frac{\partial p}{\partial y} = \frac{\partial q}{\partial x}\right) \end{aligned} \tag{1.53}$$

Now differentiate Eq. (1.40) partially with respect to x, we get

$$F_x + F_z p + F_p \frac{\partial p}{\partial x} + F_q \frac{\partial q}{\partial x} = 0$$

Along a characteristic strip, by (1.53)

$$p'(t) = -(F_x + F_z p) \tag{1.54}$$

similarly

$$q'(t) = -(F_y + F_z q) \tag{1.55}$$

So, we have from Eqs. (1.51), (1.52), (1.54), and (1.55) the following ODEs:

$$\begin{aligned} &x'(t) = F_p,\ y'(t) = F_q,\ z'(t) = pF_p + qF_q, \\ &p'(t) = -F_x - pF_z,\ q'(t) = -F_y - qF_z \end{aligned} \tag{1.56}$$

Equation (1.56) is called the characteristic equation of the differential equation (1.40) for determining the characteristic strip. The last three equations (1.56) are also called compatibility conditions.

If the functions in Eq. (1.56) satisfy a Lipschitz condition, then there exists a unique solution to the equations for every prescribed set of initial values of the variables. Thus, the characteristic strips are uniquely determined by any initial element (x_0, y_0, z_0) and any initial value t_0 for t.

The stated result is an extension of the following theorem:

Let R be the region defined by $|x-a| < l_1, |y-b| < l_2, |z-c| < l_3$. In R, let the functions $f_1(x, y, z)$ and $f_2(x, y, z)$.

(i) be continuous
(ii) satisfy a Lipschitz condition of the form

$$|f_1(x, y, z) - f_1(x, \eta, \zeta)| \leq A_1|y-\eta| + B_1|z-\zeta|$$
$$|f_2(x, y, z) - f_2(x, \eta, \zeta)| \leq A_2|y-\eta| + B_2|z-\zeta|$$

then in an appropriate interval $|x-a| < h$ there exist unique function $y(x)$ and $z(x)$ which

(a) are continuous in that interval
(b) have continuous derivatives in that interval
(c) satisfy the differential equation

$$\frac{dy}{dx} = f_1(x, y, z); \frac{dy}{dx} = f_2(x, y, z)$$

(d) satisfy $y(a) = b, z(a) = c$ where the numbers a, b, c are arbitrary.

Theorem: Along each characteristic strip of the equation $F(x, y, z, p, q) = 0$, the function $F(x, y, z, p, q)$ is a constant.

Proof: Along a characteristic strip

$$\begin{aligned}
&\frac{d}{dx}F(x(t), y(t), z(t), p(t), q(t)) \\
&\quad = F_x x' + F_y y' + F_z z' + F_p p' + F_q q' \\
&\quad = F_x F_p + F_y F_q + F_z(pF_p + qF_q) - F_p(F_x + pF_z) - F_q(F_y + qF_z) \\
&\quad = 0
\end{aligned}$$

Therefore, $F(x, y, z, p, q) = k$, a constant along the characteristic strip.

Corollary: If a characteristic strip contains at least one integral element of $F(x, y, z, p, q) = 0$, then it is an integral strip of the equation $F(x, y, z, p, q) = 0$.

Solution of Cauchy's Problem:

Suppose we want to find a solution to the nonlinear PDE. $F(x, y, z, \frac{\partial z}{\partial x}, \frac{\partial z}{\partial y}) = 0$ passing through a curve C having the freedom equations

$$x = \theta(v), y = \phi(v), z = \psi(v) \tag{1.57}$$

Let $x = x(x_0, y_0, z_0, p_0, q_0, t_0, t)$, $y = y(x_0, y_0, z_0, p_0, q_0, t_0, t)$, $z = z(x_0, y_0, z_0, p_0, q_0, t_0, t)$ be the solution of the characteristic equation (1.56). Choose $x_0 = \theta(v)$, $y_0 = \phi(v)$, $z_0 = \psi(v)$ as the initial values of x, y, z.

Find the initial values p_0, q_0 from the equations $\psi'(v) = p_0\theta'(v) + q_0\phi'(v)$(as $z' = px' + qy'$) and $F(\theta(v), \phi(v), \psi(v), p_0, q_0) = 0$.

Let t_0 be the initial value of t. Using these values of x_0, y_0, z_0, p_0, q_0 and t_0 in Eq. (1.56), we get x, y, z in terms of t and v, say

$$x = f(t, v); y = g(t, v), z = h(t, v) \tag{1.58}$$

Now eliminate t, v between the three equations (1.58). Then we get a relation of the form $H(x, y, z) = 0$ which is the equation of the integral surface of the equation $F(x, y, z, \frac{\partial z}{\partial x}, \frac{\partial z}{\partial y}) = 0$ passing through the curve C.

Example: *Solve the Cauchy problem for* $zp + q = 1$ *with initial conditions* $x_0 = \lambda, y_0 = \lambda, z_0 = \frac{\lambda}{2}, 0 \leq \lambda \leq 1$.

Solution. Given

$$f(x, y, z, p, q) \equiv zp + q - 1 = 0 \tag{1.59}$$

So, $\frac{\partial f}{\partial p} = z$, $\frac{\partial f}{\partial q} = 1$, $\frac{\partial x_0}{\partial \lambda} = 1$, $\frac{\partial y_0}{\partial \lambda} = 1$ $\frac{\partial f}{\partial q}.\frac{dx_0}{d\lambda} - \frac{\partial f}{\partial p}.\frac{dy_0}{d\lambda} = 1 - z = 1 - \frac{\lambda}{2} \neq 0$ for $0 \leq \lambda \leq 1$.

Now, we get the following ODE

$$\frac{dx}{dt} = \frac{\partial f}{\partial p}, \quad \frac{dy}{dt} = \frac{\partial f}{\partial p}$$

i.e.,

$$\frac{dx}{dt} = z, \tag{1.60}$$

$$\frac{dy}{dt} = 1 \tag{1.61}$$

Since

$$dz = \frac{\partial z}{\partial x}dx + \frac{\partial z}{\partial y}dy \Rightarrow \frac{dz}{dx} = \frac{\partial z}{\partial x}\frac{dx}{dt} + \frac{\partial z}{\partial y}\frac{dy}{dt}$$

i.e.,

$$\frac{dz}{dx} = pz + q = 1 \tag{1.62}$$

Integrating Eq. (1.61) and (1.62), we get $y = t + c_1$ and $z = t + c_2$. From given initial conditions $c_1 = \lambda$ and $c_2 = \frac{\lambda}{2}$.

Therefore,

$$y = t + \lambda, \tag{1.63}$$

$$z = t + \frac{\lambda}{2} \tag{1.64}$$

From Eq. (1.60),

$$\frac{dx}{dt} = t + \frac{\lambda}{2}$$

$$\Rightarrow \int x = \int \left(t + \frac{\lambda}{2}\right) dt$$

$$\Rightarrow x = \frac{t^2}{2} + \frac{\lambda}{2}t + c_3$$

From initial condition $c_3 = \lambda$. Hence,

$$x = \frac{t^2}{2} + \frac{\lambda}{2}t + \lambda \tag{1.65}$$

From Eq. (1.63) and (1.65), we get

$$x = \frac{t^2}{2} + \frac{t}{2}(y - t) + (y - t)$$

$$\Rightarrow t = \frac{x - y}{\frac{y}{2} - 1}$$

Now, from Eq. (1.63), we get

$$\lambda = y - t = \frac{\frac{y^2}{2} - x}{\frac{y}{2} - 1}$$

Hence, from Eq. (1.64),

$$z = \frac{x - y}{\frac{y}{2} - 1} + \frac{1}{z}\frac{\frac{y^2}{2} - x}{\frac{y}{2} - 1}$$

which is the required solution.

Compatible Systems of First-Order PDE

A pair of first-order PDEs is said to be compatible if their solutions are similar. We shall now derive the necessary and sufficient conditions for the two PDEs

$$f(x, y, z, p, q) = 0 \tag{1.66}$$

and

$$g(x, y, z, p, q) = 0 \tag{1.67}$$

to be compatible. Let

$$J = \frac{\partial(f, g)}{\partial(p, q)} \neq 0 \tag{1.68}$$

Since the solutions to Eqs. (1.66) and (1.67) are the same, we may solve them to get explicit expressions for p and q in the form

$$p = \phi(x, y, z), q = \xi(x, y, z) \tag{1.69}$$

and then, the differential relation

$$pdx + qdy = dz$$

or,

$$\phi(x, y, z)dx + \xi(x, y, z)dy = dz \tag{1.70}$$

should be integrable, for which the necessary condition is

$$\overrightarrow{X} \cdot curl\overrightarrow{X} = 0$$

where $X = \{\phi, \xi, -1\}$.

Therefore,

$$\phi(-\xi_z) + \xi(\phi_z) = \xi_x - \phi_y$$

which can be rewritten as

$$\xi_x + \phi\xi_z = \phi_y + \xi\phi_z \tag{1.71}$$

Now differentiating Eq. (1.66) with respect to x and z, we get

$$f_x + f_p\frac{\partial p}{\partial x} + f_q\frac{\partial q}{\partial x} = 0$$

$$f_z + f_p\frac{\partial p}{\partial z} + f_q\frac{\partial q}{\partial z} = 0$$

But from Eq. (1.70), we have $\frac{\partial p}{\partial x} = \frac{\partial \phi}{\partial x}, \frac{\partial q}{\partial x} = \frac{\partial \xi}{\partial x}$ and so on.

Using these results, the above equations can be recast into

$$f_x + f_p \phi_x + f_q \xi_x = 0$$
$$f_z + f_p \phi_z + f_q \xi_z = 0$$

Multiplying the second one of the above pair by ϕ and adding to the first one, we readily obtain

$$(f_x + \phi f_z) + f_p(\phi_x + \phi\phi_z) + f_q(\xi_x + \phi\xi_z) = 0$$

Similarly, from Eq. (1.67), we can deduce that

$$(g_x + \phi g_z) + g_p(\phi_x + \phi\phi_z) + g_q(\xi_x + \phi\xi_z) = 0$$

Solving the above pair of equations for $(\xi_x + \phi\xi_z)$, we have

$$\frac{\xi_x + \phi\xi_z}{f_p(g_x + \phi g_z) - g_p(f_x + \phi f_z)} = \frac{1}{f_q g_p - g_q f_p} = \frac{1}{J}$$

or,

$$\xi_x + \phi\xi_z = \frac{1}{J}[(f_p g_x - g_p f_x) + \phi(f_p g_z - g_p f_z)] = \frac{1}{J}\left[\frac{\partial(f,g)}{\partial(x,p)} + \phi\frac{\partial(f,g)}{\partial(z,p)}\right] \tag{1.72}$$

where J is defined in Eq. (1.68). Similarly, by differentiating Eq. (1.66) with respect to y and z and using Eq. (1.69), we can show that

$$\phi_y + \xi\phi_z = -\frac{1}{J}\left[\frac{\partial(f,g)}{\partial(y,q)} + \xi\frac{\partial(f,g)}{\partial(z,q)}\right] \tag{1.73}$$

Finally, substituting the values of $\xi_x + \phi\xi_z$ and $\phi_y + \xi\phi_z$ from Eqs. (1.73) and (1.74) into (1.71), we obtain

$$\frac{\partial(f,g)}{\partial(x,p)} + \phi\frac{\partial(f,g)}{\partial(z,p)} = -\left[\frac{\partial(f,g)}{\partial(y,q)} + \xi\frac{\partial(f,g)}{\partial(z,q)}\right]$$

In view of Eq. (1.69), we can replace ϕ and ζ by p and q, respectively, to get

$$\frac{\partial(f,g)}{\partial(x,p)} + p\frac{\partial(f,g)}{\partial(z,p)} + \frac{\partial(f,g)}{\partial(y,q)} + q\frac{\partial(f,g)}{\partial(z,q)} = 0 \tag{1.74}$$

This is the desired compatibility condition.

1.10 Nonlinear PDE: Charpit's Method

The method of finding the complete integral of nonlinear PDEs of the first order is partly due to the Italian mathematician Lagrange (1736–1813). Later on, it was the French mathematician Charpit who perfected it and presented it in a memoir on June 30, 1784, to the Paris Academy of Sciences. The method is based on the consideration of compatible systems of first-order equations. Let

$$f(x, y, z, p, q) = 0 \tag{1.75}$$

be a PDE of first order in two independent variables x and y. z is a function of x and y. So,

$$dz = p(x, y, z, a)dx + q(x, y, z, a)dy \tag{1.76}$$

Now, if a different relation between x, y, z, p, q can be found, such as

$$g(x, y, z, p, q) = 0 \tag{1.77}$$

and then solve Eqs. (1.75) and (1.77) for p and q and substitute in (1.76). Now, if Eq. (1.76) has a solution, it is the complete integral of Eq. (1.75).

The primary aim is to determine the second Eq. (1.77), which was already covered in the part before this one. Now you must look for an equation of the type

$$g(x, y, z, p, q) = 0$$

compatible with the given equation.

$$f(x, y, z, p, q) = 0$$

for which the necessary and sufficient condition is

$$\frac{\partial(f, g)}{\partial(x, p)} + p\frac{\partial(f, g)}{\partial(z, p)} + \frac{\partial(f, g)}{\partial(y, q)} + q\frac{\partial(f, g)}{\partial(z, q)} = 0 \tag{1.78}$$

on expansion, we have

$$\left(\frac{\partial f}{\partial x}\frac{\partial g}{\partial p} - \frac{\partial f}{\partial p}\frac{\partial g}{\partial x}\right) + p\left(\frac{\partial f}{\partial z}\frac{\partial g}{\partial p} - \frac{\partial f}{\partial p}\frac{\partial g}{\partial z}\right)$$
$$+ \left(\frac{\partial f}{\partial y}\frac{\partial g}{\partial q} - \frac{\partial f}{\partial q}\frac{\partial g}{\partial y}\right) + q\left(\frac{\partial f}{\partial z}\frac{\partial g}{\partial q} - \frac{\partial f}{\partial q}\frac{\partial g}{\partial z}\right) = 0$$

It can be transformed into

$$f_p\frac{\partial g}{\partial x} + f_q\frac{\partial g}{\partial y} + (pf_p + qf_q)\frac{\partial g}{\partial z} - (f_x + pf_z)\frac{\partial g}{\partial p} - (f_y + qf_z)\frac{\partial g}{\partial q} = 0 \tag{1.79}$$

This is a linear PDE, and we can calculate g from it. The auxiliary equation of (1.79) is

$$\frac{dx}{f_p} = \frac{dy}{f_q} = \frac{dz}{pf_p + qf_q} = \frac{dp}{-(f_x + pf_z)} = \frac{dq}{-(f_y + qf_z)} \tag{1.80}$$

These equations are known as Charpit's auxiliary equations. Any integral of (1.80) that involves p, q, or both of them can be regarded as a second relation (1.77). The desired complete integral is then obtained by integrating Eq. (1.76). It should be noted that not every Charpit's equation needs to be employed; only the simplest one will do.

Working rule:

Step 1: Write the Charpit's auxiliary equations

$$\frac{dx}{f_p} = \frac{dy}{f_q} = \frac{dz}{pf_p + qf_q} = \frac{dp}{-(f_x + pf_z)} = \frac{dq}{-(f_y + qf_z)}$$

where $f(x, y, z, p, q) = 0$ is the given equation.

Step 2: Find the simplest integral that includes p, q, or both.

Step 3: Find p and q from the given equation and the equation obtained in step 2.

Step 4: With the values of p and q taken, use $dz = pdx + qdy$.

Step 5: Integrate. This gives the complete integral.

Note: Charpit's method cannot be generalized directly to any number of independent variables. However, there is another method of solving nonlinear PDEs, which is due to the German mathematician Jacobi (1804–1851), which can be easily generalized for any number of variables.

Example: *Find a complete integral of* $z^2(p^2z^2 + q^2) = 1$

Solution. Here, $F(x, y, z, p, q) = p^2z^4 + q^2z^2 - 1 = 0$.

Charpit's auxiliary equations are

$$\frac{dp}{p(4p^2z^3 + 2zq^2)} = \frac{dq}{q(4p^2z^3 + 2zq^2)} = \frac{dz}{-2p^2z^4 - 2q^2z^2} = \frac{dx}{-2pz^4} = \frac{dy}{-2qz^2} \tag{1.81}$$

Taking the first two fractions

$$\frac{dp}{p} = dq/q$$

So that $p = aq$.

Solving for p and q, $p = \frac{a}{z\sqrt{a^2z^2+1}}$, $q = \frac{1}{z\sqrt{a^2z^2+1}}$. So that

$$dz = pdx + qdy$$
$$\Rightarrow adx + dy = z\sqrt{a^2z^2+1}dz$$

Integrating, we get

$$ax + y = \int \sqrt{(a^2z^2+1)}.zdz$$
$$ax + y + b = \frac{1}{3a^2}(a^2z^2+1)^{3/2}$$
$$9a^4(ax+y+b)^2 = (a^2z^2+1)^3$$

which is a complete integral, a and b being arbitrary constants.

Example: *Find a complete, singular and general integrals of* $(p^2+q^2)y = qz$

Solution: Here, $F(x, y, z, p, q) = (p^2+q^2)y - qz$.
Charpit's auxiliary equation are

$$\frac{dp}{-pq} = \frac{dq}{p^2+q^2-q^2} = \frac{dz}{-p(2py) - q(2qy-z)} = \frac{dx}{-2py} = \frac{dy}{-2qy+z} \tag{1.82}$$

Taking first and second parts of the equation, we get

$$pdp + qdq = 0$$

Integrating

$$p^2 + q^2 = a^2 \tag{1.83}$$

where a is an arbitrary constant. Using (1.82) and (1.83), we get $q = \frac{a^2y}{z}$. Putting the value of q in (1.83), we get $p = \frac{a}{z}\sqrt{z^2 - a^2y^2}$. Now, with these values of p, q in the equation, we get

$$dz = pdx + qdy$$

or

$$d\sqrt{z^2 - a^2y^2} = adx$$

integrating

$$z^2 - a^2y^2 = (ax+c)^2 \tag{1.84}$$

which is the required complete solution and c being an integrating constant.

To find the general integral, we write $c = \phi(a)$ in (1.84) and hence

$$z^2(ax + \phi(a))^2 + a^2y^2 \tag{1.85}$$

Differentiating (1.85) partially w.r.t. a, we get

$$0 = 2(ax + \phi(a))(x + \phi'(a)) + 2ay^2 \tag{1.86}$$

Eliminating a between (1.85) and (1.86), we get the general integral.

To find the singular integral, we differentiate (1.84) partially w.r.t a and c and get

$$-2ay^2 = 2x(ax + c) \tag{1.87}$$

$$0 = 2(ax + c) \tag{1.88}$$

Eliminating a and c from (1.84), (1.87), and (1.88), we get $z = 0$, which satisfies the given equation and hence it is the singular integral.

1.10.1 *Application of Charpits Method*

The general method given by Charpit for finding the complete integral of a nonlinear first-order PDE is usually quite lengthy and involved. However, there are some special types of first-order nonlinear PDEs whose complete integrals can be easily obtained by Charpit's method. These special types of nonlinear PDEs of first order are called standard forms of Eq. (1.75). We now take up the methods of integrating these standard forms.

Type I: Equation of the type $f(p, q) = 0$

Let the solution be $z = ax + by + c$, where $f(a, b) = 0$. Therefore, $b = \phi(a)$.

The complete solution is $z = ax + \phi(a)y + C$.

Example: *Solve* $\sqrt{p} + \sqrt{q} = 2$

Solution: Let solution be $z = ax + by + c$ where $\sqrt{a} + \sqrt{b} = 1$. Therefore, the complete solution is $z = ax + (1 - \sqrt{a})^2y + c$

Type II: Equation of the type $f(z, p, q) = 0$:

Equation not involving the independent variables, That is, equation of the type

$$f(z, p, q) = 0 \tag{1.89}$$

As a trial solution, let us assume that z is a function of $u = x + ay$, where a is an arbitrary constant. Then,

$$z = f(u) = f(x + ay) \tag{1.90}$$

$$p = \frac{\partial z}{\partial x} = \frac{dz}{du}.\frac{\partial u}{\partial x} = \frac{dz}{du}$$

$$q = \frac{\partial z}{\partial y} = \frac{dz}{du}.\frac{\partial u}{\partial y} = a\frac{dz}{du}$$

Substituting these values of p and q in the given PDE, we get

$$f\left(z, \frac{dz}{du}, a\frac{dz}{du}\right) = 0 \tag{1.91}$$

which is an ODE of first order. Solving Eq. (1.91) for $\frac{dz}{du}$, we obtain

$$\frac{dz}{du} = \phi(z, a)(say)$$

or

$$\frac{dz}{\phi(z, a)} = du$$

On integration, we find

$$\int \frac{dz}{\phi(z, a)} = u + c$$

That is,

$$F(z, a) = u + c = x + ay + c \tag{1.92}$$

which is the complete integral of the given PDE.

Example: *Find a complete integral of the equation* $zpq - p - q = 0$

Solution: Putting $p = aq$ in the above equation, we get $zaq - a - 1 = 0$, or [$p = 0, q = 0$ in which case z=constant]

If $zaq - a - 1 = 0$ then $q = \frac{1+a}{az}$ and $p = \frac{1+a}{z}$.

Now, $dz = pdx + qdy = \frac{1+a}{z}(dx + \frac{1}{a}dy)$, and hence $z^2 = \frac{2(1+a)}{a}(ax+y)+b$.

Type III: Equation of the type $f(x, p) = F(y, q)$:

An equation in which z is absent and the terms containing x and p can be separated from those containing y and q is called a separable equation.

That is, an equation of the type

$$f(x,p) = F(y,q) \tag{1.93}$$

Let's assume that as a trial solution

$$f(x,p) = F(y,q) = a(say) \tag{1.94}$$

Now, solving them for p and q, we obtain $p = \phi(x),\ q = \xi(y)$. Since

$$dz = \frac{\partial z}{\partial x}dx + \frac{\partial z}{\partial y}dy = pdx + qdy$$

or

$$dz = \phi(x)dx + \xi(y)dy$$

On integration, we get complete integral in the form

$$z = \int \phi(x)dx + \int \xi(y)dy + b$$

Example: *Solve:* $p^2 - q^2 = x - y$

Solution: Now, $p^2 - x = q^2 - y = a$.

Therefore, $p = \sqrt{a+x}$ and $q = \sqrt{a+y}$.

Now, $dz = pdx + qdy = \sqrt{a+x}dx + \sqrt{a+y}dy$.

The required solution is $z = \frac{2}{3}(a+x)^{3/2} + \frac{2}{3}(a+y)^{3/2} + c$.

Type IV: Equation of the type $z = px + qy + f(p,q)$

A first-order PDE is referred to as having Clairaut's form if it can be expressed as

$$z = px + qy + f(p,q) \tag{1.95}$$

The corresponding Charpits's equations are

$$\frac{dx}{x+f_p}\frac{dy}{y+f_q} = \frac{dz}{px+qy+pf_p+qf_q} = \frac{dp}{p-p} = \frac{dq}{q-q} \tag{1.96}$$

The integration of the last two equations of (1.96) gives as

$$p = a, \quad q = b$$

We obtain the necessary complete integral in the form of

$$z = ax + by + f(a,b) \tag{1.97}$$

by substituting these values of p and q in the provided PDE.

Example: *Find the complete integral of the PDE* $z = px + qy + 2\log pq$

Solution: The complete integral is $z = ax + by + 2\log ab$.

1.11 Second-Order PDE

Let z be a function of two variables x and y. Consider a general PDE of second order in the form

$$Rr + Ss + Tt + f(x, y, z, p, q) = 0 \tag{1.98}$$

where $p = \frac{\partial z}{\partial x}$, $q = \frac{\partial z}{\partial y}$, $r = \frac{\partial^2 z}{\partial x^2}$, $s = \frac{\partial^2 z}{\partial y^2}$, $t = \frac{\partial^2 z}{\partial x \partial y}$. R, S, and T are continuous functions of x and y and possessing partial derivatives defined in some domain D on the x-y plane. Then Eq. (1.98) is said to be

(i) Elliptic at a point (x, y) in D, if $S^2 - 4RT < 0$. The Laplace equation is the simplest illustration of an elliptic equation.
(ii) Parabolic at a point (x, y) in D, if $S^2 - 4RT = 0$. The simplest example of a parabolic equation is the heat equation.
(iii) Hyperbolic at a point (x, y) in D, if $S^2 - 4RT > 0$. The simplest example of a hyperbolic equation is the wave equation.

Example: *Classify the PDE* $\frac{\partial^2 u}{\partial x^2} + 4\frac{\partial^2 u}{\partial x \partial y} + 4\frac{\partial^2 u}{\partial y^2} = 0$

Solution: The given PDE can be written as

$$r + 4s + 4t = 0$$

Here, $R = 1$, $S = 4$, $T = 4$. So, $S^2 - 4RT = 16 - 16 = 0$.

Hence, the given equation is parabolic at all points.

Matrix Form: The general form of a second-order linear homogeneous PDE with a constant coefficient is

$$au_{xx} + 2bu_{xy} + cu_{yy} + du_x + eu_y + fu = 0 \tag{1.99}$$

Here, a, b, c, d, e, f are real constant coefficients (it's only customary to put the factor 2 in front of the b-term to improve the final PDF's appearance). The type of this equation is defined by its principal part, which consists of the highest-order terms.

$$au_{xx} + 2bu_{xy} + cu_{y}y$$

In matrix notation, this expression may be expressed as

$$\begin{pmatrix} \frac{\partial}{\partial x} & \frac{\partial}{\partial y} \end{pmatrix} \begin{pmatrix} a & b \\ b & c \end{pmatrix} \begin{pmatrix} \frac{\partial}{\partial x} \\ \frac{\partial}{\partial y} \end{pmatrix} u$$

Here, we just pretend that $\frac{\partial}{\partial x}$ and $\frac{\partial}{\partial y}$ are symbols that can be entered as components of a vector. The interpretation is that this vector is applied as the derivative operator on the function $u(x,y)$:

$$\begin{pmatrix} \frac{\partial}{\partial x} & \frac{\partial}{\partial y} \end{pmatrix} \begin{pmatrix} a & b \\ b & c \end{pmatrix} \begin{pmatrix} \frac{\partial}{\partial x} \\ \frac{\partial}{\partial y} \end{pmatrix} u = au_{xx} + 2bu_{xy} + cu_y y$$

For the last equality, we use the assumption that $u(x,y)$ is twice contentiously differentiable in order that the mixed partial derivatives are identical, that is, $u_{xy} = u_{yx}$.

The matrix $\begin{pmatrix} a & b \\ b & c \end{pmatrix}$ is said to be the coefficient matrix of the PDE, or the symbol of the PDE. PDEs are categorized according to the symbol's eigenvalues' relative sign. The classification of PDEs is based on the relative sign of the eigenvalues of the symbol. Note that the symbol is symmetric; hence, the eigenvalues are real. As in linear algebra, the determinant of this matrix is the product of the eigenvalues;

$$\lambda_1 \lambda_2 = det\ A = ac - b^2$$

which we use to define the type of the equation.

The PDE (1.99) is said to be

(i) elliptic if and only if $ac - b^2 > 0$; that is, the eigenvalues of A have the same sign and are not zero, that is, both are positive or both are negative.
(ii) parabolic if and only if $ac - b^2 = 0$; that is, at least one eigenvalue is zero.
(iii) hyperbolic if and only if $ac - b^2 < 0$; that is, the eigenvalues have opposite signs and are not zero.

Now, that we have defined our PDE kingdoms, we introduce their corresponding rulers:

(i) Laplace's equation in two dimensions is given by $u_{xx} + u_{yy} = 0$, and its symbol is $A = \begin{pmatrix} 1 & 0 \\ 0 & 1 \end{pmatrix}$ with determinant $det\ A = 1 > 0$. Hence, Laplace's equation is elliptic.
(ii) The heat equation in one (spatial) dimension is given by $u_t - ku_{xx} = 0$, and its symbol is $A = \begin{pmatrix} k & 0 \\ 0 & 0 \end{pmatrix}$ with determinant $det\ A = 0$. Hence, the heat equation is parabolic.

(iii) The wave equation in one (spatial) dimensions is given by $u_{tt} - c^2 u_{xx} = 0$, and its symbol is $A = \begin{pmatrix} -c^2 & 0 \\ 0 & 1 \end{pmatrix}$ with determinant $det\ A = -c^2 < 0$. Hence, the wave equation is hyperbolic.

Classification of PDEs if $u = \phi(x, yz)$:

We find matrix A as below

$$\begin{pmatrix} C_{xx} & C_{xy} & C_{xz} \\ C_{yx} & C_{yy} & C_{yz} \\ C_{zx} & C_{yz} & C_{zz} \end{pmatrix}$$

where C_{xx} = coefficient of u_{xx} and so on.

(a) If $|A| = 0$, then the equation is parabolic.
(b) If $|A| \neq 0$ and all coefficients are positive, then the equation is elliptic.
(c) If $|A| \neq 0$ and coefficients are positive as well as negative, then the equation is hyperbolic.

Example: *Classify the PDE* $\frac{\partial^2 u}{\partial x^2} + \frac{\partial^2 u}{\partial y^2} - \frac{\partial^2 u}{\partial z^2} = 0$.

Solution: The given PDE can be written as $u_{xx} + u_{yy} - u_{zz} = 0$.

$$A = \begin{pmatrix} 1 & 0 & 0 \\ 0 & 1 & 0 \\ 0 & 0 & -1 \end{pmatrix}$$

$|A| = -1 =\neq 0$ and coefficients are positive as well as positive. Hence, the given equation is hyperbolic type.

Characteristic Equation and Characteristic Curves:

Consider the second-order PDE (1.98). Then, the quadratic equation in λ for Eq. (1.98) is given by

$$R\lambda^2 + S\lambda + T = 0 \tag{1.100}$$

For $S^2 - 4RT \geq 0$ Eq. (1.100) has real roots. Then, the ordinary differential equations

$$\frac{dy}{dx} + \lambda = 0 \tag{1.101}$$

are called characteristic equations. The solutions of Eq. (1.101) are known as characteristic curves or characteristics of the second-order PDEs.

Case I: If $S^2 - 4RT > 0$, i.e., if Eq. (1.98) is hyperbolic, then Eq. (1.100) has two distinct roots λ_1, λ_2 (say) and corresponding to which we get two characteristic equations

$$\frac{dy}{dx} + \lambda_1 = 0, \quad \frac{dy}{dx} + \lambda_2 = 0$$

Solving these we get two distinct families of characteristics.

Case II: If $S^2 - 4RT = 0$ i.e., if Eq. (1.98) is parabolic, and we get characteristic equations

$$\frac{dy}{dx} + \lambda = 0$$

Solving it, we get only one family of characteristics.

Case III: If $S^2 - 4RT < 0$ i.e., if Eq. (1.98) is elliptic, then Eq. (1.100) has complex roots. Hence, there are no real characteristics. We get two families of complex characteristics.

Usually, boundary value problems are associated with elliptic equations and initial value problems arise in connection with hyperbolic and parabolic equations.

Canonical Forms: Consider a differential operator $L \equiv R\frac{\partial^2}{\partial x^2} + S\frac{\partial^2}{\partial x \partial y} + T\frac{\partial^2}{\partial y^2}$. The second-order PDE (1.98) can be expressed as

$$L(z) + f(x, y, z, p, q) = 0 \tag{1.102}$$

Now, change the independent variables x, y to ξ, η, where $\xi = \xi(x, y)$, $\eta = \eta(x, y)$. Also write $z(x, y)$ as $\zeta(\xi, \eta)$. Then $\frac{\partial z}{\partial x} = \frac{\partial \zeta}{\partial \xi}\frac{\partial \xi}{\partial x} + \frac{\partial \zeta}{\partial \eta}\frac{\partial \eta}{\partial x} = \frac{\partial \zeta}{\partial \xi}\xi_x + \frac{\partial \zeta}{\partial \eta}\eta_x$ and $\frac{\partial z}{\partial y} = \frac{\partial \zeta}{\partial \xi}\xi_y + \frac{\partial \zeta}{\partial \eta}\eta_y$.

Therefore,

$$\frac{\partial^2 z}{\partial x^2} = \frac{\partial^2 \zeta}{\partial \xi^2}(\xi_x)^2 + \frac{\partial^2 \zeta}{\partial \xi \partial \eta}(\xi_x \eta_x) + \frac{\partial^2 \zeta}{\partial \xi \partial \eta}(\xi_x \eta_x) + \frac{\partial^2 \zeta}{\partial \eta^2}(\eta_x)^2 \tag{1.103}$$

$$\frac{\partial^2 z}{\partial x \partial y} = \frac{\partial^2 \zeta}{\partial \xi^2}(\xi_x \xi_y) + \frac{\partial^2 \zeta}{\partial \xi \partial \eta}(\xi_y \eta_x) + \frac{\partial^2 \zeta}{\partial \xi \partial \eta}(\xi_x \eta_y) + \frac{\partial^2 \zeta}{\partial \eta^2}(\eta_x \eta_y) \tag{1.104}$$

$$\frac{\partial^2 z}{\partial y^2} = \frac{\partial^2 \zeta}{\partial \xi^2}(\xi_y)^2 + \frac{\partial^2 \zeta}{\partial \xi \partial \eta}(\xi_y \eta_y) + \frac{\partial^2 \zeta}{\partial \xi \partial \eta}(\xi_y \eta_y) + \frac{\partial^2 \zeta}{\partial \eta^2}(\eta_y)^2 \tag{1.105}$$

Multiply (1.103) by R, (1.104) by S, (1.105) by T and add. Then

$$\begin{aligned}
&R\frac{\partial^2 z}{\partial x^2} + S\frac{\partial^2 z}{\partial x \partial y} + T\frac{\partial^2 z}{\partial y^2} \\
&= L(z) \\
&= \frac{\partial^2 \zeta}{\partial \xi^2}[R\xi_x^2 + S\xi_x\xi_y + T\xi_y^2] + \frac{\partial^2 \zeta}{\partial \xi \partial \eta}[2R\xi_x\eta_x + S(\xi_x\eta_y + \xi_y\eta_x) \\
&\quad + 2T\xi_y\eta_y] + \frac{\partial^2 \zeta}{\partial \eta^2}[R\eta_x^2 + S\eta_x\eta_y + T\eta_y^2] \\
&= A(\xi_x, \xi_y)\frac{\partial^2 \zeta}{\partial \xi^2} + 2B(\xi_x, \xi_y; \eta_x, \eta_y)\frac{\partial^2 \zeta}{\partial \xi \partial \eta} + A(\eta_x, \eta_y)\frac{\partial^2 \zeta}{\partial \eta^2}
\end{aligned}$$

where $A(u, v) = Ru^2 + Suv + Tv^2$ and $B(u_1, v_1; u_2, v_2) = Ru_1u_2 + \frac{1}{2}S(u_1v_2 + u_2v_1) + Tv_1v_2$.

Equation (1.102) becomes

$$\begin{aligned}
&A(\xi_x, \xi_y)\frac{\partial^2 \zeta}{\partial \xi^2} + 2B(\xi_x, \xi_y; \eta_x, \eta_y)\frac{\partial^2 \zeta}{\partial \xi \partial \eta} + A(\eta_x, \eta_y)\frac{\partial^2 \zeta}{\partial \eta^2} \\
&\quad = F(\xi, \eta, \zeta, \zeta_\xi, \zeta_\eta) \qquad (1.106)
\end{aligned}$$

Also

$$A(\xi_x, \xi_y)A(\eta_x, \eta_y) - B^2(\xi_x, \xi_y; \eta_x, \eta_y) = \frac{1}{4}(4RT - S^2)(\xi_x\eta_y - \xi_y\eta_x)^2 \qquad (1.107)$$

on simplification. To determine ξ and η so that Eq. (1.106) is reduced to the simplest form. This we do with the help of the discriminant $S^2 - 4RT$.

Case I: Suppose $S^2 - 4RT > 0$. Then the roots α_1, α_2 of the equation

$$R\alpha^2 + S\alpha + T = 0 \qquad (1.108)$$

are real and distinct. Note that α_1, α_2 are, in general, the function of x and y. Choose ξ, η in such a way that $\xi(x, y) = c_1$ and $\eta(x, y) = c_2$ are solutions of the first-order ODE

$$\frac{dy}{dx} + \alpha_1(x, y) = 0, \quad \frac{dy}{dx} + \alpha_2(x, y) = 0 \qquad (1.109)$$

respectively. Since $\xi(x, y) = c_1$, we have $\xi_x + \xi_y\frac{dy}{dx} = 0$ i.e., $\frac{dy}{dx} = -\frac{\xi_x}{\xi_y}$. But $\xi = c_1$ is a solution of $\frac{dy}{dx} + \alpha_1 = 0$. Therefore, $\xi_x = \alpha_1\xi_y$. Similarly, $\eta_x = \alpha_2\eta_y$. Now for this choice of ξ and η, $A(\xi_x, \xi_y) = R\xi_x^2 + S\xi_x\xi_y +$

$T\xi_y^2 = (R\alpha_1^2 + S\alpha_1 + T)\xi_y^2 = 0$ (as α_1 is a root of Eq. (1.108)). Similarly, $A(\eta_x, \eta_y) = 0$. Thus, Eq. (1.106) becomes

$$2B(\xi_x, \xi_y; \eta_x, \eta_y)\frac{\partial^2 \zeta}{\partial \xi \partial \eta} = F(\xi, \eta, \zeta, \zeta_\xi, \zeta_\eta) \tag{1.110}$$

Using $A(\xi_x, \xi_y) = 0 = A(\eta_x, \eta_y)$ in Eq. (1.109), $B^2 = (S^2 - 4RT)(\xi_x\eta_y - \xi_y\eta_x)^2$ by hypothesis $S^2 - 4RT > 0$. Therefore, $B^2 > 0$ and hence $B \neq 0$. Now divide both sides of Eq. (1.111) by $2B$. Then

$$\frac{\partial^2 \zeta}{\xi \eta} = \frac{F(\xi, \eta, \zeta, \zeta_\xi, \zeta_\eta)}{2B(\xi_x, \xi_y; \eta_x, \eta_y)} = \phi(\xi, \eta, \zeta, \zeta_\xi, \zeta_\eta) \quad (say)$$

Thus, Eq. (1.102) gets reduced to the form

$$\frac{\partial^2 \zeta}{\partial \xi \partial \eta} = \phi(\xi, \eta, \zeta, \zeta_\xi, \zeta_\eta) \tag{1.111}$$

This form is known as the *canonical form* for the given second-order PDE (1.98). In this case, we say that the given differential equation is hyperbolic.

Case II: Suppose $S^2 - 4RT = 0$. Then the roots of Eq. (1.108) are equal and real (to say α_1). Choose ξ so that $\xi(x, y) = c_1$ is a solution of the equation

$$\frac{dy}{dx} + \alpha_1(x, y) = 0 \tag{1.112}$$

Then $A(\xi_x, \xi_y) = 0$. Using this and $S^2 - 4RT = 0$ in (1.107), we get $B(\xi_x, \xi_y; \eta_x, \eta_y) = 0$. Let η be any function of x, y in dependent of ξ. If $A(\eta_x, \eta_y) = 0$ then $R\eta_x^2 + S\eta_x\eta_y + T\eta_y^2 = 0$, so that $\frac{\eta_x}{\eta_y}$ will be a root of Eq. (1.112). But the roots of (1.112) are equal to $\alpha_1 = \frac{\xi_x}{\xi_y}$. Therefore, $\frac{\eta_x}{\eta_y} = \frac{\xi_x}{\xi_y}$. i.e., $\xi_x\eta_y - \xi_y\eta_x = 0$. Or $\frac{\partial(\xi,\eta)}{\partial(x,y)} = 0$ i.e., η is a function of xi, a contradiction. Thus,

$A(\eta_x, \eta_y) \neq 0$, so that Eq. (1.108) becomes

$$\frac{\partial^2 \zeta}{\eta^2} = \frac{F(\xi, \eta, \zeta, \zeta_\xi, \zeta_\eta)}{A(\eta_x, \eta_y)} = \phi(\xi, \eta, \zeta, \zeta_\xi, \zeta_\eta) \quad (say)$$

Thus, Eq. (1.102) gets reduced to the form

$$\frac{\partial^2 \zeta}{\partial \eta^2} = \phi(\xi, \eta, \zeta, \zeta_\xi, \zeta_\eta) \tag{1.113}$$

This form is known as the *canonical form* for the given second-order PDE (1.98). In this case, we say that the given differential equation is parabolic.

Case III: Suppose $S^2 - 4RT < 0$. Then the roots of Eq. (1.108) are complex conjugates. So, ξ, η are also complex conjugates. Consider the transformation $\alpha = \frac{1}{2}(\eta + \xi)$, $\beta = \frac{1}{2}i(\eta - \xi)$. Then $\alpha + i\beta = \xi$, $\alpha - i\beta = \eta$. Therefore, $\frac{\partial \alpha}{\partial \xi} = \frac{1}{2}$, $\frac{\partial \alpha}{\partial \eta} = \frac{1}{2}$, $\frac{\partial \beta}{\partial \xi} = -\frac{1}{2}i$, $\frac{\partial \beta}{\partial \eta} = \frac{1}{2}i$.

Now, $\frac{\partial \zeta}{\partial \xi} = \frac{1}{2}(\frac{\partial \zeta}{\partial \alpha} - i\frac{\partial \zeta}{\partial \beta})$. Therefore, $\frac{\partial^2 \zeta}{\partial \xi \partial \eta} = \frac{1}{4}(\frac{\partial^2 \zeta}{\partial \alpha^2} + \frac{\partial^2 \zeta}{\partial \beta^2})$.

Thus, the given second-order PDE (1.98) reduces to the Canonical form.

$$\frac{\partial^2 \zeta}{\partial \alpha^2} + \frac{\partial^2 \zeta}{\partial \beta^2} = \phi(\xi, \eta, \zeta, \zeta_\xi, \zeta_\eta) \tag{1.114}$$

In this case, we say that the given differential equation is elliptic.

1.12 Method of Separation of Variables

One method for solving several kinds of ODEs is the separation of variables method. Leibnitz discovered this method of solving equations. He also developed the process of solving the homogeneous differential equations of the first order. Additionally, he created a method to solve first-order homogeneous differential equations. Almost all linear homogeneous PDEs with constant coefficients in canonical form are amenable to the separation method. Initial boundary value problems involving linear PDEs are frequently solved using this approach in conjunction with the principle of the superposition concept. The fundamental idea behind this approach is to break down a PDE into as many ODEs as there are independent variables by expressing the solution as a product of the functions of each independent variable. After these ODEs are solved, the method reduces to solving eigenvalue problems and constructing the general solution as an eigenfunction expansion where the coefficients are evaluated by using the boundary conditions and the initial conditions. Most of the time, the solution is expressed as a set of orthogonal functions.

Wave Equation:

The wave equation is a fundamental PDE that explains how waves behave, including sound waves, light waves, and other wave-like phenomena. It is used to simulate how waves move through a medium. Typically, the one-dimensional wave equation is expressed as

$$\frac{\partial^2 u}{\partial t^2} = c^2 \frac{\partial^2 u}{\partial x^2} \tag{1.115}$$

where x stands for the spatial or "position" variable, t for the "time" variable, $u = u(x,t)$ for the unknown function, and c for a known positive constant. The vibrating string equation, often known as the wave equation, describes how a string vibrates. Physically, the "value" of a particle's normal displacement at position x and time t is represented as $u(x,t)$.

The string can be fixed at both ends, or just at one end, or we can think of an infinite string, that is not bounded at any end. Each will yield different boundary conditions for the well-posed wave equation. We can also consider the case where the string is "pushed" with an external force $h(x,t)$, or where we take into consideration the friction coefficients from the air that the string displaces. These two equations will be called "forced" and "damped," respectively. In the forced case, the wave equation is

$$\frac{\partial^2 u}{\partial t^2} = c^2\frac{\partial^2 u}{\partial x^2} + h(x,t)$$

where an example of the acting force is the gravitational force. In the damped case, the equation will look like

$$\frac{\partial^2 u}{\partial t^2} + k\frac{\partial u}{\partial t} = c^2\frac{\partial^2 u}{\partial x^2}$$

In the case of the vibrating membrane, we have two spatial variables, and the wave equation will look like

$$\frac{\partial^2 u}{\partial t^2} = c^2\left(\frac{\partial^2 u}{\partial x^2} + \frac{\partial^2 u}{\partial y^2}\right)$$

In three dimensions, the wave equation becomes

$$\frac{\partial^2 u}{\partial t^2} = c^2\left(\frac{\partial^2 u}{\partial x^2} + \frac{\partial^2 u}{\partial y^2} + \frac{\partial^2 u}{\partial z^2}\right)$$

The initial conditions for the one-dimensional wave equation (1.115) will be

$$u(x,0) = f(x), \;\; u_t(x,0) = g(x)$$

It is interesting to note here that two initial conditions should be prescribed, namely the initial displacement $u(x,0) = f(x)$ and the initial velocity $u_t(x,0) = g(x)$ that describes the initial displacement and the initial velocity at the string time $t = 0$, respectively.

Typically, we impose boundary conditions in one of the following three forms:

(i) **Dirichlet boundary conditions**: When the ends of the string are specified, we use the boundary conditions of the form

$$u(0,t) = f_1(t), \quad u(L,t) = f_2(x), \ \ t > 0$$

(ii) **Neumann boundary conditions**: The vertical forces on the string at the endpoints are given by $Tu_x(0,t)$ and $-Tu_x(0,t)$, where T is the tension in the string. By specifying these forces, we obtain Neumann boundary conditions. For example, if the ends of the string are allowed to slide vertically on frictionless sleeves, the boundary conditions become

$$u_x(0,t) = 0, \quad u_x(L,t) = 0, \ \ t > 0$$

(iii) **Robin boundary conditions**: Suppose that the ends of the string are attached to the springs. Then, the vertical forces on the string are proportional to the displacements, with a constant of proportionality given by the spring constant h. This yields the mixed boundary conditions

$$Tu_x(0,t) = hu(0,t), \quad -Tu_x(L,t) = hu(L,t), \ \ t > 0.$$

Derivation:

Let us consider a streched string of length L, which is fixed at both ends. It is assumed that (i) the string is thin and flexible. i.e., it offers no resistance to change of form except a change in length, and (ii) the tension T in the string is much larger than the force due to gravity acting on it so that the latter can be neglected. Let the string in its equilibrium state be situated along the x-axis. Let $u(x,t)$ denote the displacement of the string at time t from its equilibrium position. The shape of the string at a fixed t is represented in Fig. 1.1.

Let us further assume that the vibrations are small, which implies that the displacement $u(x,t)$ and its derivative u_x are small enough so that their square and products can be neglected. As a result of vibrations, let a segment (x_1, x_2) of the string be deformed into the AB. Then, at time t, the length of the arc $\widehat{AB}$ is given by

$$\int_{x_1}^{x_2} \sqrt{1+u_x^2}dx \approx x_2 - x_1 \tag{1.116}$$

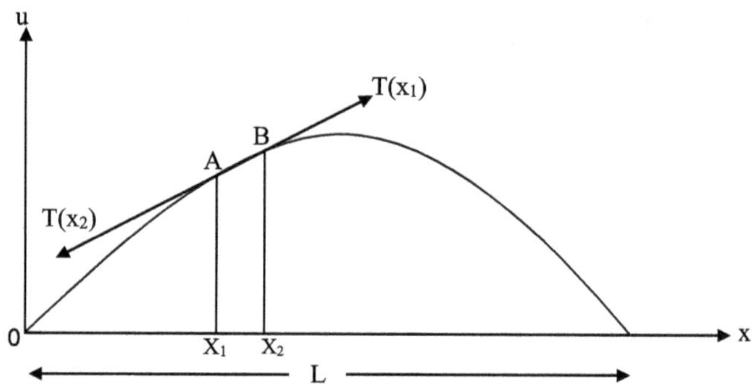

Figure 1.1. Vibrations of string of length L.

which simply means that under small vibrations, the length of the segment of the string does not change. By Hooke's law, the tension T at each point in the string is independent of t, i.e., during the motion of the string, any change in T can be neglected in comparison with the tension in equilibrium. We shall now know that the tension T is also independent of x. In fact, it is evident from Fig. 1.1 that the x-component of the resulting tension at the points A and B must be in equilibrium, i.e.,

$$T(x_1)\cos\alpha(x_1) - T(x_2)\cos\alpha(x_2) = 0$$

where $\alpha(x)$ denotes the angle between the tangent at a point x and the positive x-axis at time t. Since the vibrations are small,

$$\cos\alpha(x) = \frac{1}{\sqrt{1+\tan^2\alpha(x)}} = \frac{1}{\sqrt{1+u_x^2}} \approx 1,$$

which implies that $T(x_1) \approx T(x_2)$. Since x_1 and x_2 are arbitrary, the magnitude of T is independent of x. Hence, if T_0 denotes the tension at equilibrium and T the tension in the vibrating string, then $T \approx T_0$ for all x and t.

Now, the sum of the components of tension $T(x_1)$ at A and $T(x_2)$ at B along the u-axis must be zero, i.e.,

$$\begin{aligned}
0 &= T_0[\sin\alpha(x_2) - \sin\alpha(x_1)) \\
&= T_0\left[\frac{\tan\alpha(x_2)}{\sqrt{1+\tan^2\alpha(x_2)}} - \frac{\tan\alpha(x_1)}{\sqrt{1+\tan^2\alpha(x_1)}}\right] \\
&= T_0\left[\frac{u_{x_2}}{\sqrt{1+u_{x_2}^2}} - \frac{u_{x_1}}{\sqrt{1+u_{x_1}^2}}\right]
\end{aligned}$$

$$\begin{aligned}
&\approx T_0\left[\frac{\partial u}{\partial x_2}-\frac{\partial u}{\partial x_1}\right]\\
&=T_0\left[\left.\frac{\partial u}{\partial x}\right|_{x=x_2}-\left.\frac{\partial u}{\partial x}\right|_{x=x_1}\right]\\
&=T_0\int_{x_1}^{x_2}\frac{\partial^2 u}{\partial x^2}dx, \qquad (1.117)
\end{aligned}$$

using (1.116). Let $g(x,t)$ denote the external force per unit mass acting on the string along the u-axis. Then the component of $g(x,t)$ acting on the segment $\widehat{AB}$ along the u-axis is given by

$$\int_{x_1}^{x_2} g(x,t)dx \qquad (1.118)$$

Let ρ be the linear density of the string. Then the inertial force on the segment $\widehat{AB}$ is

$$-\int_{x_1}^{x_2} \rho(x)\frac{\partial^2 u}{\partial t^2}dx \qquad (1.119)$$

Hence, the sum of the components (1.117), (1.118), and (1.119) must be zero, i.e.,

$$\int_{x_1}^{x_2}\left[T_0\frac{\partial^2 u}{\partial x^2}+g(x,t)-\rho(x)\frac{\partial^2 u}{\partial t^2}\right]dx=0 \qquad (1.120)$$

Since x_1 and x_2 are arbitrary, it follows from (1.120) that the integrand must be zero, which gives

$$\rho(x)\frac{\partial^2 u}{\partial t^2}=T_0\frac{\partial^2 u}{\partial x^2}+g(x,t) \qquad (1.121)$$

This represents the PDE for the vibrations of the string.

If $\rho=$ constant, then (1.121) reduces to

$$\frac{\partial^2 u}{\partial t^2}=a^2\frac{\partial^2 u}{\partial x^2}+f(x,t) \qquad (1.122)$$

where $a=\sqrt{T_0/\rho}$, $f(x,t)=g(x,t)/\rho$. In the absence of external forces (1.122) becomes

$$\frac{\partial^2 u}{\partial t^2}=a^2\frac{\partial^2 u}{\partial x^2} \qquad (1.123)$$

which is the wave equation for free vibrations (oscillations) of the string.

Solve by the method of Separation:

Assume the solution of Eq. (1.115) is of the form

$$u(x,t) = X(x)T(t) \tag{1.124}$$

From (1.115), we get

$$XT'' = c^2 X'' T$$

$$\Rightarrow \frac{X''}{X} = \frac{1}{c^2}\frac{T''}{T} = \lambda \quad (say) \tag{1.125}$$

Therefore,

$$X'' - \lambda X = 0 \tag{1.126}$$

$$T'' - \lambda c^2 T = 0 \tag{1.127}$$

Case I: When $\lambda = 0$. The general solution of Eqs. (1.126) and (1.127) are

$$X = a_1 x + a_2, \; T = a_3 t + a_4$$

respectively.

Now, by using Eq. (1.124) solution of Eq. (1.115) is

$$u(x,t) = (a_1 x + a_2)(a_3 t + a_4) \tag{1.128}$$

Case II: When $\lambda > 0$. Let $\lambda = \mu^2$. Then the general solution of Eq. (1.126) and (1.127) are respectively

$$X = b_1 e^{\mu x} + b_2 e^{-\mu x}, \quad T = b_3 e^{c\mu t} + b_4 e^{-c\mu t}$$

The solution of Eq. (1.115) in this case is

$$u(x,t) = (b_1 e^{\mu x} + b_2 e^{-\mu x})(b_3 e^{c\mu t} + b_4 e^{-c\mu t}) \tag{1.129}$$

Case III: When $\lambda < 0$. Let $\lambda = -\mu^2$. Then, the general solution of Eqs. (1.126) and (1.127) are respectively

$$X = C_1 \cos(\mu x) + C_2 \sin(\mu x), \quad T = C_3 \cos(c\mu t) + C_4 \sin(c\mu t)$$

The solution of Eq. (1.115) in this case is

$$u(x,t) = (C_1 \cos(\mu x) + C_2 \sin(\mu x))(C_3 \cos(c\mu t) + C_4 \sin(c\mu t)) \tag{1.130}$$

Thus, the various possible solution of wave equation (1.126) are given, by Eqs. (1.128)–(1.130) of these three solutions, we will select the solution which is suitable as per the nature of the problem.

Heat Equation:

The diffusion equation is a PDE, which describes density fluctuations in a material undergoing diffusion. The equation can be written as

$$\frac{\partial u(\mathbf{r},t)}{\partial t} = \nabla.(D(u(\mathbf{r},t)\nabla u(\mathbf{r},t)) \tag{1.131}$$

where $u(\mathbf{r},t))$ is the density of the diffusing material at location $\mathbf{r} = (x, y, z)$ and time t. $D(u(r,t))$ denotes the collective diffusion coefficient for density u at location $\mathbf{r}$. If the diffusion coefficient does not depend on the density, i.e., D is constant, Eq. (1.131) reduces to the following linear equation

$$\frac{\partial u(\mathbf{r},t)}{\partial t} = D\nabla^2 u(\mathbf{r},t) \tag{1.132}$$

Equation (1.132) is also called the heat equation and also describes the distribution of heat in a given region over time.

Derivation: The heat equation (1.132) can be derived in a straightforward way from the continuity equation, which states that a change in density in any part of the system is due to the inflow and outflow of material into and out of that part of the system. Effectively, no material is created or destroyed.

$$\frac{\partial u}{\partial t} + \nabla.\Gamma = 0$$

where Γ is the flux of the diffusing material. Equation (1.132) can be obtained easily from the last equation when combined with the phenomenological Fick's first law, which assumes that the flux of the diffusing material in any part of the system is proportional to the local density gradient:

$$\Gamma = -D\nabla u(\mathbf{r},t)$$

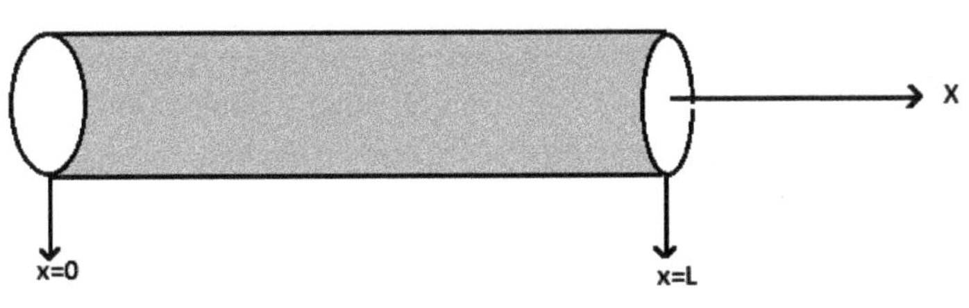

Figure 1.2. A Uniform rod of length L.

We consider the physical problem of heat conduction in a rod of length L situated on the x axis with one end at the origin and the other at $x = L$. We assume that heat is only transferred along the rod and not laterally to the surroundings, i.e., a thermally insulated rod. The rod can possibly exchange heat from its boundary points $x = 0$ and $x = L$. We also assume that the thermal properties of the rod are independent of x and t. Let $u(x,t)$ denote the temperature of the point x at time t. $u(x,t)$ satisfy the PDE

$$u_t = ku_{xx}, \quad 0 < x < L, \ t > 0$$

where k is the thermal diffusivity of the material that measures the rod ability to heat conduction.

To determine u, we must specify the temperature at every point in the rod when $t = 0$, say

$$u(x,0) = f(x), \quad 0 \leq x \leq L$$

This is the initial condition.

We must also specify boundary conditions that u must satisfy at the ends of the rod for all $t > 0$.

Dirichlet boundary conditions: One form of the boundary condition is given by the Dirichlet boundary conditions.

$$u(0,t) = a_1, \ t \geq 0; \quad u(L,t) = a_2, \ t \geq 0$$

where we assume that a_1, a_2 are constant values independent of t. If $a_1 = 0, a_2 = 0$, the heat problem is called a homogeneous problem. Note that in this case, there is no thermal source at the boundary or along the rod. Therefore, the boundary conditions $u(0,t) = 0$, $t \geq 0$; $u(L,t) = 0$, $t \geq 0$ indicate that the ends of the rod are kept at 0 temperature.

Neumann boundary condition: One form of the boundary condition is given by the Neumann boundary conditions

$$\frac{\partial u}{\partial x}(0,t) = 0, \quad \frac{\partial u}{\partial x}(L,t) = 0$$

The given boundary conditions indicate that the net heat flow through the end point is zero. Equivalently, end points can be considered insulated so that it does not allow any heat escape or enter through these points.

Robin's boundary conditions: The general form of Robin's boundary condition is

$$au(0,t) + b\frac{\partial u}{\partial x}(0,t) = 0, \quad cu(L,t) + d\frac{\partial u}{\partial x}(L,t) = 0$$

where a, b, c, d can not be simultaneously zero. Of important cases are when one side is in Dirichlet condition and the other side is in Neumann condition, like $u(x,t) = 0$ and $\frac{\partial u}{\partial x}(L,t) = 0$.

Now, consider the initial boundary value problem

$$u_t = ku_{xx}$$

with boundary condition $u(0,t) = 0,\ t \geq 0;\ u(L,t) = 0,\ t \geq 0$ and initial condition $u(x,0) = f(x),\ 0 \leq x \leq L$.

It is of interest to note that the above PDE arises in two different types, namely,

(i) **Homogeneous heat equation:** This type of equation is often given by

$$u_t = ku_{xx}, \quad 0 < x < L,\ t > 0$$

Further, a heat equation with a lateral heat loss is formally derived as a homogeneous PDE of the form

$$u_t = ku_{xx} - u, \quad 0 < x < L,\ t > 0$$

(ii) **Inhomogeneous heat equation:** This type of equation is often given by

$$u_t = ku_{xx} + g(x), \quad 0 < x < L,\ \ t > 0$$

where $g(x)$ is called the heat source which is independent of time.

Solve by the method of separation:

Let us consider one dimensional wave equation

$$\frac{\partial u}{\partial t} = k\frac{\partial^2 u}{\partial x^2} \tag{1.133}$$

Assume the solution of Eq. (1.133) is of the form

$$u(x,t) = X(x)T(t) \tag{1.134}$$

From (1.133), we get

$$XT' = kX''T$$

$$\Rightarrow \frac{X''}{X} = \frac{T'}{kT} = \lambda \quad (say) \tag{1.135}$$

Therefore,

$$X'' - \lambda X = 0 \tag{1.136}$$

$$T' - \lambda kT = 0 \tag{1.137}$$

Case I: When $\lambda = 0$. The general solution of Eqs. (1.136) and (1.137) are

$$X = a_1 x + a_2, \; T = a_3$$

respectively.

Now, by using Eq. (1.134) solution of Eq. (1.133) is

$$u(x,t) = (a_1 x + a_2)a_3 = A_1 x + A_2 \tag{1.138}$$

Case II: When $\lambda > 0$. Let $\lambda = \mu^2$. Then the general solution of Eqs. (1.136) and (1.137) are respectively

$$X = b_1 e^{\mu x} + b_2 e^{-\mu x}, \quad T = b_3 e^{k\mu^2 t}$$

The solution of Eq. (1.133) in this case is

$$u(x,t) = (B_1 e^{\mu x} + B_2 e^{-\mu x})e^{k\mu^2 t} \tag{1.139}$$

Case III: When $\lambda < 0$. Let $\lambda = -\mu^2$. Then, the general solution of Eqs. (1.136) and (1.137) are respectively

$$X = C_1 \cos(\mu x) + C_2 \sin(\mu x), \quad T = C_3 e^{-\mu^2 kt}$$

The solution of Eq. (1.133) in this case is

$$u(x,t) = (D_1 \cos(\mu x) + D_2 \sin(\mu x))e^{-\mu^2 kt} \tag{1.140}$$

Thus, the various possible solution of heat equation (1.136) are given, by Eqs. (1.138)–(1.140) of these three solutions, we will select the solution which is suitable as per the nature of the problem.

Laplace Equation:

The Laplace equation is a PDE named after the French mathematician Pierre Simon Marquis de Laplace (1749–1827). The Laplace equation, which has several applications in mathematical physics and engineering, is the most significant mathematical equation. The Laplace equation can be used to formulate the general class of issues related to the theories of gravitation, electrostatics, dielectrics, and magnetostatics in the field of mathematical physics. The accompanying fields in each of these applications can be described as a gradient of potential. The potential equation is another name for the Laplace equation. The Laplace equation is

$$\Delta u = 0$$

The operator Δ is called the Laplacian.

In two dimensions

$$\Delta u = \nabla^2 u = u_{xx} + u_{yy} = 0$$

Also in three dimensions

$$\Delta u = \nabla^2 u = u_{xx} + u_{yy} + u_{zz} = 0$$

The solutions of the Laplace equation are called harmonic functions.

The inhomogeneous Laplace equation

$$\Delta u = f$$

where f is a given function. This equation is known as the Poisson equation.

The evolution of temperature and displacement are studied, respectively, by the wave and heat equations. It is important to remember that the Laplace equation explains physical phenomena at equilibrium. Additionally, since the Laplace equation's solution is independent of time t, no initial condition is necessary, and working with the Laplace equation leaves us with a pure boundary value problem to solve. The boundary conditions associated with Laplace's equation can be identified into three types of boundary conditions, namely:

Dirichlet boundary conditions problem:

$$\Delta u = 0 \quad \text{in } \Omega$$

with boundary condition $u = f$ on $\partial\Omega$. Here, f be a continuous function on $\partial\Omega$.

Neumann boundary conditions problem:

$$\Delta u = 0 \quad \text{in } \Omega$$

with $\frac{\partial u}{\partial n} = g$ on $\partial\Omega$. Here, g be contentious function on $\partial\Omega$ and $\vec{n}$ is the outward unit normal to $\partial\Omega$. Also $\frac{\partial u}{\partial n} = \nabla u.\vec{n}$

Robin boundary conditions problem:

$$\Delta u = 0 \quad \text{in } \Omega$$

with $hu + \frac{\partial u}{\partial n} = g$ on $\partial\Omega$. Here, g and h are continuous function on $\partial\Omega$.

Derivation: By Gauss' law of electrostatics, we know that the flux of the electric vector $\mathbf{E}$ out of a surface S bounding an arbitrary volume V is 4π

times the charge contained in V. Thus, if ρ is the density of electric charge, we have

$$\int_S \mathbf{E}.dS = 4\pi \int_V \rho \, d\tau$$

Using Green's theorem in the form

$$\int_S \mathbf{E}.dS = \int_V \text{div } \mathbf{E} \, d\tau$$

and remembering that the volume V is arbitrary, we see that Gauss' law is equivalent to the equation

$$\text{div } \mathbf{E} = 4\pi\rho \tag{1.141}$$

Now, it is readily shown that the electric field is characterized by the fact that the vector $\mathbf{E}$ is derivable from a potential function ϕ by the equation

$$\mathbf{E} = -\text{grad } \phi \tag{1.142}$$

Eliminating $\mathbf{E}$ from (1.141) and (1.142), we find that ϕ satisfies the equation

$$\nabla^2\phi + 4\pi\rho = 0 \tag{1.143}$$

where we have written ∇^2 for the operator div(grad), which in rectangular Cartesian coordinates takes the form

$$\frac{\partial^2}{\partial x^2} + \frac{\partial^2}{\partial y^2} + \frac{\partial^2}{\partial z^2}$$

Equation (1.143) is known as Poisson's equation. In the absence of charges, ρ is zero, and Eq. (1.143) reduces to the simple form

$$\nabla^2\phi = 0 \tag{1.144}$$

This equation is known as a Laplace equation or a harmonic equation. If we are dealing with a problem in which the potential function ϕ does not vary in z, we then find that ∇ is replaced by

$$\frac{\partial^2}{\partial x^2} + \frac{\partial^2}{\partial y^2}$$

and that Laplace's equation becomes

$$\frac{\partial^2}{\partial x^2} + \frac{\partial^2}{\partial y^2} = 0 \tag{1.145}$$

a from which shall refer to as the two-dimensional harmonic equation.

Solve by the method of separation:

Let us consider one-dimensional wave equation

$$\frac{\partial^2 u}{\partial x^2} + \frac{\partial^2 u}{\partial y^2} = 0 \tag{1.146}$$

Assume the solution of Eq. (1.146) is of the form

$$u(x, Y) = X(x)Y(y) \tag{1.147}$$

From (1.146), we get

$$X''Y + XY'' = 0$$

$$\Rightarrow \frac{X''}{X} = -\frac{Y''}{Y} = \lambda \quad (say) \tag{1.148}$$

Therefore,

$$X'' - \lambda X = 0 \tag{1.149}$$

$$Y'' + \lambda Y = 0 \tag{1.150}$$

Case I: When $\lambda = 0$. The general solution of Eqs. (1.149) and (1.150) are

$$X = a_1 x + a_2, \ Y = a_3 y + a_4$$

respectively.

Now, by using Eq. (1.147) solution of Eq. (1.146) is

$$u(x, y) = (a_1 x + a_2)(a_3 y + a_4) \tag{1.151}$$

Case II: When $\lambda > 0$. Let $\lambda = \mu^2$. Then the general solution of Eqs. (1.149) and (1.150) are respectively

$$X = b_1 e^{\mu x} + b_2 e^{-\mu x}, \quad Y = b_3 \cos(\mu y) + b_4 \sin(\mu y)$$

Solution of Eq. (1.146) in this case is

$$u(x, y) = (b_1 e^{\mu x} + b_2 e^{-\mu x})(b_3 \cos(\mu y) + b_4 \sin(\mu y)) \tag{1.152}$$

Case III: When $\lambda < 0$. Let $\lambda = -\mu^2$. Then, the general solution of Eqs. (1.149) and (1.150) are respectively

$$X = C_1 \cos(\mu x) + C_2 \sin(\mu x), \quad Y = C_3 e^{\mu y} + C_4 e^{-\mu y}$$

The solution of Eq. (1.146) in this case is

$$u(x, y) = (C_1 \cos(\mu x) + C_2 \sin(\mu x))(C_3 e^{\mu y} + C_4 e^{-\mu y}) \tag{1.153}$$

Thus, the various possible solution of Laplace Eq. (1.149) are given, by Eqs. (1.151)–(1.153) of these three solutions, we will select the solution which is suitable as per the nature of the problem.

1.13 d'Alembert's Solution of Wave Equation

A solution to the one-dimensional equation can be found using the d'Alembert method. Consider a one-dimensional wave equation

$$\frac{\partial^2 y}{\partial x^2} = \frac{1}{c^2}\frac{\partial^2 y}{\partial t^2} \tag{1.154}$$

that describes the vibrations of a string. The general solution can be obtained by introducing new variables $\xi = x - ct$ and $\eta = x + ct$, and applying the chain rule to obtain

$$\frac{\partial}{\partial x} = \frac{\partial}{\partial \xi} + \frac{\partial}{\partial \eta}, \quad \frac{\partial}{\partial t} = -c\frac{\partial}{\partial \xi} + c\frac{\partial}{\partial \eta},$$
$$\frac{\partial^2}{\partial x^2} = \frac{\partial^2}{\partial \xi^2} + 2\frac{\partial^2}{\partial \xi \partial \eta} + \frac{\partial^2}{\partial \eta^2}, \quad \frac{\partial^2}{\partial t^2} = c^2\frac{\partial^2}{\partial \xi^2} - 2c^2\frac{\partial^2}{\partial \xi \partial \eta} + c^2\frac{\partial^2}{\partial \eta^2} \tag{1.155}$$

Using (1.155) in (1.154), we get

$$\frac{\partial^2 y}{\partial \xi \partial \eta} = 0 \tag{1.156}$$

The general solution of the PDE (1.156) is

$$y(x,t) = F(\xi) + G(\eta) = F(x - ct) + G(x + ct) \tag{1.157}$$

where F indicates a right-traveling wave and G a left-traveling wave, and F and G are arbitrary functions.

Let's assume the following initial condition as follows

$$y(x,0) = f(x), \quad y_t(x,0) = g(x), \quad -\infty < x < \infty \tag{1.158}$$

where the string's initial location is $y = f(x)$ and its initial velocity is $g(x)$ at the point x.

$$F(x) + G(x) = f(x) \tag{1.159}$$

$$-cF'(x) + cG'(x) = g(x) \tag{1.160}$$

Solving for $f(x)$ and $G(x)$, we get

$$F(x) = \frac{1}{2}f(x) - \frac{1}{2c}\int_{x_0}^{\infty} g(x)dx,$$
$$G(x) = \frac{1}{2}f(x) + \frac{1}{2c}\int_{x_0}^{\infty} g(x)dx$$

Therefore,

$$y(x,t) = \frac{1}{2}[f(x-ct) + f(x+ct)] + \frac{1}{2c}\int_{x-ct}^{x+ct} g(s)ds \qquad (1.161)$$

This is d'Alembert's solution to the wave equation. $\frac{1}{2}[f(x-ct)+f(x+ct)]$ indicates the propagation of the initial displacement without the initial velocity, i.e., $g(x) = 0$, and $\frac{1}{2c}\int_{x-ct}^{x+ct} g(s)ds$ is the initial velocity (impulse) when there is no initial displacement, i.e., when $f(x) = 0$.

In physics, a function of type $f(x-ct)$ is a propagation wave that travels forward with velocity c, and one of type $f(x+ct)$ travels backward with velocity c. Also, the wave function $f(x)$ propagates along the straight lines $y = x + ct$ and $y = x - ct$, which are referred to as characteristics.

1.14 Higher-Order Linear PDE

The PDE of higher order can be expressed as

$$F\left(\frac{\partial^n z}{\partial x^n}, \frac{\partial^n z}{\partial x^{n-1}\partial y}, \ldots, \frac{\partial^n z}{\partial y^n}, \frac{\partial^{n-1} z}{\partial x^{n-1}}, \ldots, \frac{\partial z}{\partial x}, \frac{\partial z}{\partial y}, x, y, z\right) = 0 \quad (1.162)$$

where z is the dependent variable and x, y are the two independent variables. A higher-order linear PDE is any PDE of higher order in which each partial derivative involved has a degree of one. Consider a higher-order linear PDE

$$P_0\frac{\partial^n z}{\partial x^n} + P_1\frac{\partial^n z}{\partial x^{n-1}\partial y} + \cdots + P_n\frac{\partial^n z}{\partial y^n} + Q_1\frac{\partial^{n-1} z}{\partial x^{n-1}} + \cdots + M_1\frac{\partial z}{\partial x}$$
$$+ N_1\frac{\partial z}{\partial y} + Cz = f(x,y) \qquad (1.163)$$

where P_0, $P_1, \ldots,$ M_1, N, C are functions of x and y. If, in particular, P_0, P_1, etc., are constant, then Eq. (1.163) is called a higher order linear PDE with a constant coefficient.

With the help of the symbols $D = \frac{\partial z}{\partial x}$ and $D' = \frac{\partial z}{\partial y}$, Eq. (1.163) can be expressed as

$$F(D, D') = f(x,y) \qquad (1.164)$$

where $F(D,D') = P_0D^n + P_1D^{n-1}D' + \cdots + P_nD^{'n} + \cdots + Q_1D^{n-1} + \cdots + M_1D + N_1D' + C$.

The aforementioned equation is a homogeneous PDE if $F(D, D'')$ is homogeneous in D and D''. Furthermore, the above PDE is nonhomogeneous if $F(D, D'')$ is not homogeneous in D and D''. Again, if the differential operator $F(D, D')$ can be factored into linear factors like $(\alpha D + \beta D' + \gamma)$, it is said to be reducible; if not, it is said to be irreducible.

Solution of higher-order PDE (with constant coefficients):
Consider the differential equation (1.164). The complementary function (C.F.) and particular integral (P.I.) are the two components that make up its solution, the same as an ODE. The solution of $F(D, D') = 0$ is called the complementary function, and the corresponding particular integral is given by

$$\text{P.I.} = \frac{1}{F(D, D')} f(x, y)$$

The complete solution of (1.164) is given by

$$z = \text{C.F.} + \text{P.I.}$$

Note: In the case of a nth order linear PDE, the complementary function will comprise n arbitrary functions.

1.14.1 *Working rule*

(a) C.F. of a linear homogeneous PDE with constant coefficient:
Consider the following differential equation:

$$F(D, D')z = f(x, y)$$

Calculate the linear factors of the type $(bD - aD')$ for the function $F(D, D')$, then apply the outcomes as follows:

(i) Corresponding to each non-repeated factor $(bD - aD')$, the part of the C.F. is taken as $\phi(by + ax)$.
(ii) Corresponding to each repeated factor $(bD - aD')^m$, the part of the C.F. is taken as $\phi_1(by + ax) + x\phi_2(by + ax) + x^2\phi_3(by + ax) + \cdots + x^{m-1}\phi_m(by + ax)$.
(iii) Corresponding to each non-repeated factor D, the part of the C.F. is taken as $\phi(y)$.
(iv) Corresponding to each repeated factor D^m, the part of the C.F. is taken as $\phi_1(y) + x\phi_2(y) + x^2\phi_3(y) + \cdots + x^{m-1}\phi_m(y)$.
(v) Corresponding to each non-repeated factor D', the part of the C.F. is taken as $\phi(x)$.
(vi) Corresponding to each repeated factor $(D')^m$, the part of the C.F. is taken as $\phi_1(x) + y\phi_2(x) + y^2\phi_3(x) + \cdots + y^{m-1}\phi_m(x)$.

(b) P.I. of linear homogeneous PDE with constant coefficient:

(i) When $F(a,b) \neq 0$ and $F(D,D')$ is a homogeneous function of degree n, then we have

$$\text{P.I.} = \frac{1}{F(a,b)}\phi(ax+by)$$
$$= \frac{1}{F(a,b)}\int\int\cdots\int \phi(v)dvdv\cdots dv$$

where $v = ax + by$.
Remember that the R.H.S. equation involves a multiple integral of nth order.

(ii) When $F(a,b) = 0$, we have

$$\text{P.I.} = \frac{1}{(bD - aD')}\phi(ax+by)$$
$$= \frac{x^n}{b^n n!}\phi(ax+by)$$

Example: *Solve* $(D^2 + DD' - 2D'^2)z = (2x+y)^{1/2}$

Solution: The given equation can be written as

$$(D - D')(D + 2D')z = (2x+y)^{1/2}$$

$\therefore$ C.F. $= \phi_1(y+x) + \phi_2(y-2x)$ where ϕ_1, and ϕ_2 are arbitrary functions.
Now,

$$\text{P.I.} = \frac{1}{(D^2 + DD' - 2D'^2)}(2x+y)^{1/2}$$
$$= \frac{1}{2^2 + 2.1 - 2.1^2}\int\int v^{1/2}dvdv, \textit{ where } v = 2x+y$$
$$= \frac{1}{15}(2x+y)^{5/2}$$

Hence, the general solution is

$$z = \phi_1(y+x) + \phi_2(y-2x) + \frac{1}{15}(2x+y)^{5/2}$$

Example: *Solve* $(D^2 - DD' + D' - 1)z = Cos(x+2y)$

Solution: The given equation can be written as

$$(D - D')^2 u = \tan(y+x)$$

$\therefore$ C.F. $= \phi_1(y+x) + x\phi_2(y+x)$, where ϕ_1, and ϕ_2 are arbitrary functions.

Now,

$$\text{P.I.} = \frac{1}{(D-D')^2}\tan(y+x)$$
$$= \frac{x^2}{1^1.2!}\tan(y+x)$$
$$= \frac{x^2}{2}\tan(y+x)$$

Hence, the general solution is

$$u = \phi_1(y+x) + x\phi_2(y+x) + \frac{x^2}{2}\tan(y+x)$$

(c) C.F. of linear nonhomogeneous PDE with constant coefficient:

When $F(D, D')$ can be expressed as the product of factors linear in D and D'. Then we apply the following rules:

(i) Corresponding to each non-repeated factor $(bD - aD' - c)$, the part of the C.F. is taken as $e^{\frac{cx}{b}}(by + ax)$ if $b \neq 0$.

(ii) Corresponding to each repeated factor $(bD - aD' - c)^r$, the part of the C.F. is taken as (assuming $b \neq 0$) $e^{\frac{cx}{b}}[\phi_1(by + ax) + x\phi_2(by + ax) + x^2\phi_3(by + ax) + \cdots + x^{r-1}\phi_r(by + ax)]$.

(d) P.I. of linear nonhomogeneous PDE with constant coefficient:

(i) When $f(x, y) = e^{ax+by}$ and $F(a, b) \neq 0$. Then

$$\text{P.I.} = \frac{1}{F(a,b)}e^{ax+by}$$

Example: *Solve* $(DD' + aD + bD' + ab)z = e^{px+qy}$

Solution: The given equation can be written as

$$(D + b)(D' + a)z = e^{px+qy}$$

$\therefore$ C.F. $= e^{-bx}\phi_1(y) + e^{-ay}\phi_2(x)$ where ϕ_1, ϕ_2 being arbitrary functions.

P.I. $= \frac{1}{(D+b)(D'+a)}e^{px+qy} = \frac{1}{(p+b)(q+a)}e^{px+qy}$.

Hence, the required general solution is

$$z = e^{-bx}\phi_1(y) + e^{-ay}\phi_2(x) + \frac{1}{(p+b)(q+a)}e^{px+qy}$$

(ii) When $f(x, y) = \sin(ax + by)$ or $\cos(ax + by)$. Then

$$\text{P.I.} = \frac{1}{F(D, D')} \sin(ax + by) \quad \text{or} \quad \cos(ax + by)$$

which is evaluated by putting $D^2 = -a^2, D'^2 = -b^2$ and $DD' = -ab$, provided the denominator is non-zero.

Example: *Solve* $(D^2 - DD' + D' - 1)z = \cos(x + 2y)$

Solution: The given equation can be written as

$$(D - 1)(D - D' + 1)z = \cos(x + 2y)$$

$\therefore$ C.F. $= e^x\phi_1(y) + e^{-x}\phi_2(y + x)$ where ϕ_1, ϕ_2 being arbitrary functions.
P.I. $= \frac{1}{D^2 - DD' + D' - 1} \cos(x + 2y) = \frac{1}{-1^2 - (-12) + D' - 1} \cos(x + 2y) = \frac{1}{D'} \cos(x + 2y) = \frac{1}{2} \sin(x + 2y)$

Hence, the required solution is

$$z = e^x\phi_1(y) + e^{-x}\phi_2(y + x) + \frac{1}{2} \sin(x + 2y)$$

(iii) When $f(x, y) = x^m y^n$, then

$$\text{P.I.} = \frac{1}{F(D, D')} x^m y^n = [F(D, D')]^{-1} x^m y^n$$

Example: *Solve* $r - s + p = 1$

Solution: The given equation can be written as

$$D(D - D' + 1)z = 1$$

$\therefore$ C.F. $= \phi_1(y) + e^{-x}\phi_2(y + x)$, where ϕ_1, ϕ_2 being arbitrary functions.
P.I.$= \frac{1}{D(D - D' + 1)} 1 = \frac{1}{D}(1 + D - D')^{-1} 1 = \frac{1}{D}[1 + (D - D') + \cdots] 1 = \frac{1}{D} 1 = x$

Hence, the required solution is

$$z = \phi_1(y) + e^{-x}\phi_2(y + x) + x$$

(iv) When $f(x, y) = Ve^{ax+by}$, where V is a function of x and y. Then

$$\text{P.I.} = \frac{1}{F(D, D')} Ve^{ax+by} = e^{ax+by} \frac{1}{F(D + a, D' + b)} V$$

Example: *Solve* $(D - 3D' - 2)^2 z = 2e^{2x} \sin(y + 3x)$

Solution: C.F. $= e^{2x}[\phi_1(y + 3x) + x\phi_1(y + 3x)]$
P.I. $= \frac{1}{(D - 3D' - 2)^2} 2e^{2x} \sin(y + 3x) = 2e^{2x} \frac{1}{[(D+2) - 3(D'+0) - 2]^2} \sin(y + 3x)$
$= 2e^{2x} \frac{1}{(D - 3D')^2} \sin(y + 3x) = 2e^{2x} \frac{x^2}{1^2 2!} \sin(y + 3x) = x^2 e^{2x} \sin(y + 3x)$

Hence, the required solution is

$$z = e^{2x}[\phi_1(y + 3x) + x\phi_1(y + 3x)] + x^2 e^{2x} \sin(y + 3x)$$

1.15 Non-Homogeneous PDE

A non-homogeneous PDE is a partial differential equation that contains terms dependent on different variables and functions, including an inhomogeneous term representing external influences, sources, or boundary conditions. Mathematically, a non-homogeneous PDE can be expressed as

$$L[u] = f(x_1, x_2, \ldots, x_n)$$

where $L[u]$ is a linear or nonlinear differential operator acting on the unknown function $u(x_1, x_2, \ldots, x_n)$ and $f(x_1, x_2, \ldots, x_n)$ is a given function representing external influences or sources. The term "non-homogeneous" arises because the equation is not homogeneous, meaning it does not satisfy the property of homogeneity. In a homogeneous equation, all terms are functions of the same variables, and the equation evaluates to zero when all dependent variables are set to zero. However, in a non-homogeneous equation, the presence of the $f(x_1, x_2, \ldots, x_n)$ term prevents the equation from being homogeneous.

For example, consider the wave equation

$$u_{tt} = c^2 u_{xx} + s(x,t) \tag{1.165}$$

with the boundary conditions: $u(0,t) = u(L,t) = 0$
and initial conditions: $u(x,0) = f(x), \ u_t(x,0) = g(x)$

Here, the source term $s(x,t)$ is independent of u or a derivative of u.

To solve this, we first look for a particular solution $v(x,t)$ of the PDE and boundary conditions. Then the general solution will be $u(x,t) = v(x,t) + w(x,t)$, where $w(x,t)$ is the general solution of the homogeneous PDE $u_{tt} = c^2 u_{xx}$ and boundary conditions. To satisfy our initial conditions, we must take the initial conditions for w as $w(x,0) = f(x) - v(x,0)$, $w_t(x,0) = g(x) - v_t(x,0)$.

Case 1: Steady State: If $s(x,t) = S(x)$ i.e., $s(x,t)$ does not depend on the time t, then $v(x,t) = v(x)$. The PDE (1.165) becomes $c^2 v'' + s(x) = 0$ with the boundary conditions $v(0) = v(L) = 0$. This is a standard SL system, and it can be solved.

For example, consider the PDE

$$u_{tt} = u_{xx} + x \tag{1.166}$$

boundary conditions $u(0,t) = u(1,t) = 0$
initial conditions $u(x,0) = 0, \ u_t(x,0) = 1$

The Eq. (1.166) becomes

$$v'' = -x$$

Integrating the equation, we get

$$v' = -x^2/2 + A$$

where A is a constant. Again integrating the equation, we get

$$v = -x^3/6 + Ax + B$$

Using the boundary condition, we get the particular solution

$$v(x) = (x - x^3)/6$$

The other part of the solution (1.166), $w(x,t)$, satisfies

$$u_{tt} = w_{xx}$$

boundary conditions $w(0,t) = w(1,t) = 0$
initial conditions $w(x,0) = -(x - x^3)/6,\ w_t(x,0) = 1$

Using he method of Fourier transformation, we get

$$w(x,t) = \sum_{n+1}^{\infty} \sin(n\pi x)(b_n \cos(n\pi t) + b_n^* \sin(n\pi t))$$

where $b_n = -\frac{1}{3}\int_0^1 (x - x^3)\sin(n\pi x)dx = \frac{2(-1)^n}{n^3\pi^3}$, $b_n^* = -\frac{2}{n\pi}\int_0^1 \sin(n\pi x)dx = \frac{2(1-(-1)^n)}{n^2\pi^2}$.

Therefore, the complete solution is

$$u(x,t) = \frac{x - x^3}{6} + \sum_{n=1}^{\infty} \left(\frac{2(-1)^n}{n^3\pi^3}\cos(n\pi t) + \frac{2(1-(-1)^n)}{n^2\pi^2}\sin(n\pi t) \right) \sin(n\pi x)$$

Case 2: Exponential in t: If the source term $s(x,t)$ is a function of x times an exponential in t.

For example, consider

$$u_{tt} = u_{xx} + xe^{-t} \tag{1.167}$$

boundary conditions $u(0,t) = u(1,t) = 0$
initial conditions $u(x,0) = 0,\ u_t(x,0) = 1$

Let us consider the solution $v(x,t) = V(x)e^{-t}$. Then Eq. (1.167) becomes

$$V = V'' + x$$

The general solution of this differential equation is $V(x) = x + c_1 e^x + c_2 e^{-x}$.

Using the boundary conditions, we get

$$v(x,t) = \left(x - \frac{e^{1+x}}{e^2-1} + \frac{e^{1-x}}{e^2-1}\right)e^{-t}$$

The initial conditions for $w(x,t)$ are

$$w(x,0) = -v(x,0) = -V(x) = -x + \frac{e^{1+x} - e^{1-x}}{e^2-1}$$

$$w_t(x,0) = 1 - v_t(x,0) = 1 + V(x) = 1 + x - \frac{e^{1+x} - e^{1-x}}{e^2-1}$$

Case 3: Arbitrary function of x and t: If $u(x,t)$ and $S(x,t)$ are considered arbitrary functions of x and t and it can be expanded as Fourier series through eigenfunction expansion $\phi_n(x)$, then we can handle the problem as given in the following.

$$u(x,t) = \sum_{n=1}^{\infty} b_n(t)\phi_n(x)$$

$$s(x,t) = \sum_{n=1}^{\infty} c_n(t)\phi_n(x)$$

Our PDE will give us relations between these, which will be ODEs in $b_n(t)$ for each n.

For example, let us consider the PDE

$$u_{tt} = u_{xx} + xt \tag{1.168}$$

boundary conditions $u(0,t) = u(1,t) = 0$
initial conditions $u(x,0) = 0,\ u_t(x,0) = 1$

The eigenfunctions for (1.168) are $\phi_n(x) = \sin(n\pi x)$. The Fourier sine series on the interval [0,1] can be written as

$$xt = \sum_{n=1}^{\infty} c_n \sin(n\pi t)$$

where $c_n = 2(-1)^{n+1}t/(n\pi)$. Putting these series into Eq. (1.168), we get

$$\sum_{n=1}^{\infty} b_n''(t)\sin(n\pi x) = -\sum_{n=1}^{\infty} b_n(t)(n\pi)^2\sin(n\pi x) + \sum_{n=1}^{\infty} c_n(t)\sin(n\pi x)$$

According to the uniqueness of Fourier series, the coefficients for each n must match, i.e.,

$$b_n'' = -(n\pi)^2 b_n + c_n(t) = -(n\pi)^2 b_n - \frac{2(-1)^n t}{n\pi}$$

The initial conditions for u and u_t give us initial conditions for b_n and b_n'. $u(x,0) = 0$ it imples that $b_n(0) = 0$, and $u_t(x,0){=}1{=}\sum_{n=1}^{\infty} \frac{2(1-(-1)^n)}{n\pi} \sin(n\pi x)$ implies that $b_n'(0) = \frac{2(1-(-1)^n)}{n\pi}$.

The general solution of the differential equation for b_n is

$$b_n(t) = A_n \cos(n\pi t) + B_n \sin(n\pi t) - \frac{2(-1)^n t}{(n\pi)^3}$$

From the initial conditions, we get $b_n(0) = 0 = A_n$ and $b_n'(0) = \frac{2(1-(-1)^n)}{n\pi} = n\pi B_n \frac{2(-1)^n}{(n\pi)^3}$, so $B_n = \frac{2(-1)^n}{(n\pi)^4} + \frac{2(1-(-1)^n)}{(n\pi)^2}$.

The complete solution is

$$u(x,t) = \sum_{n=1}^{\infty} \left(\left(\frac{2(-1)^n}{(n\pi)^4} + \frac{2(1-(-1)^n}{(n\pi)^2} \right) \sin(n\pi t) - \frac{2(-1)^n t}{(n\pi)^3} \right) \sin(n\pi t)$$

1.16 Solving Non-homogeneous PDEs by Separation of Variable

To solve a homogeneous PDE, we can directly apply the separation of variables method. But this can be generalized to non-homogeneous PDEs with homogeneous boundary conditions by solving non-homogeneous ODEs in time. Let us consider a non-homogeneous PDE

$$u_t = (p(x)u_x)_x - q(x)u + f(x,t), \; 0 < x < L \tag{1.169}$$

with

$$\alpha_1 u(0,t) + \beta_1 u_x(0,t) = 0, \quad \beta_1 u(L,t) + \beta_2 u_x(L,t) = 0, \; t > 0$$

$$u(x,0) = \phi(x), \quad 0 < x < L$$

At first, let us consider $v(x,t) = X(x)S(t)$ is a solution of the homogeneous PDE

$$v_t = (p(x)v')' - q(x)v, \quad 0 < x < L \tag{1.170}$$

Substituting into the PDE (1.170), we get

$$S'(t)X(x) = S(t)(p(x)X'(x))' - q(x)S(t)X(x)$$

Therefore,

$$\frac{S'(t)}{S(t)} = \frac{(p(x)X'(x))' - q(x)X(x)}{X(x)} = -\lambda$$

where λ is a constant.

So, we find the ODE

$$S'(t) + \lambda S(t) = 0$$
$$-(p(x)X'(x))' + q(x)X(x) = \lambda X(x)$$

This is a Sturm–Liouville problem. It has eigenvalues $\{\lambda_n\}_{n=1}^{\infty}$ and corresponding eigenfunctions $\{X_n(x)\}_{n=1}^{\infty}$ that satisfy the boundary conditions. Moreover, these eigenfunctions are orthogonal.

i.e.,

$$\int_0^L X_n(x)X_m(x)dx = 0, \ m \neq n$$

They also form a basis for the function space consisting of functions that satisfy the boundary conditions, meaning that any such function can be expressed as a linear combination of these eigenfunctions. The solution $u(x,t)$ and the external source term $f(x,t)$ in (1.169) can be expressed as

$$u(x,t) = \sum_{n=1}^{\infty} u_n(t)X_n(x), \ f(x,t) = \sum_{n=1}^{\infty} f_n(t)X_n(x)$$

where $u_n(t) = \frac{\int_0^L u(x,t)X_n(x)dx}{\int_0^L X_n(x)^2}dx, \ f_n(t) = \frac{\int_0^L f(x,t)X_n(x)dx}{\int_0^L X_n(x)^2}dx.$

To solve the PDE (1.169), we multiply both sides by $X_m(x)$ and integrate from 0 to L. Resulting in

$$\int_0^L u_t(x,t)X_m(x)dx = \int_0^L (p(x)u_x(x,t))_x - q(x)u(x,t)X_m(x)dx + \int_0^L f(x,t)X_m(x)dx$$

$X_m(x)$ is an eigenfunction of the spatial ODE. So, we get

$$u'_m(t) = -\lambda_m u_m(t) + f_m(t). \tag{1.171}$$

Also, from the initial condition, we obtain $u_m(0) = a_m = \frac{\int_0^L \phi(x)X_m(x)dx}{\int_0^L X_m(x)^2dx}$ The non-homogeneous ODE (1.171) can be solved using an integrating factor $\mu(t) = e^{\lambda_m t}$. We then have

$$u_m(t) = e^{-\lambda t}a_m + \int_0^t e^{-\lambda_m(t-s)} f_m(s)ds$$

Finally, we obtain the solution of (1.171) as

$$u(x,t) = \sum_{n=1}^{\infty} u_n(t)X_n(x)$$

where $X_n(x)$ is the solution of $-(p(x)X_n'(x))' + q(x)X_n(x) = \lambda_n X_n(x)$.

This solution satisfies the boundary conditions, and the coefficients $u_n(t)$ of the eigenfunction expansion of u(x; t) can be expressed as

$$u_n(t) = e^{-\lambda_n t}a_n + \int_0^t e^{-\lambda_n(t-s)} f_n(s)ds$$

where $a_n = \frac{\int_0^L \phi(x)X_n(x)dx}{\int_0^L X_n(x)^2 dx}$, $f_n(t) = \frac{\int_0^L f(x,t)X_n(x)dx}{\int_0^L X_n(x)^2 dx}$. Through this process we can find $u(x,t)$, which is known as the method of eigenfunction expansions.

Example: *We will use the method of eigenfunction expansions to solve the* initial boundary value problem (IBVP)

$$\begin{aligned} u_t &= \alpha^2 u_{xx} + \sin(3\pi x), \quad 0 < x < 1 \\ u(0,t) &= 0, \quad u(1,t) = 0, \ t > 0 \\ u(x,0) &= \sin(\pi x), \quad 0 < x < 1 \end{aligned}$$

Solve: At first applying the separation of variables to the corresponding homogeneous PDE,

$$u_t = \alpha^2 u_{xx}$$

We get the ODE

$$\begin{aligned} T' + \alpha^2 \lambda T &= 0 \\ X'' + \lambda X &= 0 \end{aligned}$$

with boundary conditions $X(0) = 0$, $X(1) = 0$. The spatial ODE has the solutions or eigenfunctions,

$$X_n(x) = \sin(n\pi x), \ n = 1, 2, \ldots$$

with corresponding eigenvalues $\lambda_n = (n\pi)^2$, $n = 1, 2, \ldots$. Expanding the source term, initial data, and solution in the basis of eigenfunctions, we get

$u(x,t) = \sum_{n=1}^{\infty} u_n(t)\sin(n\pi x)$, $\sin(\pi x) = \sum_{n=1}^{\infty} a_n \sin(n\pi x)$, $\sin(3\pi x) = \sum_{n=1}^{\infty} f_n \sin(n\pi x)$.

Because of the orthogonality of the eigenfunctions, $a_1 = 1,\ f_3 = 1$, and $a_n = 0$ for all $n \neq 1$ and $f_n = 0$ for all $n \neq 3$.

From the formula

$$u_n(t) = e^{-(n\pi\alpha)^2 t} a_n + \int_0^t e^{-(n\pi\alpha)^2(t-s)} f_n ds$$

we obtain $u_1 = e^{-(\pi\alpha)^2 t} a_1 = e^{-(\pi\alpha)^2 t}$, $u_3 = e^{-(3\pi\alpha)^2 t} a_3 + f_3 \int_0^t e^{-(3\pi\alpha)^2(t-s)} ds = \frac{1}{(3\pi\alpha)^2} e^{-(3\pi\alpha)^2 t} e^{(3\pi\alpha)^2 s}|_0^t = \frac{1}{(3\pi\alpha)^2}[1 - e^{-(3\pi\alpha)^2 t}]$ and $u_n(t) = 0$ for all $n \neq= 1, 3$.

Therefore, the solution of the IBVP is

$$u(x,t) = e^{-(\pi\alpha)^2 t}\sin(\pi x) + \frac{1}{(3\pi\alpha)^2}[1 - e^{-(3\pi\alpha)^2 t}]\sin(3\pi x)$$

When $t \to \infty$, we obtain the steady-state solution $\frac{1}{(3\pi\alpha)^2}\sin(3\pi x)$.

Note: In Chapter 3, we shall find out the approximate solution of non-homogeneous NLPDEs in detail.

1.17 Equations in Physics

Today, PDEs are used frequently across a wide range of academic disciplines. A dynamical viewpoint on the difficulties of physical models underpins the significance of learning and using such equations. PDEs were first applied by Euler, Lagrange, and d'Alembert to describe the mechanics of continuum and serve as a foundation for analytic model studies in physical research. Therefore, the primary goal of this project is to investigate a certain class of PDEs that have traveling wave solutions.

The fundamental concepts and numerous significant nonlinear model equations that appear in a wide range of physical issues are covered in this chapter. Included are one-dimensional wave, Klein–Gordon, sine-Gordon, Burgers, Fisher, Korteweg–de Vries, Boussinesq, modified KdV, nonlinear Schrödinger, Benjamin–Bona–Mahony, Burgers–Huxley, and Kadomtsev Petviashvili and many other equations.

The most general first-order nonlinear PDE in two independent variables x and y has the form

$$F(x, y, u, u_x, u_y) = 0 \tag{1.172}$$

The most general second-order nonlinear PDE in two independent variables x and y has the form

$$F(x, y, u, u_x, u_y, u_{xx}, u_{xy}, u_{yy}, u_{yy}) = 0 \tag{1.173}$$

More formally, it is possible to write these equations in the operator form

$$L_x u(x) = f(x) \tag{1.174}$$

where L_x is a partial differential operator and $f(x)$ is a given function of two or more independent variables $x = (x, y, \ldots)$. Here, L_x is not a linear operator, (1.174) is called a nonlinear PDE. Equation (1.174) is called an inhomogeneous nonlinear equation if $f(x) \neq 0$. On the other hand, (1.174) is called a homogeneous nonlinear equation if $f(x) = 0$.

In general, the linear superposition principle can be applied to linear PDE if certain convergence requirements are satisfied. This principle is usually used to find a new solution as a linear combination of a given set of solutions. For nonlinear PDE, however, the linear superposition principle cannot be applied to generate a new solution. So, because most solution methods for linear equations cannot be applied to nonlinear equations, there is no general method of finding analytical solutions of nonlinear PDE, and numerical techniques are usually required for their solution. Sometimes a nonlinear equation can be solved by finding a transformation of the variables that turns it into a linear equation, or by using some other improvised technique. In fact, new methods are usually required for finding solutions of nonlinear equations. The theory of nonlinear PDE encompasses more than just nonlinear equation solution techniques. The existence, uniqueness, and stability of nonlinear PDE solutions are fundamentally important, just like with linear equations. These and other characteristics of nonlinear equations have made the field one of the most active and diverse in contemporary mathematics.

1.18 Some Nonlinear Model Equations

The formulation of fundamental natural laws and the mathematical investigation of a wide range of physical issues usually involve nonlinear PDEs. Here are a few key model equations of greatest relevance.

1.18.1 *Kinematic wave equation*

The kinetic wave equation was introduced by Sir James Lighthill and Gerald B. Whitham in their paper "On Kinematic Waves: II. A Theory of Traffic

Flow on Long Crowded Roads," which was published in the *Proceedings of the Royal Society of London* in 1955. The simplest nonlinear wave equation is the kinematic wave equation, which is expressed as

$$\frac{\partial u}{\partial t} + c(u)\frac{\partial u}{\partial x} = 0, \quad -\infty < x < \infty, \ t > 0 \tag{1.175}$$

where $c(u)$ is a given function of u. When the viscocity $\nu = 0$, then this equation becomes a special case of the Burgers' equation. It then discusses the propagation of nonlinear waves, e.g., waves in traffic flow on highways, shock waves, flood waves, and waves in plasmas. Moreover, this equation also explains the propagation of chemical exchange processes in chromatography, and sediment transport in rivers.

1.18.2 *Reaction–diffusion equation*

In 1952, Alan Turing, a British mathematician, first established the reaction-diffusion equation in his seminal paper, "The Chemical Basis of Morphogenesis." Turing's work discovered a mathematical framework to explain how patterns and structures can arise in biological systems through the interaction of two fundamental processes: reaction and diffusion. This research helps us to investigate pattern formation in many biological and chemical situations, profoundly influencing subjects such as developmental biology, chemistry, physics, and computer science.

It is a mathematical model that describes how the concentration of one or more substances varies over time as a result of chemical reactions and diffusion. The reaction-diffusion equation is

$$\frac{\partial u}{\partial t} = D\nabla^2 u + f(u). \tag{1.176}$$

where u, t, D, and $f(u)$ represent the concentration of the substance(s) being modeled, time, diffusion coefficient, and reaction term, respectively. $f(u)$ can have many different forms depending on the chemical interactions being modeled.

1.18.3 *Klein–Gordon equation*

In 1927, physicists O. Klein and W. Gordon introduced an equation that is known as the Klein–Gordon equation (or Klein–Gordon–Fock equation). This equation describes a relativistic quantum mechanical wave equation for particles with no internal degree of freedom. According to Schrödinger, the Klein–Gordon equation was a quantum wave equation. Later in 1926, Fock, in an article, discussed the generalization for the case of magnetic fields and derived this equation independently. The nonlinear Klein–Gordon equation is

$$\frac{\partial^2 u}{\partial t^2} - c^2\nabla^2 u + V'(u) = 0 \tag{1.177}$$

where c is a constant, and $V'(u)$ is a nonlinear function of u generally selected as the derivative of the potential energy $V(u)$.

Particularly when the value of $c = 1$, $V' = \sin u$ and $\frac{\partial}{\partial y} = \frac{\partial}{\partial z} = 0$, then we get the Sine-Gordon equation

$$\frac{\partial^2 u}{\partial t^2} - \frac{\partial^2 u}{\partial x^2} + \sin u = 0. \tag{1.178}$$

Especially, when the particle is massless, the Klein–Gordon equation (1.177) reduces to the relativistic wave equation. The Klein–Gordon equations are important for many scientific purposes, e.g., nonlinear dispersion, solid-state physical problems, nonlinear optics, quantum field theory, and nonlinear meson theory, etc. It is born out of numerous physical problems like nonlinear dispersion and nonlinear meson theory.

Several methods like the finite difference method, the finite element method, the collocation method, the Bäcklund transformation, and the inverse scattering method have been extensively used to study the Klein–Gordon equation.

1.18.4 *Sine-Gordon equation*

The Sine-Gordon equation appears in differential geometry and relativistic field theory. It is denominated according to its similar form to the Klein–Gordon equation. The Sine-Gordon equation has a long history that began in the 19th century in the course of the study of surfaces of constant negative curvature. The sine-Gordon equation is one of the basic nonlinear evolution equations that describes various important nonlinear physical phenomena. This equation has a wide range of applications in mathematics and physics, e.g., in differential geometry, relativistic field theory, solid-state physics, nonlinear optics, etc. It has arisen classically in the study of differential geometry and in the propagation of a 'slip' dislocation in crystals. More recently, it has arisen in a wide variety of physical problems, including the propagation of magnetic flux in Josephson-type superconducting tunnel junctions, the phase jump of the wave function of superconducting electrons along long Josephson junctions (a junction between two superconductors), the motion of rigid pendular attached to a stretched wire, the propagation of short optical pulses in resonant laser media, the stability of fluid motions in ferromagnetism and ferroelectric materials, in the dynamics of certain molecular chains such as DNA, in elementary particle physics, and in weakly unstable baroclinic wave packets in a two-layer fluid.

The standard form of this equation is

$$\frac{\partial^2 u}{\partial t^2} - \frac{\partial^2 u}{\partial x^2} = \sin u \tag{1.179}$$

where $u(x, t)$ is a real-valued function representing the wave or soliton. The equation, as well as several solution techniques, was known in the 19th century, but the equation grew greatly in importance in the 1970s when it was first realized that it led to solutions (so-called "kink" and "antikink") with the collisional properties of solitons. Various methods for finding particular solutions to this equation were developed at that time. One of the methods is known as the Bäcklund transformation, which will be considered in a subsequent section. Other methods include traveling wave solutions, the similarity method, the inverse scattering method, and the method of separation of variables, which deals with the representation of solutions as functions of independent variables. All of these methods will be discussed in subsequent sections.

1.18.5 *Burgers' equation*

In fluid mechanics, Burgers' equation is a basic PDE. The modeling of traffic flow and gas dynamics are just two examples of how it can be found in diverse applications of mathematics. It also arises in a variety of physical issues, such as sound waves in a viscous medium, waves in fluid-filled viscous elastic tubes, and magnetohydrodynamic waves in a medium with limited electrical conductivity. In honor of Johannes Martinus Burgers (1895–1981), it bears his name. The generic version of Burgers' equation (also known as the viscous Burgers' equation) is

$$\frac{\partial u}{\partial t} + u\frac{\partial u}{\partial x} = \nu\frac{\partial^2 u}{\partial x^2} \tag{1.180}$$

for a given velocity u and viscosity coefficient ν. It is the important nonlinear model equation representing phenomena described by a balance between time evolution, nonlinearity, and diffusion.

When $\nu = 0$, Burgers' equation transforms into the inviscid Burgers' equation

$$\frac{\partial u}{\partial t} + u\frac{\partial u}{\partial x} = 0 \tag{1.181}$$

a model for equations whose solutions might exhibit discontinuities (shock waves). A specific instance of the kinematic wave equation ($c(u) = u$) is the inviscid Burgers' equation, often known as the Hopf equation. The inviscid Burgers' equation is hyperbolic, whereas the Burgers' equation is parabolic. The characteristics of the parabolic equation's solution are very different from those of the hyperbolic equation.

The Cole–Hopf Transformation

The Cole–Hopf transformation is a mathematical technique used in the theory of PDEs, particularly in the context of soliton solutions. It is named after J. Cole and E. Hopf, who developed this transformation in the mid-20th century. The Cole–Hopf transformation is most commonly applied to nonlinear PDEs, such as the Korteweg–de Vries (KdV) equation. The KdV equation is a well-known PDE that describes the behavior of long waves in shallow water. It has soliton solutions, which are stable, localized waveforms that do not disperse or change shape as they propagate. The transformation involves converting a given nonlinear PDE into a linear one through a change of variables. In the context of the KdV equation, the transformation is often used to convert the equation into a linear Schrödinger-like equation. This linearized form makes it easier to find solutions and analyze the behavior of solitons. The Cole–Hopf transformation is a powerful tool in the study of solitons and nonlinear wave equations, enabling researchers to find exact solutions and understand the underlying physics of these complex systems.

Eberhard Hopf (1950) and Julian Cole (1951) independently introduced a transformation to convert the Burgers' equation into the linear heat equation and solved it exactly for any arbitrary initial condition. This transformation is known as the Cole–Hopf transformation.

Solve the initial value problem for Burger's equation

$$\frac{\partial u}{\partial t} + u\frac{\partial u}{\partial x} = \epsilon\frac{\partial^2 u}{\partial x^2}, \quad -\infty < x < \infty \tag{1.182}$$

$$u(x,0) = f(x) \tag{1.183}$$

First, we write (1.182) in a form similar to a conservation law,

$$\frac{\partial u}{\partial t} + \frac{\partial}{\partial x}\left(\frac{1}{2}u^2 - \epsilon\frac{\partial u}{\partial x}\right) = 0 \tag{1.184}$$

This can be regarded as the compatibility condition for a function ψ to exist, such that

$$u = \frac{\partial \psi}{\partial x}, \tag{1.185}$$

$$\epsilon\frac{\partial u}{\partial x} - \frac{1}{2}u^2 = \frac{\partial \psi}{\partial t} \tag{1.186}$$

We substitute the value of u from (1.185) in (1.186) to obtain

$$\epsilon\frac{\partial^2 \psi}{\partial x^2} - \frac{1}{2}\left(\frac{\partial \psi}{\partial x}\right)^2 = \frac{\partial \psi}{\partial t} \tag{1.187}$$

Next, we introduce $\psi = -2\epsilon\log v$ so that

$$u = \frac{\partial \psi}{\partial x} = -2\epsilon\frac{\frac{\partial v}{\partial x}}{v} \tag{1.188}$$

This called the *Cole–Hopf transformation*, which, by differentiating, gives

$$\frac{\partial^2 \psi}{\partial x^2} = 2\epsilon\left(\frac{\frac{\partial v}{\partial x}}{v}\right)^2 - \frac{2\epsilon}{v}\frac{\partial^2 v}{\partial x^2}$$

$$\frac{\partial \psi}{\partial t} = -2\epsilon\left(\frac{\frac{\partial v}{\partial t}}{v}\right)$$

Consequently, (1.187) reduces to the linear diffusion equation

$$\frac{\partial v}{\partial t} = \epsilon\frac{\partial^2 v}{\partial x^2} \tag{1.189}$$

Many solutions of this equation are well known in the literature. We substitute the given solution for v to find solutions of the Burger's equation.

At $t = 0$, $u(x,0) = f(x)$. The new variable must initially satisfy $f(x) = -2\epsilon\frac{\frac{\partial}{\partial x}v(x,0)}{v(x,0)}$, i.e., $\frac{\partial}{\partial x}v(x,0) = -\frac{1}{2\epsilon}f(x)v(x,0)$. This is a linear first-order ODE for $v(x,0)$ and integrating factor is $e^{\frac{1}{2\epsilon}\int_0^x f(s)ds}$ and the general solution is

$$v(x,0) = \alpha e^{-\frac{1}{2\epsilon}\int_0^x f(s)ds} = \alpha g(x)$$

where α is an integrating constant and $g(x) = e^{-\frac{1}{2\epsilon}\int_0^x f(s)ds}$ exists. Thus, for a given $f(x)$, we compute $g(x)$ by quadrature of course. It is understood that the integral $\int_0^x f(s)ds$ exists. So, we need to solve the following linear problem for $v(x,0)$, from Eq. (1.189)

$$\frac{\partial v}{\partial t} = \epsilon \frac{\partial^2 v}{\partial x^2}, \quad -\infty < x < \infty \tag{1.190}$$

$$v(x,0) = \alpha g(x) \tag{1.191}$$

Now, taking Fourier transformation of (1.190) and assuming $v, v_x \to 0$ as $|x| \to \infty$, we get

$$\int_{-\infty}^{\infty} \frac{\partial v}{\partial t} e^{ipx} dx = \int_{-\infty}^{\infty} \epsilon \frac{\partial^2 v}{\partial x^2} e^{ipx} dx,$$

$$\Rightarrow \frac{d}{dt} \int_{-\infty}^{\infty} v e^{ipx} dx = \epsilon \left[\frac{\partial v}{\partial x} e^{ipx} \right]_{-\infty}^{\infty} - \epsilon ip \int_{-\infty}^{\infty} \frac{\partial v}{\partial x} e^{ipx} dx,$$

$$\Rightarrow \frac{dV}{dt} + \epsilon p^2 V = 0$$

where $V = \int_{-\infty}^{\infty} v e^{ipx} dx$. This is a linear first order ODE for V and the general solution is $V = A(p) e^{-\epsilon p^2 t}$.

Now, $V(x,0) = \int_{-\infty}^{\infty} v(x,0) e^{ipx} dx = \int_{-\infty}^{\infty} \alpha g(x) e^{ipx} = F(p)$.

Therefore, $V = F(p) e^{-\epsilon p^2 t}$.

Taking inverse Fourier transformation, we have

$$\begin{aligned} v(x,t) &= \frac{1}{2\pi} \int_{-\infty}^{\infty} F(p) e^{-\epsilon p^2 t} e^{ipx} dp, \\ &= \frac{1}{2\pi} \int_{-\infty}^{\infty} e^{-\epsilon p^2 t} \left[\int_{-\infty}^{\infty} \alpha g(\xi) e^{ip\xi} d\xi \right] e^{ipx} dp, \\ &= \frac{\alpha}{2\pi} \int_{-\infty}^{\infty} g(\xi) \left[\int_{-\infty}^{\infty} e^{-\epsilon p^2 t + ip\xi - ipx} dp \right] d\xi \end{aligned} \tag{1.192}$$

changing the order of integration. Now,

$$\int_{-\infty}^{\infty} e^{-\epsilon t (p^2 + \frac{(x-\xi)}{\epsilon t}) ip} dp = \frac{(x-\xi)^2}{4\epsilon t} \sqrt{\frac{\pi}{\epsilon t}}.$$

From (1.192), we have

$$v(x,t) = \frac{\alpha}{2\sqrt{\pi \epsilon t}} \int_{-\infty}^{\infty} g(\xi) e^{-\frac{(x-\xi)^2}{4\epsilon t}} d\xi \tag{1.193}$$

Then follows that

$$\frac{\partial}{\partial x} v(x,t) = \frac{-\alpha}{2\sqrt{\pi \epsilon t}} \int_{-\infty}^{\infty} \frac{(x-\xi) g(\xi)}{2\epsilon t} e^{-\frac{(x-\xi)^2}{4\epsilon t}} d\xi. \tag{1.194}$$

Therefore, $u(x,t)$ is given by

$$u(x,t) = -2\epsilon \frac{\frac{\partial v}{\partial x}}{v} = \frac{\int_{-\infty}^{\infty} \frac{(x-\xi)g(\xi)}{t} e^{-\frac{(x-\xi)^2}{4\epsilon t}} \, d\xi}{\int_{-\infty}^{\infty} g(\xi) e^{-\frac{(x-\xi)^2}{4\epsilon t}} \, d\xi}. \tag{1.195}$$

1.18.6 *Fisher equation*

Diffusion and reaction interact in a process that is described by the Fisher equation. R. A. Fisher developed the Fisher equation in 1936 to examine the wave propagation of a gene in a population and logistic growth-diffusion phenomena. This equation explains wave propagation phenomena in a variety of chemical and biological systems, including the theory of combustion, diffusion and mass transfer, nonlinear diffusion, chemical kinetics, ecology, chemical wave propagation, neutron population in a nuclear reactor, etc. The Fisher equation is

$$\frac{\partial u}{\partial t} - \frac{\partial^2 u}{\partial x^2} - u(1-u) = 0 \tag{1.196}$$

Using $\xi = x - ct$ in Eq. (1.196), it will be converted into ODE

$$-cu' - u'' - u(1-u) = 0 \tag{1.197}$$

Applying the tanh–coth method (balancing u'' with u^2 gives $M = 2$), we get the kink solution

$$u(x,t) = \frac{1}{4}\left(1 - \tanh\left[\frac{1}{2\sqrt{6}} 9x - \frac{5}{\sqrt{6}} t)\right]\right)^2$$

1.18.7 *Boussinesq equation*

J. V. Boussinesq first proposed the Boussinesq equation in 1872, and it is used to explain a wide variety of physical phenomena and scientific applications, including the propagation of long waves in shallow water, nonlinear lattice waves, iron sound waves in plasma, and vibrations in a nonlinear string. Being completely integrable, the Boussinesq equation admits an infinite number of conservation laws and N-soliton solutions. The fourth-order nonlinear Boussinesq equation is

$$\frac{\partial^2 u}{\partial t^2} - \frac{\partial^2 u}{\partial x^2} - \frac{\partial^2}{\partial x^2}(3u^2) + \lambda \frac{\partial^4 u}{\partial x^4} = 0 \tag{1.198}$$

Water waves that are dispersive and propagate in both positive and negative x-directions are described by this equation. The first two terms define the equation as a wave equation; the third term is a nonlinearity resulting from physical limitations and expansion; and the fourth term is a dispersive term. Our long-wave assumptions lead us to infer that the first term of this equation is negligibly small. Boussinesq equations come in two different varieties. Due to the challenges in obtaining descriptive solutions, the equation with $\lambda = -1$ is known as the bad Boussinesq equation or the ill-posed classical equation, whereas the equation with $\lambda = 1$ is called the good Boussinesq equation or the well-posed equation.

The various ansatz approaches, including the tanh, sech, Exp-function, (G/G')-Expansion, and Jacobi elliptic function methods, are used to generate precise solutions for various equations. These techniques lead to a variety of Boussinesq equation solutions. A lot of focus has been placed on modified Boussinesq equations in recent years that take into account a velocity potential at an arbitrary height and the free surface displacement of water, as well as novel forms of Boussinesq equations that deal with the propagation of water waves.

Let us consider the bad Boussinesq equation as

$$\frac{\partial^2 u}{\partial t^2} - \frac{\partial^2 u}{\partial x^2} + \frac{\partial^2}{\partial x^2}(3u^2) - \frac{\partial^4 u}{\partial x^4} = 0 \tag{1.199}$$

using $u = f(z)$, $z = x - ct$ into (1.199) and after some calculation we get the exact solution of the equation (1.199)

$$u(x,t) = \frac{c}{2}\operatorname{sech}^2\left[\frac{\sqrt{c}}{2}x + \frac{\sqrt{c(1+ct)}}{2}\right. \tag{1.200}$$

1.18.8 *Korteweg–de Vries equation*

John Scott Russell made the initial observation of the lone wave in shallow water, and in 1895, Diederik Jonannes Kortewag (1848–1941) and de Vries later modeled it. Although Boussinesq first presented this equation in 1877, it wasn't until the renowned finding made by Zabusky and Kruskal in 1965 that the study of the KdV equation gained widespread attention. They have demonstrated that these waves are essentially unaffected in shape by passing through one another, and as a result, the name "soliton" is used to describe these waves. This exceptional characteristic spurred research into numerous KdV equations, which were then studied in relation to shallow water waves, long internal waves in a concept, acoustic waves in plasma, and acoustic crystal tails. This equation has a number of intriguing characteristics that have also been revealed through various investigations.

For the KdV equation, Miura discovered conservation principles in 1968, and Tabor demonstrated in 1989 that an infinite number of conserved values would be possible for the KdV equation. It displays a nonlinear wave's stable form. The isospectral integrability requirement for pairs of linear operators, among other things, is identical to the KdV equation, as P. Lax demonstrated in 1968.

The KdV equation can be written as

$$\frac{\partial \phi}{\partial t} + a\phi \frac{\partial \phi}{\partial x} + b\frac{\partial^3 \phi}{\partial x^3} = 0. \tag{1.201}$$

where $\phi(x, t)$ represents the amplitude of the wave. The KdV equation is a useful approximation in the modeling of many physical systems. Let x and t be the two independent variables, and a and b be two non-zero constants. Equation (1.201) is nonlinear to the convective term $\phi\frac{\partial \phi}{\partial x}$ and dispersive term $\frac{\partial^3 \phi}{\partial x^3}$. When $\frac{\partial^3 \phi}{\partial x^3} = 0$, that is the presence of only the nonlinear terms, and in the absence of dispersion, the breaking of waves occurs.

The KdV equation is particularly interesting because it exhibits a soliton solution. The KdV equation has applications in various fields such as fluid dynamics, plasma physics, nonlinear optics, and even in some models in biology and economics. It's a fundamental equation in the study of integrable systems and nonlinear phenomena.

1.18.9 *Modified KdV equation*

The popularity of the modified KdV equation is due to its unique solution behavior, or breathers. The parameter of the medium on which it appears determines the coefficient of the KdV equation. The medium's nonlinearity is no longer included in the equation if the coefficient of the nonlinear term disappears. Instead of using quadratic nonlinearity in this case, the researchers strive to discover an equivalent equation that uses cubic nonlinearity. Modified KdV (mKdV) equation is the name given to such an equation. Electrodynamics, nonlinear optics, and other fields make use of this equation. Let us consider an mKdV equation as

$$\frac{\partial \phi}{\partial t} + a\phi^2 \frac{\partial \phi}{\partial x} + b\frac{\partial^3 \phi}{\partial x^3} = 0. \tag{1.202}$$

Alfven waves in a collisionless plasma and nonlinear acoustic waves in a harmonic lattice are both described by the mKdV equation. It also occurs in a variety of other bodily circumstances. Various methods have been used to construct solutions to the mKdV equation. Hirota constructed the exact solution for the mKdV equation in 1972. In the same year, Tanaka was the first to solve the mKdV equation by using the inverse scattering technique. Wadati, using this same method, succeeded in obtaining the exact N-soliton solution for the mKdV equation. In 1976, rational solutions to the mKdV equation were derived by Ono by using the Bäcklund transformation. A limiting procedure based on bilinear results gave other type of solutions in 1978.

In 1983, rational solutions to the mKdV equation were determined by recurrence relations by Tanaka. Solitary wave solutions and cnoidal wave solutions were given in 1994. Some classes of periodic solutions of mKdV equation were given in 2004. More recently, in 2012, solutions to the mKdV equation were constructed via a bilinear Bäcklund transformation, and rational solutions in terms of Wronskians were obtained. In 2016, periodic solutions and rational solutions of first and second order were presented using a Darboux transformation were constructed.

Miura Transformation

In 1968, Miura, Gardner, and Kruskal developed a method for determining an infinite number of conservation laws by introducing the Miura transformation $u = v^2 + v_x$, which establishes a connection between the modified KdV equation and the KdV equation, i.e., between two nonlinear PDEs. The Miura transformation is a non-invertible mapping.

The KdV equation is defined as

$$u_t - 6uu_x + u_{xxx} = 0 \tag{1.203}$$

and we desire to obtain a solution to this equation by application of the Miura transformation, given by

$$u = v^2 + v_x \tag{1.204}$$

where $u = u(x,t)$ and $v = v(x,t)$. This equation, when substituted directly into Eq. (1.203) yields

$$\left(2v + \frac{\partial}{\partial x}\right)(v_t - 6v^2 v_x + v_{xxx}) = 0 \tag{1.205}$$

Thus, If v is a solution to,

$$v_t - 6v^2 v_x + v_{xxx} = 0 \tag{1.206}$$

which is known as the mKdV equation.

1.18.10 *Third-order generalized KdV equation*

We consider the nonlinear family of third-order KdV equation of the form

$$\frac{\partial u}{\partial t} + P(u)\frac{\partial u}{\partial x} + \frac{\partial^3 u}{\partial x^3} = 0 \tag{1.207}$$

where $u(x,t)$ is a function of space x and time t.

When $P(u) = au$, Eq. (1.207) becomes the standard KdV equation as

$$\frac{\partial u}{\partial t} + au\frac{\partial u}{\partial x} + \frac{\partial^3 u}{\partial x^3} = 0$$

When $p(u) = au^2$, Eq. (1.207) becomes mKdV equation as

$$\frac{\partial u}{\partial t} + au^2\frac{\partial u}{\partial x} + \frac{\partial^3 u}{\partial x^3} = 0$$

For $p(u) = au^n$, Eq. (1.207) is called the generalized KdV equation. This equation can be written as

$$\frac{\partial u}{\partial t} + au^n\frac{\partial u}{\partial x} + \frac{\partial^3 u}{\partial x^3} = 0$$

Unlike the KdV equation and the mKdV equation, the generalized KdV equation is not integrable for $n \geq l3$, and therefore does not give multiple soliton solutions.

The Gardner equation is obtained when $p(u) = 2au - 3bu^2$, $a > 0, b > 0$. This equation takes the form:

$$\frac{\partial u}{\partial t} + (2au - 3bu^2)\frac{\partial u}{\partial x} + \frac{\partial^3 u}{\partial x^3} = 0$$

If $p(u) = u_x$, then Eq. (1.207) is called the potential KdV (pKdV) equation which is of the form

$$\frac{\partial u}{\partial t} + \left(\frac{\partial u}{\partial x}\right)^2 + \frac{\partial^3 u}{\partial x^3} = 0$$

When $p(u) = au^n - bu^{2n}$, we get the generalized KdV equation with two power nonlinearities. This equation takes the form

$$\frac{\partial u}{\partial t} + (au^n - bu^{2n})\frac{\partial u}{\partial x} + \frac{\partial^3 u}{\partial x^3} = 0 \tag{1.208}$$

The Gardner equation can be obtained by setting $n = 1$ in (1.208).

1.18.11 *Fifth-order KdV equation*

In the context of nonlinear fifth-order KdV-type equations, studies are flourishing because these equations are able to describe the real features in a variety of scientific applications and engineering areas and would have much practical and physical meaning. Fifth-order KdV-type equations take the form

$$u_t = u_{xxxxx} + f(x, t, u_x, u_{xx}, u_{xxx}) \tag{1.209}$$

arising naturally in modeling many different wave phenomena such as gravity-capillary waves, the propagation of shallow water waves over a flat surface and magnetosound propagation in plasmas. The fifth-order KdV-type equations appear in the literature in three main families, given as follows:

1. The first family reads

$$u_t + auu_x + bu_{xxx} - ru_{xxxxx} = 0 \tag{1.210}$$

where a, b, r are constants. This equation is called the Kawahara equation which models the dynamics of long waves in a viscous fluid.

2. The second family of nonlinear fifth-order KdV equations in the standard form reads:

$$u_t + \alpha u^2 u_x + \beta u_x u_{xx} + \gamma u u_{xxx} + u_{xxxxx} = 0 \tag{1.211}$$

where α, β, γ are arbitrary nonzero and real parameters, and $u = u(x, t)$ is a smooth function. The fifth-order KdV Eq. (1.211) involve the linear dispersive term u_{xxxxx} in addition to three nonlinear terms. Because the parameters α, β, γ are arbitrary and take different values, this will drastically change the characteristics of the fifth-order KdV Eq. (1.211). The most well-known fifth-order KdV equations that have been thoroughly examined are the Sawada–Kotera (SK) equation, the Caudrey–Dodd–Gibbon equation, the Lax equation, the Kaup–Kuperschmidt equation, and the Ito equation. These forms are:

(i) the Sawada–Kotera (SK) equation is given by

$$u_t + 5u^2 u_x + 5u_x u_{xx} + 5uu_{xxx} + u_{xxxxx} = 0 \tag{1.212}$$

(ii) the Caudrey–Dodd–Gibbon equation is given by

$$u_t + 180u^2 u_x + 30u_x u_{xx} + 30uu_{xxx} + u_{xxxxx} = 0 \tag{1.213}$$

It is the same as (1.212) under the transformation $u \to \frac{1}{6}u$.

(iii) the Lax equation reads

$$u_t + 30u^2u_x + 20u_xu_{xx} + 10uu_{xxx} + u_{xxxxx} = 0 \quad (1.214)$$

(iv) the Kaup–Kuperschmidt (KK) equation reads

$$u_t + 20u^2u_x + 25u_xu_{xx} + 10uu_{xxx} + u_{xxxxx} = 0 \quad (1.215)$$

and is characterized by $\beta = \frac{5}{2}\gamma$, $\alpha = \frac{1}{5}\gamma^2$. The KK equation is known to be integrable and possess bilinear representations.

(v) the Ito equation

$$u_t + 2u^2u_x + 6u_xu_{xx} + 3uu_{xxx} + u_{xxxxx} = 0 \quad (1.216)$$

It was found that the SK, CDG, Lax, and KK equations belong to the completely integrable hierarchy of higher-order KdV equations. These four equations have infinite sets of conservation laws, and therefore these equations have N-soliton solutions. However, the Ito equation is not completely integrable but has a limited number of special conservation laws.

3. The third family reads

$$u_{ttt} - u_{txxxx} - \alpha(u_tu_x)_{xx} - \beta(u_xu_{xt})_x = 0 \quad (1.217)$$

or equivalently

$$u_{ttt} - u_{txxxx} - \alpha(u_tu_x)_{xx} - \beta(u_xu_{xx})_t = 0 \quad (1.218)$$

The third family is integrable only if $\alpha = \beta$.

1.18.12 *Seventh-order KdV equation*

The history of the seventh-order Korteweg–de Vries (KdV) equation is relatively niche compared to its more well-known counterparts, like the classic third-order KdV equation or even the fifth-order KdV equation. One notable aspect of the seventh-order KdV equation is its increased complexity compared to lower-order variants. The seventh-order KdV equation is a PDE that describes the propagation of long waves in certain physical systems. It is an extension of the well-known KdV equation, which is a third-order nonlinear PDE. The seventh-order KdV equation is a nonlinear PDE, which means that it can exhibit soliton solutions, much like the original KdV equation. The seventh-order KdV equation arises in various physical contexts, such as fluid dynamics, plasma physics, and nonlinear optics. It's a mathematical model used to describe the behavior of certain types of waves and is a higher-order extension of the KdV equation, which itself is a fundamental equation in the study of solitons and wave phenomena.

$$u_t + u_{xxxxxxx} + \alpha u u_x + \beta u_{xxx} = 0. \tag{1.219}$$

Here, u is the wave amplitude.

1.18.13 *Generalized KdV equation*

In studying long waves, tides, solitary waves, and related phenomena, one is led to an equation of the form

$$u_t + (m+1)(m+2)u^m u_x + u_{xxx} = f(x,t) \tag{1.220}$$

where $f(x,t)$ is a given function and $m = 1, 2, \ldots$, with $u,\ u_x, u_{u_{xx}} \to 0$ as $|x| \to \infty$. This equation is referred to as a generalized KdV equation. If $m = 0$, $m = 1$, and $m = 2$, (1.220) becomes a linearized KdV equation, respectively.

1.18.14 *Gardner equation*

Long solitary waves are frequently observed in density-layered shallow water, according to remote sensing and in situ studies. Despite its efficiency in representing solitary waves across a large range of parameters, the KdV model fails in some cases, such as when certain key parameter values cause the nonlinear term coefficient in the equation to be dismissed. To overcome such issues, the KdV equation must be extended to include higher-order nonlinearities in addition to quadratic nonlinearities. These extended models frequently find value in numerous sectors of application.

When dual nonlinearity is taken into account in the KdV model, the Gardner equation emerges, which claims validity for the larger parametric zone for internal wave motion in many physical settings. Clifford Gardner, a notable mathematician, introduced this equation in 1968. There is some great research on Gardner's mode for modeling internal waves in the extended parametric domain. The Gardner equation is essentially a combination of the KdV equation and the mKdV equation. The Gardner equation is also called the combined KdV–mKdV equation. The Gardner equation is

$$\frac{\partial u}{\partial t} + au\frac{\partial u}{\partial x} + bu^2\frac{\partial u}{\partial x} + c\frac{\partial^3 u}{\partial x^3} = 0. \tag{1.221}$$

where $u(x,t)$ is the amplitude of the wave. Also, a and b are the coefficients of the quadratic and cubic nonlinear terms, respectively, whereas c presents a dispersive effect.

1.18.15 *Geophysical KdV equation*

The geophysical KdV equation is a modification of the KdV equation that is used to simulate certain waves and phenomena in geophysics, particularly oceanography and atmospheric sciences. It takes into account elements such as water density, depth, Coriolis force, and atmospheric properties. The equation investigates internal waves, Rossby waves, and wave interactions, which aid in weather forecasting, climate modeling, and ocean circulation research. The simplified form of the geophysical KdV equation is

$$\frac{\partial u}{\partial t} + a\frac{\partial u}{\partial x} + bu\frac{\partial u}{\partial x} + c\frac{\partial^3 u}{\partial x^3} = 0. \tag{1.222}$$

where u, a, b, and c represent the wave amplitude, advection (the horizontal transport of a quantity), nonlinearity, and dispersion respectively.

1.18.16 *Potential KdV equation*

The potential KdV equation is a significant variation of the traditional KdV equation that is frequently applied to soliton theory and integrable systems. The term "potential" in the potential KdV equation refers to the relationship it has with a certain potential function. It is derived from the Lax pair formulation, a mathematical framework for studying integrable systems that includes a spectral issue and a compatibility requirement.

The potential KdV equation has a history that dates back to the 1960s and 1970s when there were considerable advances in our understanding of solitons and integrable systems. Peter Lax and his associates made a significant contribution by developing the Lax pair formulation for some nonlinear PDEs, including the KdV problem. In the context of the KdV equation, the Lax pair is made up of two linear differential equations involving a spectral parameter and a potential function. The nonlinear KdV equation is obtained by applying the compatibility condition to these equations. When specific potential function options are selected, the potential KdV equation emerges. The potential KdV equation is

$$\frac{\partial u}{\partial t} + a\left(\frac{\partial u}{\partial x}\right)^2 + b\frac{\partial^3 u}{\partial x^3} = 0. \tag{1.223}$$

where $u(x,t)$ is the wave amplitude. The second and third terms of the equation are the potential term and the dispersion term, respectively. The parameters a, b are the strengths of the nonlinear potential and dispersion, respectively.

1.18.17 *Damped forced KdV equation*

The damped-forced KdV equation is an extension of the KdV equation that incorporates damping and external forcing effects. It is used to simulate specific sorts of waves and events in nonlinear systems that are sensitive to damping and external pressures. The simplified form of the damped-forced KdV equation is given by

$$\frac{\partial u}{\partial t} + au\frac{\partial u}{\partial x} + b\frac{\partial^3 u}{\partial x^3} + cu = f(x,t). \tag{1.224}$$

where $u(x,t)$ is the wave amplitude. The terms auu_x, bu_{xxx}, and cu are the nonlinear advection term, the dispersion term, and the damping term, respectively. The damping term dissipates energy from the system. $f(x,t)$ is an external forcing term that may rely on both space and time, introducing external impacts on wave dynamics. The parameters a, b, and c govern the strengths of the nonlinear advection, dispersion, and damping effects, respectively.

The damped forced KdV equation is relevant to the study of wave propagation in dissipative and driven nonlinear systems. It has applications in a variety of domains, including fluid dynamics, plasma physics, nonlinear optics, and geophysical sciences, where waves are affected by damping and external forces. Solving this equation enables academics to better understand wave behavior under different situations and forecast how nonlinear systems will respond to external pressures.

1.18.18 *Cylindrical KdV equation*

The cylindrical KdV (cKdV) equation is an nonlinear PDE that represents wave evolution in cylindrical coordinates. It extends the KdV equation, which describes waves in one spatial dimension, to the case of cylindrical symmetry. The simplified form of the cKdV equation is

$$\frac{\partial u}{\partial t} + au\frac{\partial u}{\partial x} + b\frac{\partial^3 u}{\partial x^3} + \frac{1}{2t}u = 0 \tag{1.225}$$

where $u(x,t)$ is the wave amplitude. The second, third, and fourth terms of Eq. (1.225) represent, respectively, the nonlinear advection term, dispersion term, and cylindrical term. Also, the parameters a and b are the strengths of the nonlinear advection and dispersion, respectively.

1.18.19 *Spherical KdV equation*

The spherical KdV (sKdV) equation, an nonlinear PDE, describes the evolution of waves in cylindrical coordinates. It is an extension of the KdV equation, which describes waves in one spatial dimension, to cylindrical symmetry. The sKdV equation is given by

$$\frac{\partial u}{\partial t} + au\frac{\partial u}{\partial x} + b\frac{\partial^3 u}{\partial x^3} + \frac{1}{t}u = 0 \tag{1.226}$$

where $u(x, t)$ is the wave amplitude. The second, third, and fourth terms of (1.226) respectively represent nonlinear advection terms, dispersion terms, and spherical terms. Here, the parameters a, b are the strengths of the nonlinear advection, dispersion, respectively.

1.18.20 *Schamel–KdV (SKdV) equation*

The Schamel equation (S-equation) is an NLPDE. This equation, similar to the KdV, describes how a localized, coherent wave structure propagates in a nonlinear dispersive medium. Hans Schamel developed it in 1973 to describe the effects of electron trapping in the potential trough of a solitary electrostatic wave structure moving at ion acoustic speed in a two-component plasma. The Schamel–KdV equation is given by

$$\frac{\partial u}{\partial t} + a\sqrt{u}\frac{\partial u}{\partial x} + bu\frac{\partial u}{\partial x} + c\frac{\partial^3 u}{\partial x^3} = 0, \tag{1.227}$$

where $u(x, t)$ is the wave amplitude. The coefficients a, b, and c are the activation of trapping, convection, and dispersion. in the solitary waves, respectively. The SKdV equation, which balances the effects of nonlinear convection and linear diffusion, is the simplest second-order nonlinear equation to formulate.

1.18.21 *Benjamin–Bona–Mahony equation*

For the purpose of analyzing long wave propagation (where nonlinear dispersion is incorporated), T. B. Benjamin, J. L. Bona, and J. J. Mahony developed the Benjamin–Bona–Mahony (BBM) equation in 1972. In order to describe the unidirectional propagation of weakly long dispersive waves over inviscid fluids as an alternative to the KdV equation, the BBM equation was developed. As a result, the KdV-type equation family includes the BBM equation. The BBM equation is a better model for long waves than the KdV equation, and the regularity results for its solutions were utilized to make use of the boundedness of the BBM equation's dispersion relation. The KdV equation is a model for the propagation of one-dimensional, small-amplitude, weakly dispersive waves. In contrast to the KdV equation, which can be solved entirely using the inverse scattering method, the BBM problem is discovered to not be entirely integrable, and as a result, it lacks N-soliton solutions. The BBM equation has a stable solitary wave solution and dispersive properties. Both the BBM and KdV equations are applicable for studying shallow water waves, surface waves of long wavelengths in liquids, acoustic-gravity waves in compressible fluids, hydromagnetic waves in cold plasma, acoustic waves in anharmonic crystals, etc.

The BBM equation is given by

$$u_t + u_x + uu_x - u_{xxt} = 0, \; x \in R, \; t > 0 \tag{1.228}$$

Also, the KdV equation can be written as

$$u_t + u_x + uu_x - u_{xxx} = 0, \; x \in R, \; t > 0 \tag{1.229}$$

The approximate dispersion relations for the two linearized equations can be compared to quickly identify the fundamental mathematical differences between the BBM and KdV equations. We are looking for a plane wave solution to both linearized equations of the form

$$u(x,t) \sim exp[i(\omega t - kx)].$$

The dispersion relation of the linearized KdV equation, the phase velocity, and the group velocity are respectively given by

$$\omega = k - k^3, \; C_p = \frac{\omega}{k} = 1 - k^2, \; C_g = \frac{d\omega}{dk} = 1 - 3k^2$$

In the case of $k^2 > 1$, the phase velocity and group velocity both become negative. This indicates that, contrary to the original assumption that waves only move in the positive x-direction, all waves with large wavenumbers (small wavelengths) propagate in the negative x-direction. This is a problematic aspect of the KdV equation's physical structure. Equation (1.228) attempts to do away with this unrealistic aspect of the KdV equation. In the linearized form of (1.228), the dispersion relation, phase velocity, and group velocity are respectively

$$\omega = \frac{k}{(1+k^2)}, \; C_p = \frac{\omega}{k} = (1+k^2)^{-1}, \; C_g = (1-k^2)(1+k^2)^{-2}$$

Short waves cannot propagate because both C_p and C_g trend to zero as $k \to \infty$. The BBM model roughly exhibits the property of significantly responding primarily to short-wave components included in the initial wave form. The BBM equation appears to be a better model as a result. However, it has not yet been clearly established that the BBM model is superior to the KdV model.

1.18.22 *Burgers–Huxley equation*

The Burgers–Huxley (BH) equation is an extended form of the famous Burgers–Huxley equation. The BH equation is a prototype model for describing the interaction between reaction mechanisms, convection effects, and diffusion transport. It finds its application in many fields, such as chemistry, biology, mathematics, engineering, and nonlinear acoustics. The generalized form of the BH equation is given as

$$u_t + \alpha u^\delta u_x - u_{xx} = \beta(1-u^\delta)(u^\delta - \gamma)u, \quad 0 \le x \le 1, \ t \ge 0 \tag{1.230}$$

where $\alpha(>0)$ is the advection coefficient, $\beta(\ge 0)$ is the reaction coefficient, and γ and $\delta(>0)$ are real coefficients with $0 < \gamma < 1$. Here, u_{xx} is the diffusive term, $u^\delta u_x$ is the advection term, and $(1-u^\delta)(u^\delta-\gamma)u$ is the reaction term.

When $\alpha \neq 0$, $\beta \neq 0$, $\delta = 1$, Eq. (1.230) becomes the BH equation, which is of the form

$$u_t + \alpha u u_x - u_{xx} = \beta u(1-u)(u-\gamma) \tag{1.231}$$

When $\alpha = 0$, $\delta = 1$, Eq. (1.230) reduced to the Huxley equation, which is given as

$$u_t - u_{xx} = \beta u(1-u)(u-\gamma) \tag{1.232}$$

This equation describes nerve pulse propagation in nerve fibers and wall motion in liquid crystals.

Also, when $\beta = 0$, $\delta = 1$, Eq. (1.230) becomes the following Burgers equation:

$$u_t + \alpha u u_x - u_{xx} = 0. \tag{1.233}$$

1.18.23 *Kadomtsev–Petviashvili equation*

The Kadomtsev–Petviashvili (KP) equation, a nonlinear PDE with dimension (2+1), describes the wave processes in various nonlinear mediums.

This equation can also be termed the (2+1)-dimensional KdV equation. This equation was first presented in 1970 by two Soviet scientists, Boris B. Kadomtsev (1928–1998) and Vladimir I. Petviashvili (1936–1993). This equation is also modeled as acoustic waves in plasma under the effects of dissipation at long transverse disturbances, just like the KdV equation. So, the KdV equation's application to two special dimensions follows naturally. In nonlinear optics and other related sciences, it is also applicable. The KP equation can also be used to model waves in ferromagnetic media, as well as two-dimensional matter-wave pulses in Bose–Einstein condensates. The KP equation is of the form

$$(u_t + Auu_x + Bu_{xxx})_x + Cu_{yy} = 0 \tag{1.234}$$

where $u = u(x, y, t)$ is a real-valued function of two spatial variables, x, y, and the time variable t, A, B, and C are real constants. The KP equation is reduced to the KdV equation when $C = 0$. In the case, where $C < 0$, the equation is referred to as the KP-I equation, which is a good model when the surface tension is strong. The KP-II equation, which is a good model when surface tension is weak or nonexistent, is the equation that is used when $C > 0$.

1.18.24 *Zakharov–Kuznetsov equation*

In 1974, Zakharov and Kuznetsov derived a nonlinear equation named the ZK equation as a nonlinear evolution equation to study IAWs in a strongly magnetized plasma in the z-dimension. The dynamics of IASWs are the silent character in the nonlinear phenomena of modern plasma physics. Nonlinear differential equations, such as the KdV equation and the mKdV equation, describe the different properties of nonlinear plasma waves in one dimension. The Zakharov–Kuznetsov (ZK) equation and the KP equation describe the different properties of nonlinear plasma waves in more than one dimension. Zakharov and Kuznetsov were derived and used to study three-dimensional acoustic ion waves in magnetized plasma. This equation regulates the nature of weakly nonlinear waves in a plasma that consists of cold ions and hot super-thermal electrons in a magnetic field.

The general form of the ZK equation is

$$u_t + auu_x + bu_{xxx} + c(u_{yy})_x = 0. \tag{1.235}$$

Here, the electrostatic wave potential is represented by $u(x, y, t)$, which is the function of spatial variables x, y and temporal variables t. a, b, and c, respectively, denote the coefficients of the nonlinearity, dispersion, and disturbed wave velocity in the direction of y. The ZK equation is extensively utilized in describing the weakly nonlinear ion-acoustic waves in a strongly magnetized lossless plasma comprised of cold ions and hot isothermal electrons. This equation is a two-dimensional generalization of the well-known KdV equation, which is spatially limited as the 1D model of weakly nonlinear waves in shallow water. The 2D quadratic ZK equation governs, for example, weakly nonlinear ion-acoustic waves in a plasma comprising cold ions and hot isothermal electrons in the presence of a uniform magnetic field (Monro and Parkes, 1999). Melkonian and Maslowe (1989), this equation appears as the amplitude equation for two-dimensional long waves on the free surface of a thin film flowing down a vertical plane with moderate values of the fluid surface tension and large viscosity. While originally the equation was proposed by Zakharov and Kuznetsov in the 3D setting, see Zakharov and Kuznetsov (1974); the first rigorous derivation was done in Lannes (2013) from the Euler–Poisson system.

1.18.25 *KdV–Burgers' equation*

The KdV–Burgers' equation combines the KdV equation with the Burgers' equation. It appears in fluid dynamics and nonlinear wave theory, and it depicts a system with both dispersive and dissipative characteristics. The KdV–Burgers' equation is

$$\frac{\partial u}{\partial t} + au\frac{\partial u}{\partial x} + b\frac{\partial^2 u}{\partial x^2} - c\frac{\partial^3 u}{\partial x^3} = 0. \tag{1.236}$$

where a, b, and c are the strengths of advection, dispersion, and higher-order dispersion effects, respectively. The KdV–Burgers' equation appears in a variety of physical systems, including fluid dynamics, nonlinear optics, and plasma physics, where both dispersive and dissipative effects are important. Solving this equation allows researchers to better understand the behavior of waves in these systems and forecast phenomena such as soliton production, wave breaking, and turbulence.

1.18.26 *KP–Burgers' equation*

The KP–Burgers' equation is an NLPDE that incorporates features from both the KP and Burgers' equations. This equation appears in the study of physical systems that exhibit both dispersive and dissipative effects. The KP equation is an integrable PDE that describes the development of two-dimensional, weakly nonlinear waves, specifically shallow water waves. It is a crucial model for studying soliton theory and integrable systems. The Burgers' equation, on the other hand, represents the dynamics of viscous fluid flow by including both nonlinear convection and viscous diffusion factors. The KP–Burgers' equation is of the form

$$\frac{\partial}{\partial x}\left(\frac{\partial u}{\partial t} + au\frac{\partial u}{\partial x} + b\frac{\partial^2 u}{\partial x^2} + c\frac{\partial^3 u}{\partial x^3}\right) + d\frac{\partial^2 u}{\partial y^2} = 0. \tag{1.237}$$

where a, b, c, and d are arbitrary constants. The KP–Burgers' equation is being studied because it has applications in a variety of physical systems, including fluid dynamics, plasma physics, and nonlinear optics. It sheds light on the complex dynamics of wave propagation in dispersive and dissipative media, with applications in a wide range of science and engineering disciplines. The evolution of the KP–Burgers' equation is inextricably linked to the advancement of integrable systems, nonlinear wave equations, and the study of complex wave phenomena. Its examination is still ongoing, with academics looking into its features, solutions, and uses in a variety of settings.

1.18.27 *ZK–Burgers' equation*

The ZK–Burgers' equation incorporates parts from both the ZK and Burgers' equations. The ZK equation is an NPDE that characterizes wave propagation in dispersive media, whereas Burgers' equation represents the dynamics of viscous fluid flow. The ZK–Burger equation is of the form

$$\frac{\partial u}{\partial t} + au\frac{\partial u}{\partial x} + b\frac{\partial^2 u}{\partial x^2} + c\frac{\partial^3 u}{\partial x^3} + d\frac{\partial}{\partial x}\left(\frac{\partial^2 u}{\partial y^2}\right) = 0. \tag{1.238}$$

Here, b is the kinematic viscosity, and a and c are coefficients determining the strength of the nonlinear and dispersive effects, respectively. d is the disturbed wave velocity in the direction of y. The ZK–Burgers' equation describes the combined effects of dispersion, nonlinearity, and viscosity in wave propagation phenomena. It occurs in many physical systems where these effects are considerable, including fluid dynamics, plasma physics, and nonlinear optics. The ZK–Burgers' equation is being studied because of its importance in understanding the dynamics of wave propagation in complex media, as well as its applicability in a wide range of science and engineering disciplines. Researchers research its features, solutions, and behaviors in order to gain insight into the underlying physical processes and create predictive models for practical use.

1.18.28 *Navier–Stokes equation*

The Navier–Stokes equations are fundamental PDEs used to describe the motion of viscous fluids. They are named after Claude–Louis Navier and George Gabriel Stokes, who each contributed to the development of fluid dynamics in the 19th century. The Navier–Stokes equations for an incompressible fluid in three-dimensional Cartesian coordinates take the following form:

$$\frac{\partial \mathbf{v}}{\partial t} + (\mathbf{v}.\nabla)\mathbf{v} = -\frac{1}{\rho}\nabla \mathbf{p} + \nu\nabla^2\mathbf{v} + \mathbf{f}. \tag{1.239}$$

where $\mathbf{v}$, t, $\mathbf{p}$, ρ, ν, and $\mathbf{f}$ are the velocity vector fields of the fluid, time, the pressure, the fluid density, the kinematic viscosity, and the external body forces per unit volume acting on the fluid (e.g., gravity), respectively. Also, ∇, ∇^2, and ∇ denote the gradient operator, the Laplacian operator, and the divergence operator, respectively.

The first equation is the momentum equation, which describes the conservation of momentum for the fluid. It states that the change in velocity with respect to time is equal to the sum of convective acceleration, pressure gradient force, viscous force, and external body forces.

The second equation is the continuity equation, which expresses the conservation of mass for the fluid. It states that the divergence of the velocity field is zero, indicating that the fluid is incompressible.

The Navier–Stokes equations are highly nonlinear and exhibit turbulent behavior in many practical situations. They are of paramount importance in various fields of science and engineering, including fluid mechanics, aerodynamics, weather prediction, and oceanography. However, despite their importance, a complete mathematical understanding of the Navier–Stokes equations, particularly their behavior in turbulent flow, remains one of the most unsolved problems in mathematics and physics.

1.18.29 *Schrödinger equation*

In quantum physics, the Schrödinger equation describes how a physical system's quantum state changes over time. It was conceived by Austrian physicist Erwin Schrödinger in 1925. A single non-relativistic particle's time-dependent Schrödinger equation is as follows:

$$i\hbar\frac{\partial}{\partial t}\psi(\mathbf{r},t) = -\frac{\hbar^2}{2m}\nabla^2\psi(\mathbf{r},t) + \mathbf{V}(\mathbf{r},t)\psi(\mathbf{r},t). \tag{1.240}$$

Here, $\psi(\mathbf{r},t)$, t, and $\mathbf{r}$ are the wave function of the particle, time, and the position vector in three-dimensional space. Also ι, $\hbar = \frac{h}{2\pi}$, and m is the imaginary unit, the reduced Planck constant, and the mass of the particle. $\mathbf{V}(\mathbf{r},t)$ is the potential energy function, which depends on the position and time. The Schrödinger equation describes the evolution of the wave function $\psi(\mathbf{r},t)$ in time. The term on the right-hand side represents the total energy of the particle, which is the sum of its kinetic energy $(-\frac{\hbar^2}{2m}\nabla^2\psi(\mathbf{r},t))$ and potential energy $(\mathbf{V}(\mathbf{r},t)\psi(\mathbf{r},t))$.

The Schrödinger equation has versions for different physical conditions, such as the time-independent Schrödinger equation for stationary states and the relativistic Dirac equation. The Schrödinger equation is applied in many domains, including atomic and molecular physics, solid-state physics, quantum chemistry, theoretical physics, quantum optics and photonics, nanotechnology, quantum computing, and particle physics.

1.18.30 *Nonlinear Schrödinger equation*

When leading-order nonlinear and dispersive effects are taken into account in a dispersive system, the envelope of a small-amplitude narrow-wave pulse may emerge, satisfying the nonlinear Schrödinger equation (NLSE). In certain situations, this problem can be solved using envelope soliton. Envelope soliton refers to a set of waves that propagate without changing their form. Envelope soliton endures when higher-order effects are taken into account. There are two varieties of envelope soliton: bright and dark. Ion acoustic dark or bright envelope solitons are created to modulate both stable and unstable plasma waves.

The standard form of the NLSE is

$$i\frac{\partial\psi}{\partial\tau} + P\frac{\partial^2\psi}{\partial\xi^2} + Q|\psi|^2\psi = 0 \tag{1.241}$$

where ψ denotes the electric potential correction ϕ_1^1. The group velocity dispersion coefficient $P(=\frac{1}{2}\frac{d^2\omega}{dk^2})$ is a real-valued parameter, and Q is the cubic nonlinear (local) coefficient. The coefficient P is proportional to the group dispersion $\frac{dv_g}{dK}$, and the coefficient Q is proportional to the nonlinear frequency shift.

If the amplitude of the plasma wave is not small, nonlinearities give rise to a modulation of the wave amplitude. When the amplitude varies slowly throughout oscillation, the equation describing the evolution of the wave amplitude (in certain cases) becomes the nonlinear Schrödinger equation.

In plasma, it is found that under certain conditions, the conversion of an initially uniform wave train into a spatially modulated wave proves to be energetically favorable. This effect is known as *modulationally instability.* The plane wave solutions of Eq. (1.241) are modulationally unstable if $PQ > 0$. Let us rewrite Eq. (1.241) as

$$i\frac{\partial \psi}{\partial \tau'} = -\frac{\partial^2 \psi}{\partial \xi^2} - \frac{Q}{P}|\psi|^2\psi \tag{1.242}$$

where $\tau' = P\tau$. The Schrödinger equation for quasiparticles, the wave function of which is given by ψ, is trapped by a self-generated nonlinear potential $V = -\frac{Q}{P}|\psi|^2$. For $PQ > 0$, this potential will have an attractive sign. If somehow the quasiparticle density $|\psi|^2$ increases, the depth of this potential increases; more particles are attracted, leading to a further increase in the potential depth. In this respect, the instability may be the result of the self-trapping of the quasiparticles.

For electron plasma waves, the modulational instability can be easily understood in terms of the pondermotive force. When an electromagnetic wave of vary high intensity interacts with plasma particles in a plasma nonlinear process, the force due to radiation pressure is coupled to the plasma particles and is called the pondermotive force. Suppose we have a plasma wave with a ripple on the envelope. The gradient in wave intensity causes a pondermotive force, which pushes plasma out of the region with maximum intensity towards the region with minimum intensity.

It causes a ripple, in a density depression in the region of maximum intensity. The dispersion relation is

$$\omega^2 = \omega_{pe}^2 + \frac{3}{2}k^2V_{Th}^2 \tag{1.243}$$

It indicates that waves of large k can exist only in a region of small ω_{pe} or low density. Thus, the density of the region's depression can trap plasma waves, leading to further enhancement of the local field and, hence, instability.

During the nonlinear evolution of the wave, when the nonlinear effects are balanced by the dispersion effect, a stable nonlinear wave structure is formed, called the envelope soliton. When both P and Q are positive, Eq. (1.241) has a solitary wave solution.

1.19 Dispersive Waves

Wave components do not need to travel at the same phase speed. Non-dispersive waves are those with the same phase speed. Dispersive waves occur when wave components do not have the same phase speed. Wave energy in dispersive waves travels at a group speed that differs from the speed of the wave component crests (phase speed). Wave energy in non-dispersive waves moves at phase speed.

Dispersive effects can be classified as either linear or nonlinear dispersive waves, depending on the circumstances. Linear dispersive waves have a linear dispersion between frequency and wavenumber. In a linear dispersive system, the wave shape changes over time as different frequencies propagate at different rates. However, the wave's amplitude remains constant throughout its transmission. In contrast, nonlinear dispersive waves have a nonlinear dispersion relation.

In the past, Stokes' groundbreaking research on water waves in 1847 served as the impetus for the study of nonlinear dispersive waves. The existence of periodic wave trains in nonlinear dispersive wave systems was initially demonstrated by Stokes. Additionally, he discovered that nonlinear wave behavior undergoes notable qualitative changes as a result of the dispersion relation on amplitude. Beyond just correcting linear results, it also presents a number of novel phenomena in the theory of dispersive waves. The field of nonlinear dispersive wave phenomena in general and nonlinear water waves in particular has greatly benefited from these basic concepts and Stokes' findings. Considered to be the origin of the modern theory of nonlinear dispersive waves, Stokes conducted extensive research on water waves. Water wave research really served as the inspiration for many of the basic ideas and findings surrounding nonlinear dispersive waves. Over the past three decades, research on nonlinear dispersive waves has advanced remarkably quickly.

1.19.1 *Linear dispersive waves*

We consider a dynamical problem so that a small disturbance $\phi(x,t)$ with reference to the undisturbed stable state is governed by a linear PDE with constant coefficients:

$$P\left(\frac{\partial}{\partial t}, \frac{\partial}{\partial x_1}, \frac{\partial}{\partial x_2}, \frac{\partial}{\partial x_3}\right)\phi(x,t) = F(x,t) \tag{1.244}$$

where P is a polynomial in partial derivatives, t and $x = (x_1, x_2, x_3)$ are independent time and space variables. The term $F(x,t)$ represents the action of external forces on the dynamical system and is usually referred to as a given forcing term.

Let us consider a general linear PDE in x and t with constant coefficients

$$P\left(\frac{\partial}{\partial t}, \frac{\partial}{\partial x_1}, \frac{\partial}{\partial x_2}, \frac{\partial}{\partial x_3}\right)\phi(x,t) = 0 \tag{1.245}$$

where P is a polynomial.

The elementary plane wave solution of (1.245) is

$$\phi(x,t) = Ae^{i(\kappa\cdot x - \omega t)}, \tag{1.246}$$

where A, ω, $\kappa = (k_1, k_2, k_3)$ represents the amplitude, frequency and wavenumber vector respectively.

Replacing $\frac{\partial}{\partial t} \leftrightarrow -i\omega$, $\frac{\partial}{\partial x_j} \leftrightarrow ik_j$ in (1.245), we get

$$P(-i\omega, ik_1, ik_2, ik_3)A = 0 \tag{1.247}$$

If this has a solution for real ω and κ, then (1.245) is dispersive, and the dispersion relation must be

$$D(\omega, k) \equiv P(-i\omega, ik_1, ik_2, ik_3) = 0 \tag{1.248}$$

D contains all of the information about the problem. By reversing the above argument, the original equation can be reconstructed from the dispersion relation. If, in particular, the dispersion relation (1.247) can be solved explicitly in terms of real roots given by

$$\omega = \Omega(\kappa) = \Omega(k_i), \tag{1.249}$$

then there may be several possible roots for the frequency, corresponding to different modes of propagation. A significant quantity associated with each mode is the group velocity. The phase velocity of the waves is the velocity at which a surface of a constant phase moves. It is defined by the relation

$$c = \frac{\omega}{\kappa}\hat{\kappa}, \tag{1.250}$$

where $\hat{\kappa}$ is the unit vector in the direction of the wavenumber vector κ. Thus, for any particular wave mode $\omega = \Omega(\kappa)$, the phase velocity depends on the wave numbers k_1, k_2, and k_3. In other words, different waves propagate with different phase velocities, and such waves are called dispersive provided the determinant $|\frac{\partial^2\omega}{\partial k_i \partial k_j}| \neq 0$. On the other hand, waves are called non-dispersive if the phase velocity does not depend on the wave number.

For one-dimensional dispersive waves, $\omega''(k) \neq 0$, i.e., $\omega'(k)$ is not constant. So, the group velocity is not constant.

However, in general, the governing dispersion relation for one-dimensional waves is $\omega = \Omega(\kappa)$, which may give a complex ω for a real

κ. In such a case, the phase velocity depends not only on the wavenumber but also on the amplitude of the waves. This means that the amplitude will grow or decay in time accordingly, as $Im(\omega) > 0$ or < 0. So, the former case leads to instability.

Note: (i) For the three-dimensional wave motion, the group velocity is $C(\kappa) = \nabla_\kappa \omega = (\frac{\partial\omega}{\partial k_1}, \frac{\partial\omega}{\partial k_2}, \frac{\partial\omega}{\partial k_3})$, where $k = (k_1, k_2, k_3)$ is the wave number vector.

(ii) For one-dimensional waves, the group velocity $C(\kappa) = \frac{d\omega}{d\kappa}$.

(iii) The group velocity plays a fundamental role in the theory of dispersive waves and is much more important than the phase velocity.

The dispersion relation of waves is given in Table 1.1.

The phase and the group velocities associated with these examples can readily be calculated. For dispersive waves, the former are different from the latter. The general solution for any dispersive wave system can then be obtained by the Fourier superposition of the plane wave solution (1.246) for different wavenumbers (k_1, k_2, k_3) with the corresponding frequencies ω to satisfy the dispersion relation (1.249). The solution is represented by

$$\phi(x,t) = \int_{-\infty}^{\infty}\int_{-\infty}^{\infty}\int_{-\infty}^{\infty} F(\kappa)exp[i(\kappa.x - \Omega(\kappa)t)]d\kappa \tag{1.251}$$

where the spectrum function $F(\kappa)$ is determined from appropriate initial or boundary conditions.

Although (1.251) is the exact solution, the principal features of the dispersive waves cannot be described without an exact or approximate evaluation of the integral in (1.251). We shall discuss this point in the next section, which is concerned with initial-value problems. It is relevant to mention here that, in general, the governing equations for dispersive waves are inherently nonlinear, and the corresponding dispersion relation is also nonlinear in the sense that the frequency ω is not only a function of κ but also of other parameters, such as amplitude and local properties of the medium. The theory of water waves provides an excellent example of nonlinear dispersion. In the study of linear dispersive systems, the preceding parameters are assumed to be small, and the governing equations for such systems are obtained by linearizing the original nonlinear equations. The classical theory of water waves satisfying Laplace's equation and the linearized free surface conditions can again be cited as excellent examples of the process of linearization. However, in his pioneering work based upon averaged variational principles, Whitham (1974) gave a new description of nonlinear water waves and a completely different approach to nonlinear

Table 1.1. Linear dispersive waves.

	Equation	Dispersion Relation
(i)	Surface waves on water of constant depth h	$\omega^2 = gk \tanh kh$
(ii)	The KdV equation for long water waves $\phi_t + c_0\phi_x + \alpha\phi_{xxx} = 0$	$\omega = c_0 k - \alpha k^3$, $c_0^2 = gh$
(iii)	The Boussinesq equation for long water waves $\phi_t t - c_0^2\phi_x x - \beta^2\phi_{xxtt} = 0$	$\omega^2 = c_0^2 k^2(1+\beta^2 k^2)^{-1}$
(iv)	The BBM equation $\phi_t - c_0\phi_x - \alpha\phi_{xxt} = 0$	$\omega = -c_0 k(1+\alpha k^2)^{-1}$
(v)	The Klein–Gordon equation $\phi_t t - c^2\nabla^2\phi + d^2\phi = 0$	$\omega = \pm(c^2\kappa^2 + d^2)^{1/2}$
(vi)	The internal waves in a stratified ocean, governed by the equation $(\frac{\partial}{\partial t} + U\frac{\partial}{\partial x})^2(\frac{\partial\psi}{\partial x^2} + \frac{\partial\psi}{\partial z^2}) + N^2\frac{\partial\psi}{\partial x^2} = 0$, where $\psi(x,z,t)$ represents the stream function	$\omega = Uk \pm \frac{Nk}{\sqrt{k^2+m^2}}$, $N = (-\frac{g}{\rho_0}\frac{\partial\bar{\rho}}{\partial z})^{1/2}$
(vii)	The inertial waves in a rotating liquid, governed by the equation $(\frac{\partial}{\partial t} + U\frac{\partial}{\partial z})\nabla^2\chi + f^2\frac{\partial\chi}{\partial z^2} = 0$, $(f = 2\Omega)$	$\omega = iUm \pm \frac{fm}{\kappa}$
(viii)	The internal-inertial waves in a rotating stratified ocean, governed by the equation $(\frac{\partial^2}{\partial t^2} + N^2)\nabla_h^2\chi + (\frac{\partial^2}{\partial t^2} + f^2)\chi_{zz} = 0$, where $\nabla_h^2 = \frac{\partial^2}{\partial x^2} + \frac{\partial^2}{\partial y^2}$	$\omega^2 = \frac{f^2m^2 + N^2(k^2+l^2)}{\kappa^2}$
(ix)	The Rossby waves in a β-plane ocean, governed by the equation $(\nabla_h^2 - a^2)\psi_t + \beta\psi_x = 0$ $(a^2 = \frac{f^2}{c_0^2})$	$\omega = -\frac{\beta k}{k^2+l^2+a^2}$.
(x)	The Alfvén gravity waves in an electrically conducting liquid of constant depth h	$\omega^2 = (gk + a^2k^2)\tanh kh$
(xi)	Electromagnetic waves in plasmas	$\omega^2 = \omega_p^2 + c^2k^2$
(xii)	The Schrödinger equation and de Broglie waves $i\hbar\phi_t = V\psi - \frac{\hbar^2}{2m}\nabla^2\psi$, where $h(=2\pi\hbar)$ is the Planck constant and V is a constant potential	$\omega = \frac{\hbar k^2}{2m} + \frac{V}{\hbar}$

dispersive waves in general. It has become clear from Whitham's theory that nonlinear dispersion is much more important than the corresponding linearized concept and plays a significant role in the general theory of dispersive wave systems.

1.19.2 *Nonlinear dispersive waves*

To describe a slowly evolving nonlinear and non-uniform oscillatory wave-train in a dispersive medium, we assume the existence of a one-dimensional solution in the form, so that

$$\phi(x,t) = a(x,t)exp(i\theta(x,t)) + c.c., \tag{1.252}$$

where $a(x,t)$ is the complex amplitude, with $n = 1$, and *c.c.* denotes the complex conjugate. The phase function $\theta(x,t)$ is given by

$$\phi(x,t) = xk(x,t) = t\omega(x,t) \tag{1.253}$$

k, ω, and a are slowly varying functions of space variable x and time t. Considering that k and ω vary slowly, it is plausible to infer that they still satisfy the dispersion relation.

$$\omega = W(k) \tag{1.254}$$

Differentiating (1.253) with respect to x and t, respectively, we obtain

$$\theta_x = k + (x - tW'(k))k_x \tag{1.255}$$

$$\theta_t = -W(k) + (x - tW'(k))k_t \tag{1.256}$$

$W'(k) = (x/t) > 0$ defines stationary locations, and these equations become

$$\theta_x = k(x,t), \quad \theta_t = -\omega(x,t) \tag{1.257}$$

This is the local wavenumber and local frequency of a slowly varying nonlinear wavetrain.

In view of (1.257), relation (1.254) gives a nonlinear PDE for the phase $\theta(x,t)$ in the form

$$\frac{\partial \theta}{\partial t} + W\left(\frac{\partial \theta}{\partial x}\right) = 0 \tag{1.258}$$

The geometry of the wave pattern is found by solving this equation. We eliminate θ from (1.257) to obtain the equation

$$\frac{\partial k}{\partial t} + \frac{\partial \omega}{\partial x} = 0 \tag{1.259}$$

The *Whitham equation* for the conservation of waves is this one, in which ω is the wave flux and k is the wave density.

Using (1.254), Eq. (1.259) gives

$$\frac{\partial k}{\partial t} + C(k)\frac{\partial k}{\partial x} = 0 \tag{1.260}$$

where $C(k) = W'(k)$ is the group velocity. This is the basic nonlinear wave (hyperbolic) equation for wavenumber k propagating at group velocity $C(k)$.

To find the general solution of (1.260) with the initial condition $k = f(x)$ at $t = 0$. In this case, the solution takes the form

$$k(x,t) = f(\xi), \quad x = \xi + tF(\xi) \tag{1.261}$$

where $F(\xi) = C(f(\xi))$. This further confirms the propagation of wavenumber k with group velocity C. Equations (1.260) and (1.254) reveal that ω also satisfies the first-order, nonlinear wave (hyperbolic) equation

$$\frac{\partial \omega}{\partial t} + W'(k)\frac{\partial \omega}{\partial x} = 0 \tag{1.262}$$

It follows from Eqs. (1.260) and (1.262) that both k and ω remain constant on the characteristic curves defined by

$$\frac{dx}{dt} = W'(k) = C(k) \tag{1.263}$$

in the (x,t)-plane. The characteristics are straight lines with slope $C(k)$ since k or ω is constant on each characteristic. The solution for $k(x,t)$ is given by (1.261). Finally, it follows from the preceding analysis that any constant value of the phase ϕ propagates according to $\theta(x,t)$=constant, and hence,

$$\theta_t + \left(\frac{dx}{dt}\right)\theta_x = 0 \tag{1.264}$$

which gives, by (1.257),

$$\frac{dx}{dt} = -\frac{\theta_t}{\theta_x} = \frac{\omega}{k} = c \tag{1.265}$$

Thus, the phase of the waves propagates with the phase speed c. On the other hand, Eq. (1.260) ensures that the wavenumber k propagates with group velocity $C(k) = (d\omega/dk) = W'(k)$.

Next, we study the wave energy propagation via a dispersive medium. We examine the integral that involves the square of the energy or wave amplitude between any two points $x = x_1$ and $x = x_2 (0 < x1 < x2)$:

$$Q(t) = \int_{x_1}^{x_2} |A|^2 dx = \int_{x_1}^{x_2} AA^* dx \tag{1.266}$$

$$= 2\pi \int_{x_1}^{x_2} \frac{F(k)F^*(k)}{t|W''(k)|} dx, \tag{1.267}$$

which, due to a change of variable $x = tW'(k)$, is

$$= 2\pi \int_{k1}^{k2} F(k)F^*(k)dk \tag{1.268}$$

where $x_r = tW'(k_r),\ r = 1, 2$.

When k_r is kept fixed as t varies, $Q(t)$ remains constant so that

$$0 = \frac{dQ}{dt} = \frac{d}{dt}\int_{x_1}^{x_2} |A|^2 dx = \int_{x_1}^{x_2} \frac{\partial}{\partial t}|A|^2 dx + |A|_2^2 W'(k_2) - |A|_1^2 W'(k_1) \tag{1.269}$$

In the limit, as $x_2 - x \to 0$, this result reduces to the first-order, PDE

$$\frac{\partial}{\partial t}|A|^2 + \frac{\partial}{\partial x}[W'(k)|A|^2] = 0 \tag{1.270}$$

This is the wave energy conservation equation, where the energy density and energy flux are denoted by $|A|^2$ and $|A|^2 W'(k)$, respectively. Consequently, the energy propagates with a group velocity of $W'(k)$. It is demonstrated that the group velocity is also propagated by the wavenumber k. The group velocity thus serves a dual purpose. Another significant finding of the previous analysis is that, for the three functions k, ω, and A, (1.254), (1.259), and (1.270) form a closed set of equations. These are, in fact, *Whitham's equations*, the basic formulas for nonlinear dispersive waves.

References

[1] I. N. Sneddon, *Elements of Partial Differential Equations*, Dover Publications, INC, Mineola, New York, 2006.

[2] B. B. Kadomtsev and V. I. Petviashvili, *Sov. Phys. Dokl.* 15, 539–541 1970s.

[3] A. Mukherjee and N.K. Bej, *Ordinary and Partial Differential Equations*, Shreedhar prakashani, Kolkata, India, 2011.

[4] R. K. Ghosh and K. C. maity, *An Introduction to Differential Equations,* New Central book agency (P) LTD, Kolkata, India, 2007.

[5] S. L. Ross, *Differential Equation*, John Wiley & Sons (Asia) Pte. Ltd., Singapore, 2004.

[6] P. Chatterjee, K. Roy, and U. N. Ghosh, *Waves and Wave Interaction in Plasmas*, World Scientific Publishing Co. Pte. Ltd., Singapore, 2023.

[7] Tyn Myint-U and Lokenath Debnath, *Linear Partial Differential Equations for Scientists and Engineers*, Birkhauser, Boston, 2007.

[8] G. B. Whitham, *Linear and Nonlinear Waves*, John Wiley & Sons, New York, 1973.

[9] Abdul-Majid Wazwaz, *Partial Differential Equation and Solitary Waves Theory*, Springer, Higher Education Press, Beijing, 2009.
[10] I. Shingareva C. and Lizarraga-celaya, *Solving Nonlinear Partial Differential Equations with Maple and Mathematica*, Springerwien New York, 2011.
[11] P. K. Kythe, P. Puri, Michael and R. Schaferkotter, *Partial Differential Equations and Mathematica*, CRC Press, New York, 1997.
[12] L. Debnath, *Nonlinear Partial Differential Equations for Scientists and Engineers*, Birkhauser, Boston, 2011.
[13] M. E. Taylor, *Partial Differential Equations III: Nonlinear Equations: 117 (Applied Mathematical Sciences)*, Springer International Publishing AG, Switzerland, 2023.
[14] R. A. Struble, *Nonlinear Differential Equations*, Dover Publications Inc., Moneola, New York, 2018.
[15] G. Baumann, *Mathematica for Theoretical Physics*, Springer, New York, 2005.

Chapter 2

Traveling Wave Solution Method

2.1 Introduction

The partial differential equations (PDEs) and their analytical study were stated in the 18th century because of a large set of physical situations that arose due to the works of Euler, Cauchy, d'Alembert, Hamilton, Jacobi, Lagrange, Laplace, Monge, and others. Since the middle of the 19th century, PDEs became an essential tool to study different branches of science and mathematics. Then the equation becomes omnipresent in mathematical modeling. To model any physical, biological, and social phenomena that depend on both space and time, they need to be explained through PDEs.

S. Lie developed an important method to determine explicit solutions of some nonlinear partial differential equations (NLPDEs), accordingly which is named after him as Lie symmetries. Accordingly, PDE theory plays a crucial role in the advancement of mathematics. They help us to describe the evolution of many phenomena in various fields. Since the 20th century, the investigation of NLPDEs has emerged as an independent field, expanding into numerous research directions.

Some analytic methods are developed to solve some NLPDEs, but generally, there is no central theory for finding the solution of NLPDEs. Till today there is no unified method that can be applied for solving all types of NLPDEs. The "nonlinearity" makes each problem unique and complicated. One needs to discover new methods for solving at least for a class of NLPDEs. Moreover, the functions and data in NLPDE problems are frequently defined in discrete points. Therefore, we need to develop numerical methods for NLPDEs through different approaches. In this chapter, we study traveling wave approaches to solve NLPDEs and nonlinear systems with the aid of computer algebra systems (CAS). We distinguish the approaches, in which it is very useful to apply computer algebra for solving NLPDEs and their systems.

In the upcoming chapters, we choose the most important and recently developed methods which allow us to construct solutions for NLPDEs (e.g., transformations methods, traveling wave and self-similarity methods, ansatz methods, method of separation of variables and its generalizations, qualitative methods, Painlevé test methods, truncated expansion methods, Hirota method and its generalizations, Adomian decomposition method and its generalizations, perturbation methods, finite difference methods, method of lines, etc.).

2.2 Solutions of NLPDEs

A classical solution to an NLPDE

$$F(x_1, x_2, \ldots, u, u_{x_1}, u_{x_2}, \ldots, u_{x_1x_1}, u_{x_1x_2}, \ldots, u_{x_ix_j}, \ldots) = 0 \qquad (2.1)$$

is a function $u = u(x_1, \ldots, x_n)$ defined in a domain D. This function is continuously differentiable and has all partial derivatives involved in the equation that satisfy Eq. (2.1). The concept of classical solution can be extended by adding the notion of weak solution (also known as generalized solution, which will not be discussed here), to encompass discontinuous and non-differentiable functions.

On the general solution of an NLPDE, we define an explicit closed-form expression that may contain movable critical singularities. A general solution (or general integral) to a first-order PDE is an equation of the form $f(\phi, \psi) = 0$, where f is an arbitrary function of the functions $\phi = \phi(x, y)$, $\psi = \psi(x, y)$ and offers a solution to this PDE.

An exact solution of an NLPDE is one that is defined in the entire domain of the PDE's definition and can be represented in closed form, i.e., as a finite expression (infinite functional series and products are not included). The exact solution of an NLPDE is not novel if it can be reduced to a known exact solution. An NLPDE's exact solution is redundant if there are more general solutions, and thus redundant solutions can be considered a special case of more general solutions.

A traveling wave is a wave of fixed shape that moves with a consistent velocity. Using the Galilean type transformation $\xi = x - ct$, $u(x,t) = u(\xi)$ (where c is the wave velocity), it is feasible to transform the PDE (in x, t) into an ODE (in ξ) that may be solved by appropriate methods. In other words, a traveling wave solution for a given NLPDE is a solution of the equation after reduction $\xi = x - ct$ if it exists. Periodic solutions are traveling wave solutions that are periodic, such as $\cos(x-t)$. Kink solutions are traveling wave solutions that rise or descend from one asymptotic state to the next; near infinity, kink solutions approach a constant. Standing wave solutions are two traveling wave solutions with the same amplitude and speed (but in opposite directions). Standing waves have a time-varying amplitude that does not travel spatially.

A solitary wave solution of an NLPDE is a traveling wave that obeys specific decreasing conditions as $\xi \to \pm\infty$ ($\xi = x - ct$). Solitary waves are confined traveling waves, asymptotically zero at long distances. Solitons are a unique type of solitary wave. The soliton solution is a spatially localized solution, i.e., $u'(\xi) \to 0$, $u''(\xi) \to 0$, $u'''(\xi) \to 0$ as $\xi \to \pm\infty$ ($\xi = x - ct$). Solitons' key trait is that they maintain their identity when interacting with other solitons.

The Bäacklund transformation, among other ways, can be used to obtain a one-soliton solution from the solution's first iteration. Similarly, an N-soliton solution, where N is any positive integer, is the Nth iteration of the vacuum solution using the Bäcklund transformation. A soliton equation is an NLPDE that admits a N-soliton solution, where N is any positive integer. A similarity or invariant solution is a solution to a PDE that results from invariance under a one-parameter Lie group of transformations. A self-similar or automodel solution is an invariant solution that results from invariance under a one-parameter Lie group of scaling.

2.3 Traveling Wave Solution

Wave propagation is a crucial concept in the study of both linear and nonlinear PDEs. It plays a crucial role in a variety of domains, including fluid mechanics (water waves, aerodynamics), acoustics (sound waves in the atmosphere), elasticity (stress waves, earthquakes), biology (epizootic waves), chemistry (combustion and detonation waves), and physics. Nerve fiber impulses are represented mathematically as traveling waves. Traveling waves, also known as shock profiles, comply with fluid dynamics conservation laws. A *traveling wave* is a wave with a defined shape and constant speed while moving in a specific direction.

In mathematics, traveling wave solutions are stated as $u(x,t) = f(x - ct)$, where x, t, and c denote the wave's spatial and time domains, as well as its velocity. At $t = 0$, the wave has the form $f(x)$, which is the initial wave profile. Thus, clearly, $u(x,t) = f(x - ct)$ depicts a wave traveling in the proper direction of the origin with a speed $c > 0$. Similarly, $u(x,t) = f(x + ct)$ denotes a wave traveling to the left at a speed $c > 0$. These waves travel undistorted along the lines $x \pm ct =$ constant.

If $c = 0$, the resultant wave is stationary. These non-propagating waves are typically seen when a fixed boundary is induced. Actually, there are forms that we can assign to traveling waves depending on specific characteristics. Wave fronts are formed when a traveling wave approaches the constant states $u(-\infty, t) = u_l$ and $u(\infty, t) = u_r$, where $u_l \neq u_r$. However, if the constant states are equivalent, $u_l = u_r$, the resulting wave is referred to as a pulse wave. A wave is considered spatially periodic if $f(x - ct + \lambda) = f(x - ct)$, where $\lambda > 0$. Geometric representations of such solutions will be discussed in order to provide more information about the many forms of traveling waves.

An NPDE describes the dynamical evolution of the waveform $u(x, t)$ and serves as the starting point. The following set of processes is then performed to obtain the solutions.

For (1+1) dimensional PDE:

Let us start with PDEs that have two independent variables: x for space and t for time.

Step 1: Conversion of PDE to ODE

A traveling wave solution for these types of PDEs requires a single coordinate: $\xi = c(x - vt)$. We write $u(x, t) = U(\xi)$, where $U(\xi)$ represents the traveling wave solution that moves at a speed of v. This is an example of a stationary wave with $L = c^{-1}$ characteristic width. Without loss of generality, let us define $c > 0$. As a result, the derivatives change as

$$\frac{\partial}{\partial t} \to -cv\frac{d}{d\xi}, \quad \frac{\partial}{\partial x} \to c\frac{d}{d\xi}\ \frac{\partial^2}{\partial x^2} \to c\frac{d^2}{d\xi^2} \quad \text{and so on.}$$

Accordingly, the PDE is converted into an ODE.

Step 2: Integration

Then the ODE will be successively integrated. This procedure comes to an end when at least one term contains no derivatives. The associated integration constants are taken to be zero in view of the localized solutions.

For (1+3) dimensional PDE:

Let us consider PDEs that have four independent variables: x, y, z for space and t for time.

Step 1: Conversion of PDE to ODE

For this type of PDE, the traveling wave solution requires one coordinate $\xi = lx + my + nz - vt$ and $u(x, y, z, t) = U(\xi)$, where $U(\xi)$ represents the wave solution, which travels with speed v, and l, m, and n represent the direction cosines of the wave along the space coordinates x, y, and z, respectively. Here also, we consider $c > 0$. Consequently, the derivatives undergo a change.

$$\frac{\partial}{\partial t} \to -cv\frac{d}{d\xi}, \quad \frac{\partial}{\partial x} \to l\frac{d}{d\xi}, \quad \frac{\partial}{\partial y} \to m\frac{d^2}{d\xi}, \quad \frac{\partial}{\partial z} \to n\frac{d}{d\xi} \text{ and so on.}$$

So, the PDE is converted into an ODE.

Step 2: Integration

The ODE will be integrated if all terms contain derivatives. This procedure comes to an end when at least one of the terms contains no derivatives. The associated integration constants are also taken to be zero.

Note: For simplicity, we use the notation $u(\xi)$.

2.4 Solution of KdV Equation and KdV-like Equations

2.4.1 *KdV equation*

To find the solution of the standard KdV equation

$$\frac{\partial u}{\partial t} + u\frac{\partial u}{\partial x} + \frac{\partial^3 u}{\partial x^3} = 0, \tag{2.2}$$

We follow steps 1 and 2 of Section 2.3.

Step 1: To convert the KdV equation to the following ODE, apply the Galilean transformation $\xi = x - vt$, where v is the right-ward-moving wave's velocity and obtained

$$\frac{d^3 u}{d\xi^3} + u\frac{du}{d\xi} - v\frac{du}{d\xi} = 0. \tag{2.3}$$

with $u \equiv u(\xi) \equiv u(x - ct)$.

Step 2: Integrating both sides of Eq. (2.3) and taking the integrating constant as zero, we obtain

$$\frac{d^2 u}{d\xi^2} + u\left(\frac{u}{2} - v\right) = 0. \tag{2.4}$$

Multiplying both sides of Eq. (2.4) by $2\frac{du}{d\xi}$ and after integration, setting the integrating constant to zero, we get

$$\left(\frac{du}{d\xi}\right)^2 + \frac{u^3}{3} - vu^2 = 0.$$

Therefore,

$$\frac{du}{d\xi} = \frac{u}{\sqrt{3}}\sqrt{(3v-u)}$$

$$\Longrightarrow \int \frac{du}{u\sqrt{(3v-u)}} = \int \frac{d\xi}{\sqrt{3}} \tag{2.5}$$

If we substitute $u = 3v\text{sech}^2 p$ in Eq. (2.5), then we get

$$\int \frac{-6v\text{sech}^2 p \tanh p}{3v\sqrt{(3v)}\text{sech}^2 p \tanh p} dp = \int \frac{d\xi}{\sqrt{3}}$$

$$\Rightarrow u = 3v\text{sech}^2\left[\frac{(x-vt)\sqrt{v}}{2}\right] \tag{2.6}$$

The soliton solution of the KdV equation (2.2) is represented by Eq. (2.6) and shown in Fig. (2.1). If we express it as $u = u_0\text{sech}^2(\frac{x-vt}{\delta})$, where δ denotes the solitary wave's width, v represents its velocity, and u_0 is its amplitude. Here, the velocity is denoted by v, the amplitude by $3v$, and the width of the solitary wave solution of the given KdV equation is indicated by $\frac{2}{\sqrt{v}}$. In this case, velocity is inversely proportional to the square root of width but proportional to the amplitude.

If we substitute $u = 3v\text{cosech}^2 p$ in Eq. (2.5) and proceed in the above manner, we get another solution of the KdV equation (2.2) of the form

$$u = 3v cosch^2\left[\frac{(x-vt)\sqrt{v}}{2}\right]. \tag{2.7}$$

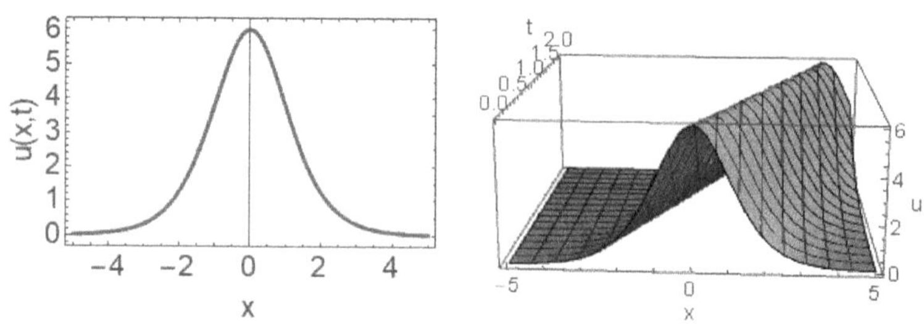

Figure 2.1. Regular soliton solution of Eq. (2.2) (given in Eq. (2.6)).

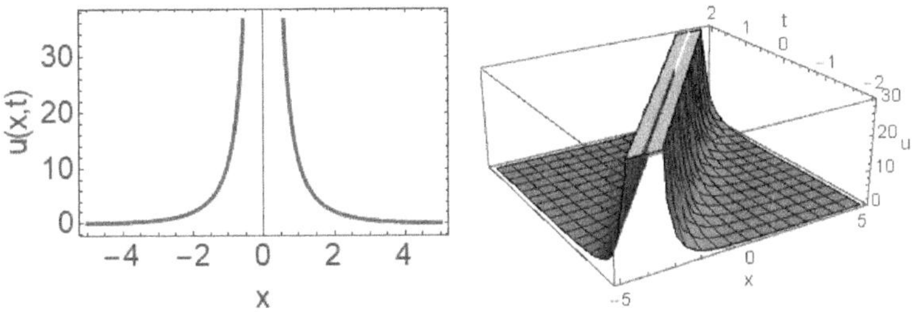

Figure 2.2. Singular soliton solution of Eq. (2.2) (given in Eq. (2.6)).

This is known as the singular soliton solution of the KdV equation, as seen in Fig. 2.2.

2.4.2 *mKdV equation:*

The analytical solution of the mKdV equation

$$\frac{\partial u}{\partial t} + u^2 \frac{\partial u}{\partial x} + \frac{\partial^3 u}{\partial x^3} = 0 \tag{2.8}$$

is obtained by applying step 1 and step 2 of Section 2.3.

Step 1: Take the traveling wave transformation $\xi = x - vt$ to convert the mKdV Eq. (2.8) to the following ODE

$$-v\frac{du}{d\xi} + u^2 \frac{du}{d\xi} + \frac{d^3 u}{d\xi^3} = 0 \tag{2.9}$$

Step 2: Integrating Eq. (2.9) and setting the integrating constant to zero, we have

$$\frac{d^2 u}{d\xi^2} + u\left(\frac{u^2}{3} - v\right) = 0 \tag{2.10}$$

Multiplying both sides of Eq. (2.10) by $2\frac{du}{d\xi}$ and integrating and setting the integrating constant to zero, we get

$$\left(\frac{du}{d\xi}\right)^2 + \frac{u^4}{6} - vu^2 = 0$$

$$\Rightarrow \frac{du}{d\xi} = \frac{u}{\sqrt{6}}\sqrt{(6v - u^2)}$$

$$\Rightarrow \int \frac{du}{u\sqrt{(6v - u^2)}} = \int \frac{d\xi}{\sqrt{6}}. \tag{2.11}$$

Now, if we take $u = \sqrt{(6v)}sechp$, then from Eq. (2.11), we obtained

$$\int \frac{-\sqrt{(6v)}sechp\tanh p}{6vsechp\tanh p}dp = \int \frac{d\xi}{\sqrt{6}}$$

$$\text{or} \quad p = -\sqrt{v}\xi$$

$$\therefore \quad u = \sqrt{(6v)}sech[(x - vt)\sqrt{v}] \tag{2.12}$$

It represents the soliton solution of the mKdV equation. It can be written in the form $u = u_0\text{sech}(\frac{x-vt}{\delta})$, shown in Fig. (2.3) where u_0 is the amplitude, v is velocity, and δ is the width of the solitary wave. So, here it is seen that v is the velocity, $\sqrt{6v}$ is the amplitude, and $\frac{1}{\sqrt{v}}$ is the width of the solitary wave solution of the given mKdV equation. Here, velocity is proportional to the square root of the amplitude but inversely proportional to the square root of the width of the solitary wave solution of the mKdV equation.

If we take $u = \sqrt{(6v)}cschp$ and put this value in Eq. (2.11), we get the singular soliton solution of the mKdV equation as (see Fig. 2.4).

$$u = -\sqrt{(6v)}csch[(x - vt)\sqrt{v}] \tag{2.13}$$

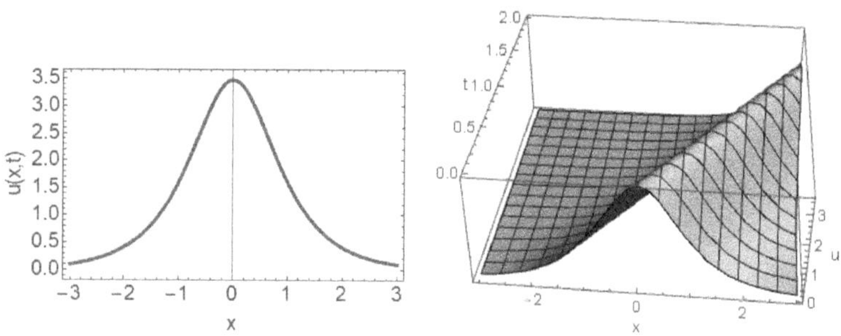

Figure 2.3. Regular soliton solution of Eq. 2.8 (given in Eq. (2.12)).

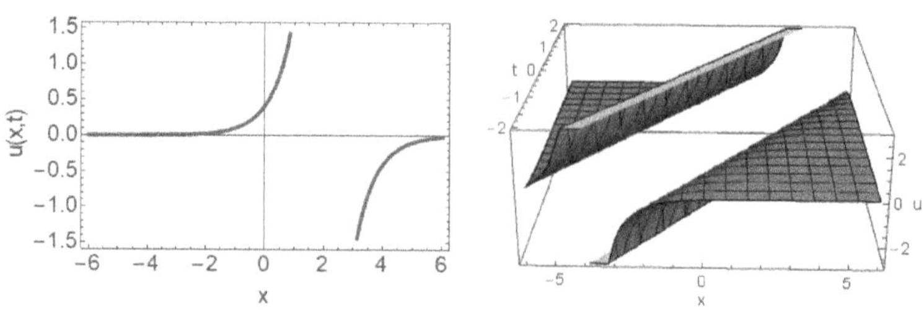

Figure 2.4. Singular soliton solution of Eq. 2.8 (given in Eq. (2.13)).

2.4.3 *Schamel KdV (SKdV) equation*

The analytical solution of the Schamel KdV (SKdV) equation

$$\frac{\partial u}{\partial t} + \sqrt(u)\frac{\partial u}{\partial x} + \frac{\partial^3 u}{\partial x^3} = 0 \tag{2.14}$$

is obtained by applying steps 1 and 2 of Section 2.3. Following the same way as it is done for the KdV & modified KdV equation and following steps 1 and 2, we get

$$\frac{d^2u}{d\xi^2} + \frac{2}{3}u^{\frac{3}{2}} - vu = 0 \tag{2.15}$$

where $\xi = x - vt$. Multiplying both sides of Eq. (2.15) by $2\frac{du}{d\xi}$ and integrating and setting the integrating constant to zero, we get

$$\left(\frac{du}{d\xi}\right)^2 + \frac{8}{15}u^{\frac{5}{2}} - vu^2 = 0$$

$$\Rightarrow \frac{du}{d\xi} = \frac{u}{\sqrt{15}}\sqrt(15v - 8\sqrt{u})$$

$$\Rightarrow \int \frac{du}{u\sqrt(15v - 8\sqrt{u})} = \int \frac{d\xi}{\sqrt{15}}. \tag{2.16}$$

Now, if we consider $u = \frac{225}{64}v^2\text{sech}^4(p)$ and put this value in (2.16), we get

$$\int \frac{-\frac{225}{64}(4v^2)\text{sech}^4(p)\tanh(p)}{\frac{225}{64}v^2\sqrt(15v)\text{sech}^4(p)\tanh(p)}dp = \int \frac{d\xi}{\sqrt{15}}$$

$$\therefore u = \frac{225}{64}v^2\text{sech}^4\left(\frac{(x - vt)\sqrt{v}}{4}\right). \tag{2.17}$$

Equation (2.17) represents the soliton solution of Eq. (2.14) shown in Fig. (2.5). Here, v is the velocity, $\frac{225}{64}v^2$ is the amplitude, and $\frac{4}{\sqrt{v}}$ is the

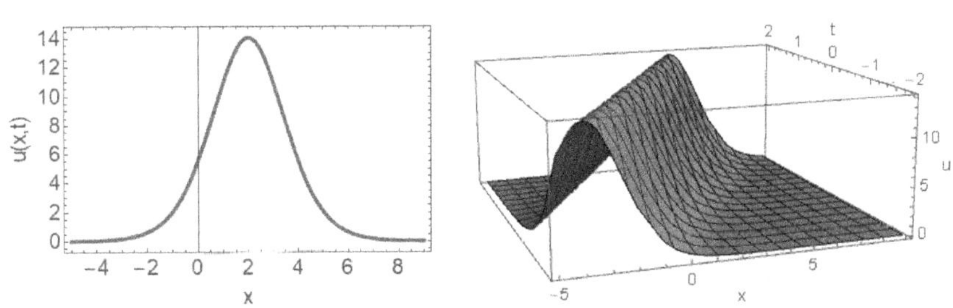

Figure 2.5. Regular soliton solution of Eq. (2.14) (given in Eq. (2.17)).

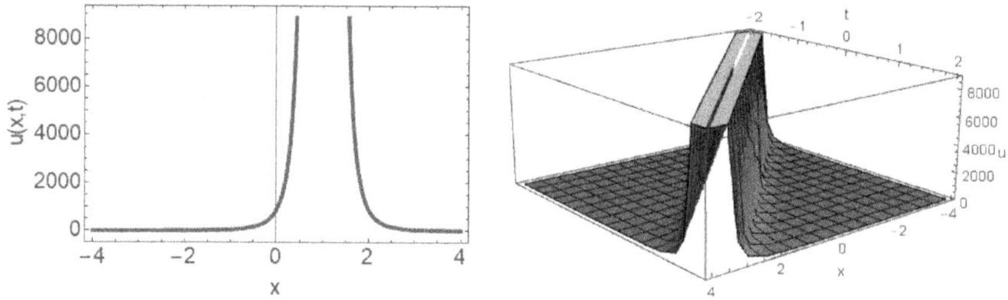

Figure 2.6. Singular soliton solution of Eq. (2.14) (given in Eq. (2.18)) shown in Fig. (2.5).

width of the solitary wave solution of the given SKdV equation. Here, the velocity is proportional to the square of the amplitude but inversely proportional to the square root of the width of the solitary wave solution of the SKdV equation.

If we substitute $u = \frac{225}{64}v^2 csch^4 p$ in Eq. (2.16) and proceed in the above manner, we get another solution of the SKdV equation (2.14) of the form

$$u = \frac{225}{64}v^2 csch^4\left(\frac{(x-vt)\sqrt{v}}{4}\right), \tag{2.18}$$

which is known as the singular soliton solution (see Fig. 2.6) of the Schamel KdV equation (2.14).

2.4.4 *KP equation*

The analytical solution of the KP equation

$$\frac{\partial}{\partial x}\left(\frac{\partial u}{\partial t} + u\frac{\partial u}{\partial x} + \frac{\partial^3 u}{\partial x^3}\right) + \frac{\partial^2 u}{\partial y^2} = 0 \tag{2.19}$$

is obtained by applying steps 1 and 2 of Section 2.3. This is (2 + 1) dimensional NLPDE, where the independent space variables are x & y and the time variable is t.

Step 1: The traveling wave transformation $\xi = Lx + My - vt$ converts the KP Eq. (2.19) to the following ODE

$$L\frac{d}{d\xi}\left(-v\frac{du}{d\xi} + uL\frac{du}{d\xi} + L^3\frac{d^3u}{d\xi^3}\right) + M^2\frac{d^2u}{d\xi^2} = 0 \tag{2.20}$$

Step 2: Integrating twice Eq. (2.20) and setting the integrating constant to zero, we have

$$(M^2 - Lv)u + \frac{1}{2}L^2u^2 + L^4\frac{d^2u}{d\xi^2} = 0$$

$$\Rightarrow \frac{d^2u}{d\xi^2} = \frac{h}{L^4}u - \frac{u^2}{2L^2}, \quad \text{where } [h = Lv - M^2] \tag{2.21}$$

Multiplying both sides of Eq. (2.21) by $2\frac{du}{d\xi}$ and after integration, taking the constant to be zero, we get

$$\left(\frac{du}{d\xi}\right)^2 = \frac{h}{L^4}u^2 - \frac{u^3}{3L^2},$$

$$\Rightarrow \frac{du}{d\xi} = u\sqrt{(A - Bu)}, \left(\text{where } A = \frac{h}{L^4} \quad \text{and} \quad B = \frac{1}{3L^2}\right),$$

$$\Rightarrow \int \frac{du}{u\sqrt{(A - Bu)}} = \int d\xi. \tag{2.22}$$

Now, if we consider $u = \frac{A}{B}\text{sech}^2 p$ and put this value in (2.22), we get

$$u = \frac{A}{B}\text{sech}^2\left(\frac{\xi\sqrt{A}}{2}\right). \tag{2.23}$$

Equation (2.23) represents the soliton solution of the KP equation shown in Fig. (2.7). If we write in the form $u = u_0\text{sech}^2(\frac{Lx+My-vt}{\delta})$, where u_0, v, and δ represent the amplitude, velocity, and width of the solitary wave, respectively, it is seen that v is the velocity, $\frac{3(Lv-M^2)}{L^2}$ is the amplitude, and $\frac{2L^2}{\sqrt{(Lv-M^2)}}$ is the width of the solitary wave solution of the given KP equation. Here, the amplitude of the solitary wave solution is proportional

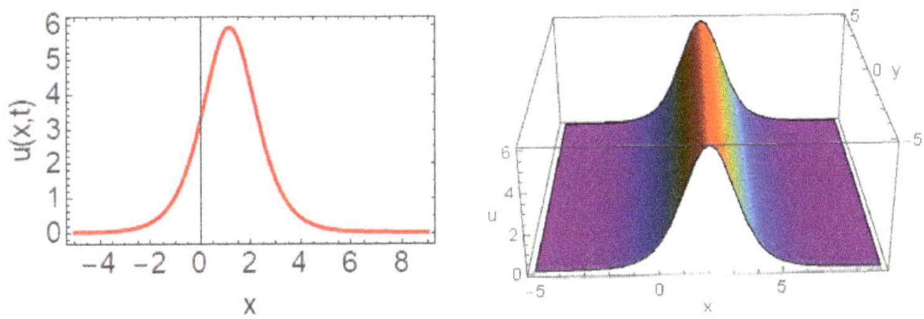

Figure 2.7. Regular soliton solution of Eq. (2.19) (given in Eq. (2.23)).

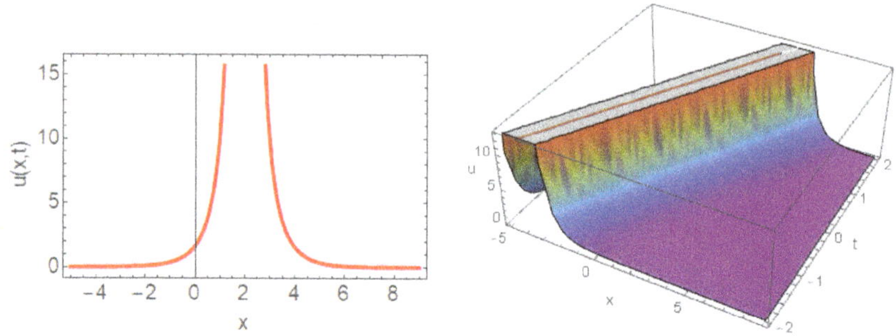

Figure 2.8. Singular soliton solution of Eq. (2.19) (given in Eq. (2.24)).

to the velocity, but the width of the wave is inversely proportional to the square root of the velocity of the wave.

If we consider $u = \frac{A}{B} csch^2 p$ and substitute it in Eq. (2.22) and proceed in the above manner, we get another singular soliton solution (see Fig. 2.8)) of the KP Eq. (2.19) of the form

$$u = \frac{3(Lv - M^2)}{L^2} csch^2 \left[\frac{(Lx + My - vt)\sqrt{(}Lv - M^2)}{2L^2} \right] \tag{2.24}$$

2.4.5 *Modified KP equation*

To obtain the analytical solution of the modified KP equation

$$\frac{\partial}{\partial x}\left(\frac{\partial u}{\partial t} + u^2 \frac{\partial u}{\partial x} + \frac{\partial^3 u}{\partial x^3} \right) + \frac{\partial^2 u}{\partial y^2} = 0 \tag{2.25}$$

We take the traveling wave transformation $\xi = Lx + My - vt$ and following the steps (1 & 2), we get

$$(M^2 - Lv)\frac{du}{d\xi} + L^2 u^2 \frac{du}{d\xi} + L^4 \frac{d^3 u}{d\xi^3} = 0 \tag{2.26}$$

Integrating this equation and setting the integrating constant to zero, we have

$$(M^2 - Lv)u + \frac{1}{3}L^2 u^3 + L^4 \frac{d^2 u}{d\xi^2} = 0,$$

$$\text{or} \quad \frac{d^2 u}{d\xi^2} = \frac{h}{L^4} u - \frac{u^3}{3L^2} (\text{where } h = Lv - M^2) \tag{2.27}$$

Multiplying both sides of Eq. (2.27) by $2\frac{du}{d\xi}$ and after integration set constant to be zero, we obtain

$$\left(\frac{du}{d\xi}\right)^2 = \frac{h}{L^4}u^2 - \frac{u^4}{6L^2}$$

$$\text{or} \quad \frac{du}{d\xi} = u\sqrt{(A - Bu^2)}, \quad \text{where } A = \frac{h}{L^4} \quad \text{and} \quad B = \frac{1}{6L^2},$$

$$\text{or} \quad \int \frac{du}{u\sqrt{(A - Bu^2)}} = \int d\xi \tag{2.28}$$

Now, if we consider $u = \sqrt{(\frac{A}{B})}sech(p)$ and put this value in (2.28), and after integration, we get

$$u = \sqrt{\left(\frac{6(Lv - M^2)}{L^2}\right)} sech\left(\frac{(Lx + My - vt)\sqrt{(Lv - M^2)}}{L^2}\right) \tag{2.29}$$

It represents the soliton solution of the modified KP equation shown in Fig. (2.7), where it is seen that v is the velocity, $\sqrt{\frac{6(Lv-M^2)}{L^2}}$ is the amplitude, and $\frac{L^2}{\sqrt{(Lv-M^2)}}$ is the width of the solitary wave solution of the given modified KP equation. Here, the amplitude of the solitary wave solution is proportional to the square root of velocity, but the width of the wave is inversely proportional to the square root of the velocity of the wave.

If we consider $u = \sqrt{(\frac{A}{B})}csch(p)$ and substitute it in Eq. (2.28) and proceed in the above manner, we get another solution (see Fig. (2.10)) of the modified KP Eq. (2.25) of the form

$$u = -\sqrt{\left(\frac{6(Lv - M^2)}{L^2}\right)} csch\left(\frac{(Lx + My - vt)\sqrt{(Lv - M^2)}}{L^2}\right) \tag{2.30}$$

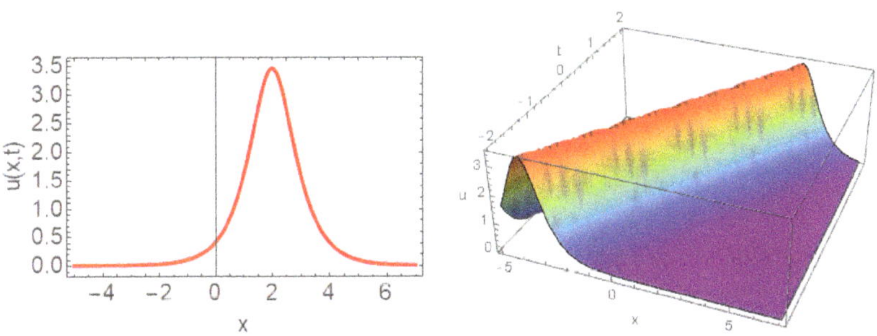

Figure 2.9. Regular soliton solution of Eq. (2.25) (given in Eq. (2.29)).

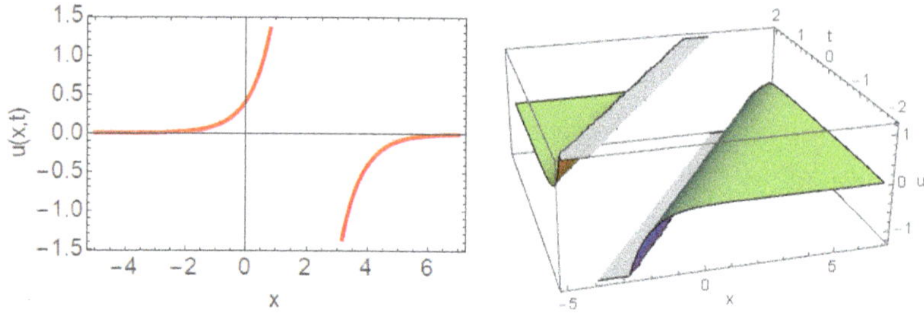

Figure 2.10. Singular soliton solution of Eq. (2.25) (given in Eq. (2.30)).

which is known as the singular soliton solution of modified KP equation (2.25).

2.4.6 *Schamel KP (SKP) equation*

To obtained the analytical solution of the Schamel KP equation

$$\frac{\partial}{\partial x}\left(\frac{\partial u}{\partial t}+\sqrt{u}\frac{\partial u}{\partial x}+\frac{\partial^3 u}{\partial x^3}\right)+\frac{\partial^2 u}{\partial y^2}=0 \tag{2.31}$$

we take the traveling wave transformation $\xi = Lx + My - vt$ and following the steps (1 & 2), we get

$$(M^2 - Lv)u + \frac{2}{3}L^2u^{\frac{3}{2}} + L^4\frac{d^2u}{d\xi^2} = 0$$

$$\Rightarrow \frac{d^2u}{d\xi^2} = \frac{h}{L^4}u - \frac{2u^{\frac{3}{2}}}{3L^2}\ \ [\text{where}\ \ h = Lv - M^2] \tag{2.32}$$

Multiplying both sides of (2.32) by $2\frac{du}{d\xi}$ and after integration taking constant to zero gives

$$\left(\frac{du}{d\xi}\right)^2 = \frac{h}{L^4}u^2 - \frac{8u^{\frac{5}{2}}}{15L^2}$$

$$\text{or}\quad \int\frac{du}{u\sqrt{(A-B\sqrt{u})}} = \int d\xi\ \ \left[A = \frac{h}{L^4}\ \ \&\ \ B = \frac{8}{15L^2}\right] \tag{2.33}$$

Now, if we consider $u = \frac{A^2}{B^2}\text{sech}^4(p)$ and put this value in (2.33), and after integration, we get

$$u = \frac{225(Lv-M^2)^2}{64L^4}\text{sech}^4\left(\frac{(Lx+My-vt)\sqrt{(Lv-M^2)}}{4L^2}\right). \tag{2.34}$$

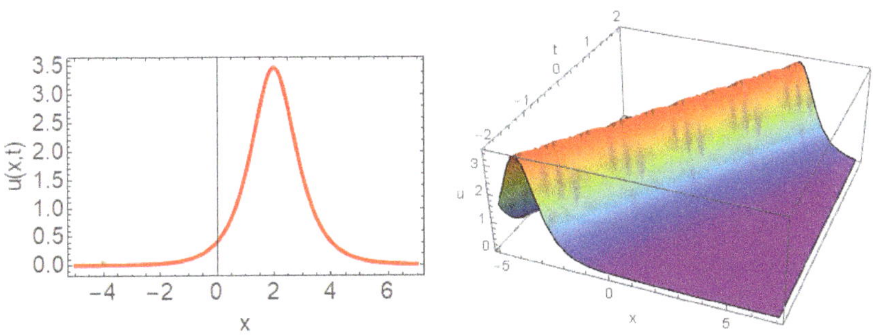

Figure 2.11. Regular soliton solution of Eq. (2.31) (given in Eq. (2.34)).

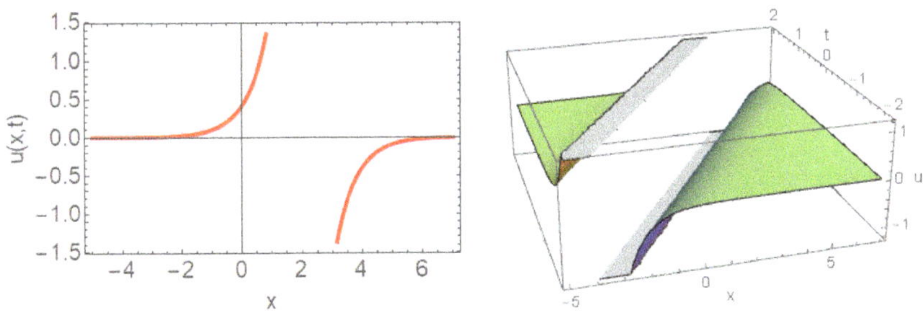

Figure 2.12. Singular soliton solution of Eq. (2.31) (given in Eq. (2.35)).

Equation (2.34) represents the soliton solution of the SKP equation (2.31) shown in Fig. (2.11), where $\frac{225(Lv-M^2)^2}{64L^4}$, v and $\frac{4L^2}{\sqrt{(Lv-M^2)}}$ are the amplitude, velocity and width of the solitary wave respectively. Here, the amplitude of the solitary wave solution is proportional top the square of the velocity and width is inversely proportional to the square root of the velocity of the solitary wave.

Again, if we consider $u = \frac{A^2}{B^2} csch^4(p)$ and substitute it in Eq.(2.33) and proceed in the above manner, we found another solution (see Fig. (2.12)) of the SKP equation (2.31) of the form

$$u = \frac{225(Lv - M^2)^2}{64L^4} csch^4 \left(\frac{(Lx + My - vt)\sqrt{(}Lv - M^2)}{4L^2} \right), \qquad (2.35)$$

which is known as the singular soliton solution of the SKP equation (2.31).

2.4.7 *ZK equation*

The analytical solution of the ZK equation

$$\frac{\partial u}{\partial t}+u\frac{\partial u}{\partial x}+\frac{\partial^3 u}{\partial x^3}+\frac{\partial}{\partial y}\left(\frac{\partial^2 u}{\partial x^2}\right)=0 \tag{2.36}$$

is obtained by virtue of steps 1 and 2 of Section 2.3. This is also a $(2+1)$ dimensional NLPDE. By taking a traveling wave transformation $\xi = Lx + My - vt$ and following steps (1 & 2), we get

$$\frac{d^2u}{d\xi^2}=\left(\frac{v}{L^3+ML^2}\right)u-\frac{1}{2(L^2+ML)}u^2. \tag{2.37}$$

Multiplying both sides of Eq. (2.37) by $2\frac{du}{d\xi}$ and after integration, set all the constants to zero. Then we get

$$\left(\frac{du}{d\xi}\right)^2=\left(\frac{v}{L^3+ML^2}\right)u^2-\frac{1}{3(L^2+ML)}u^3$$

$$\Rightarrow \frac{du}{d\xi}=u\sqrt{(A-Bu)}\quad\left[\text{set}\ \ A=\frac{v}{L^3+ML^2}\ \ \text{and}\ \ B=\frac{1}{3(L^2+ML)}\right]$$

$$\Rightarrow \int\frac{du}{u\sqrt{(A-Bu)}}=\int d\xi \tag{2.38}$$

Now, if we consider $u=\frac{A}{B}\text{sech}^2(p)$ and put this value in (2.38), and after integration, we get

$$u=\frac{3v}{L}\text{sech}^2\left[\frac{(Lx+My-vt)\sqrt{v}}{2L\sqrt{(L+M)}}\right]. \tag{2.39}$$

Equation (2.39) represents the soliton solution of the ZK equation (2.36) shown in Fig. 2.13, where $\frac{3v}{L}$, v, and $\frac{2L\sqrt{(L+M)}}{\sqrt{v}}$ represents the amplitude, velocity, and width of the solitary wave respectively. Here, the amplitude is proportional to the velocity and the width is inversely proportional to the square root of the velocity of the solitary wave solution.

Again, if we consider $u=\frac{A}{B}csch^2p$ and substitute it in Eq. (2.38) and proceed in the above manner, we find another solution [see Fig. (2.14)] of

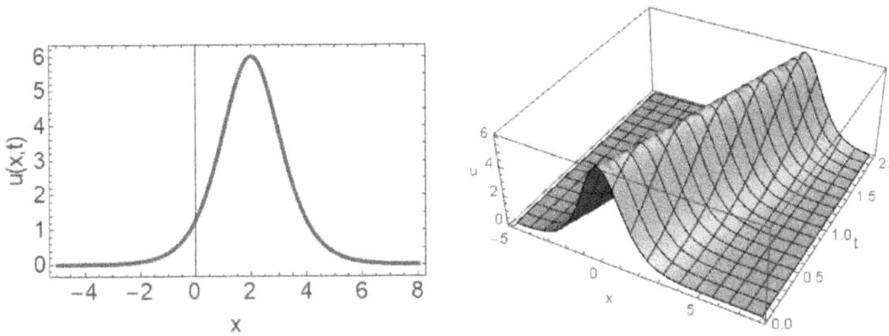

Figure 2.13. Regular soliton solution of Eq. (2.36) (given in Eq. (stzksol)).

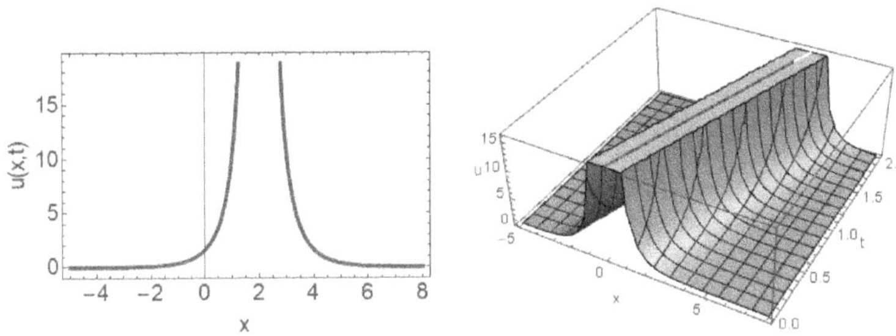

Figure 2.14. Singular soliton solution of Eq. (2.36) (given in Eq. (2.40)).

the ZK equation (2.36) of the form

$$u = \frac{3v}{L} csch^2 \left[\frac{(Lx + My - vt)\sqrt{v}}{2L\sqrt{(L+M)}} \right] \tag{2.40}$$

which is known as the singular soliton solution of ZK equation (2.36).

2.4.8 *Modified ZK (mZK) equation*

The analytical solution of the modified ZK equation

$$\frac{\partial u}{\partial t} + u^2 \frac{\partial u}{\partial x} + \frac{\partial^3 u}{\partial x^3} + \frac{\partial}{\partial y}\left(\frac{\partial^2 u}{\partial x^2}\right) = 0 \tag{2.41}$$

is obtained by virtue of step 1 and step 2 of Section 2.3. This is also a (2 + 1) dimensional NLPDE. By taking a traveling wave transformation $\xi = Lx + My - vt$ and following steps (1 & 2), we get

$$\frac{d^2 u}{d\xi^2} = \left(\frac{v}{L^3 + ML^2}\right) u - \frac{1}{3(L^2 + ML)} u^3. \tag{2.42}$$

Multiplying both side by $2\frac{du}{d\xi}$ and integrate and setting the integrating constant to be zero, we have

$$\left(\frac{du}{d\xi}\right)^2 = \left(\frac{v}{L^3 + ML^2}\right)u^2 - \frac{1}{6(L^2+ML)}u^4$$

$$\Rightarrow \frac{du}{d\xi} = u\sqrt{(A - Bu^2)}, \quad \left[\mathrm{A} = \frac{v}{L^3 + ML^2} \text{ and } \mathrm{B} = \frac{1}{6(L^2+ML)}\right]$$

$$\Rightarrow \int \frac{du}{u\sqrt{(A - Bu^2)}} = \int d\xi \tag{2.43}$$

If we consider $u = \sqrt{(\frac{A}{B})}sech(p)$ and substitute it in (2.43), and after integration, we obtain

$$u = \sqrt{\left(\frac{6v}{L}\right)}sech\left(\frac{(Lx + My - vt)\sqrt{v}}{L\sqrt{(L+M)}}\right). \tag{2.44}$$

Equation (2.44) represents the soliton solution of the mZK equation (2.41) shown in Fig. 2.15, where $\sqrt{\frac{6v}{L}}$, v and $\frac{L\sqrt{(L+M)}}{\sqrt{v}}$ represents the amplitude, velocity and width of the solitary wave respectively. Here, amplitude of the wave is proportional to the square root of the velocity but the width of the wave is inversely proportional to the velocity of the solitary wave solution.

Again if we consider $u = \sqrt{(\frac{A}{B})}csch(p)$ and substitute it in Eq. (2.43) and proceed in the above manner, we found another solution of the mZK Eq. (2.41) of the form

$$u = -\sqrt{\left(\frac{6v}{L}\right)}csch\left(\frac{(Lx + My - vt)\sqrt{v}}{L\sqrt{(L+M)}}\right) \tag{2.45}$$

which is known as the singular soliton solution (see Fig. 2.16).

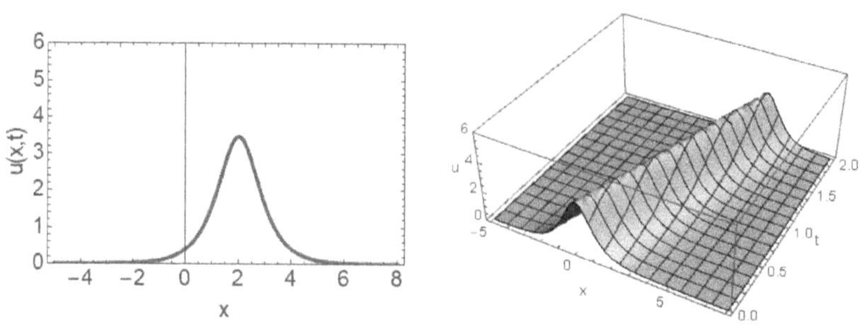

Figure 2.15. Regular soliton solution of Eq. (2.41) (given in Eq. (2.44)).

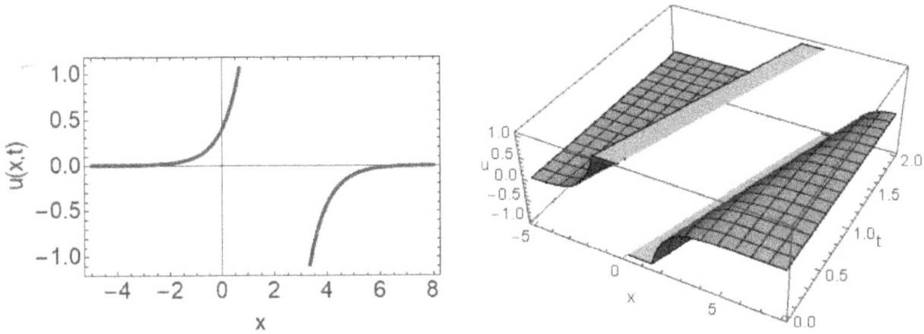

Figure 2.16. Singular soliton solution of Eq. (2.41) (given in Eq. (2.45)).

2.4.9 *Schamel ZK (SZK) equation*

The analytical solution of the Schamel ZK equation

$$\frac{\partial u}{\partial t} + \sqrt{u}\frac{\partial u}{\partial x} + \frac{\partial^3 u}{\partial x^3} + \frac{\partial}{\partial y}\left(\frac{\partial^2 u}{\partial x^2}\right) = 0 \tag{2.46}$$

is obtained by virtue of steps 1 and 2 of Section 2.3. This is also a $(2+1)$ dimensional NLPDE. By taking a traveling wave transformation $\xi = Lx + My - vt$ and following steps (1 & 2), we get

$$\frac{d^2u}{d\xi^2} = \left(\frac{v}{L^3 + ML^2}\right)u - \frac{2}{3(L^2+ML)}u^{\frac{3}{2}}. \tag{2.47}$$

Multiplying both sides by $2\frac{du}{d\xi}$ and after integration, setting the integrating constant to zero, we have

$$\left(\frac{du}{d\xi}\right)^2 = \left(\frac{v}{L^3+ML^2}\right)u^2 - \frac{8}{15(L^2+ML)}u^{\frac{5}{2}}$$

$$\Rightarrow \frac{du}{d\xi} = u\sqrt{(A - B\sqrt{u})}\left[\text{where } A = \frac{v}{L^3+ML^2} \text{ and } B = \frac{8}{15(L^2+ML)}\right]$$

$$\therefore \int \frac{du}{u\sqrt{(A-B\sqrt{u})}} = \int d\xi \tag{2.48}$$

If we consider $u = \frac{A^2}{B^2}\text{sech}^4(p)$ and substitute it in (2.48), then after integration, we get

$$u = \frac{225v^2}{64L^2}\text{sech}^4\left[\frac{(Lx+My-vt)\sqrt{v}}{4L\sqrt{(L+M)}}\right] \tag{2.49}$$

Equation (2.49) represents the soliton solution of the SZK equation (2.46) shown in Fig. 2.17, where $\frac{225v^2}{64L^2}$, v and $\frac{4L\sqrt{(L+M)}}{\sqrt{v}}$ represent the amplitude, velocity, and width of the solitary wave respectively.

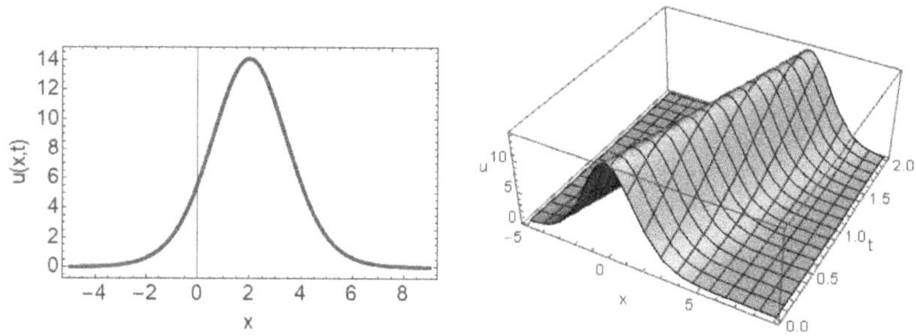

Figure 2.17. Regular soliton solution of Eq. (2.46) (given in Eq. (2.49)).

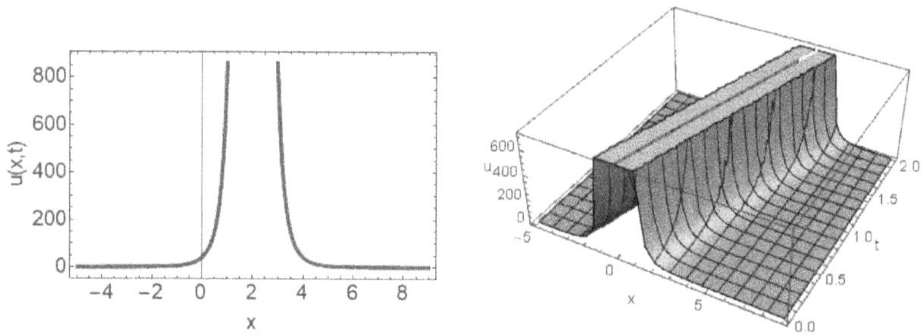

Figure 2.18. Singular soliton solution of Eq. (2.46) (given in Eq. (2.50)).

Again, if we consider $u = \frac{A^2}{B^2} csch^4(p)$ and substitute it in Eq. (2.46) and proceed in the above manner, we find another solution of the SZK Eq. (2.46) of the form

$$u = \frac{225v^2}{64L^2} csch^4 \left[\frac{(Lx + My - vt)\sqrt{v}}{4L\sqrt{(L+M)}} \right], \tag{2.50}$$

known as the singular soliton solution (see Fig. 2.18) of the SZK equation (2.46).

2.5 Hyperbolic Tangent (Tanh) Method

Recently, traveling wave solutions to nonlinear wave equations were found with the help of the tanh function. This method was developed to study solitons and shocks in different areas of physics. This technique is based on the prior assumption that the solutions can be written in terms of tanh functions. We have here systematized an approach that can be written in terms of the hyperbolic tangent function. First, a schematic outline is presented to demonstrate the main features of this technique. In this section, we will discuss a few examples. The starting point is an NLPDE.

$$P(u, u_t, u_x, u_{xx}, u_{xxx}, \ldots) = 0 \tag{2.51}$$

which describes the dynamical evolution of the wave form $u(x, t)$. To obtain the solution, the following five steps are then applied; the solution when the first two steps are the same as discussed in Section 2.3.

Step 1: Conversion of a PDE to an ODE.

Step 2: Integration until a non-derivative term appears.

Step 3: Change of an independent variable.

After Step 2, an ODE is formed, where ξ is the independent variable. Now, we introduce a new independent variable $Y = \tanh(\xi)$. Then the dependent variable is transformed from $u(\xi)$ to $S(Y)$, and the corresponding derivatives are changed as follows:

$$\frac{d}{d\xi} = \frac{\partial Y}{\partial \xi}\frac{d}{dY} = \operatorname{sech}^2(\xi)\frac{d}{dY} = (1 - Y^2)\frac{d}{dY}$$

Therefore,

$$\frac{d}{d\xi} = (1 - Y^2)\frac{d}{dY} \tag{2.52}$$

$$\begin{aligned}
\frac{d^2}{d\xi^2} &= \frac{d}{d\xi}\left(\frac{d}{d\xi}\right) = (1 - Y^2)\frac{d}{dY}\left[(1 - Y^2)\frac{d}{dY}\right] \\
&= (1 - Y^2)\left[-2Y\frac{d}{dY} + (1 - Y^2)\frac{d^2}{dY^2}\right]
\end{aligned} \tag{2.53}$$

$$\begin{aligned}
\frac{d^3}{d\xi^3} &= \frac{d}{d\xi}\left(\frac{d^2}{d\xi^2}\right) \\
&= (1 - Y^2)\frac{d}{dY}\left[(1 - Y^2)\left\{-2Y\frac{d}{dY} + (1 - Y^2)\frac{d^2}{dY^2}\right\}\right] \\
&= -2Y(1 - Y^2)\left[-2Y\frac{d}{dY} + (1 - Y^2)\frac{d^2}{dY^2}\right] + (1 - Y^2)^2 \\
&\quad \times \left(-2\frac{d}{dY} - 4\frac{d^2}{dY^2} + (1 - Y^2)\frac{d^3}{dY^3}\right)
\end{aligned} \tag{2.54}$$

Higher-order derivatives can be found accordingly.

Step 4: Homogeneous Balancing.

The solution we are looking for must be expressed in terms of Y. There is no general procedure at this stage. Often, the following finite term series expansion is mostly accepted:

$$S(Y) = \sum_{m=0}^{M} a_m Y^m \tag{2.55}$$

The parameter M will be found by balancing the linear term(s) of highest order with the nonlinear term(s). The linear term of the highest order is contained in the highest derivative when applied to $S(Y)$. This is readily seen using the set of relations (2.52)–(2.54). The Y^2 component of Eq. (2.52) leads to the order $(2 + M - 1) = M + 1$, and the Y^3 component of the first and the Y^4 component of the second derivative of Eq. (2.53) both lead to the order $M + 2$. On the other hand, the nonlinear terms yield multipliers of results. Balancing the order of the nonlinear term with the highest or the derivative, the value of the parameter M will be found.

Step 5: Obtaining the solution

Normally, M will be a positive integer so that a closed analytical solution can be obtained. Finally, the series expansion (2.55) or (2.56) is substituted into the relevant equation, and recursion relations appear because the coefficients a_m $(m = 0, 1, 2, \ldots, M)$ are then calculated. However, we may find examples where M takes a negative value. To avoid singularities for $Y \to 0$ in that case, $(-1 \leq Y \leq 1)$, the series expansion (2.53) must be changed. As an obvious choice, we introduce

$$\left(\sum_{m=0}^{M} a_m Y^m\right)^{-1} \quad \text{with } M' = -M(> 0) \tag{2.56}$$

In principle, an infinite value for M is also allowed. It can be used in those cases where a finite value of M does not lead to a solution. But such cases are not treated here.

A few examples are given as follows.

Burgers' Equation

Let us derive the solution to the Burgers' equation

$$\frac{\partial u}{\partial t} + u\frac{\partial u}{\partial x} - \frac{\partial^2 u}{\partial x^2} = 0 \tag{2.57}$$

by Tanh method.

Step 1: Take the transformation $\xi = c(x - vt)$, so $u(x,t) = U(\xi)$ and the Burgers' equation overtly to the following ODE

$$-cv\frac{dU(\xi)}{d\xi} + cU(\xi)\frac{dU(\xi)}{d\xi} - c^2\frac{d^2U(\xi)}{d\xi^2} = 0 \tag{2.58}$$

Step 2: Taking integration of Eq. (2.58) and setting the integrating constant to zero, we get

$$-vU(\xi) + \frac{1}{2}U^2(\xi) - c\frac{dU(\xi)}{d\xi} = 0 \tag{2.59}$$

Step 3: Introducing the new independent variable $Y = \tanh(\xi)$ and substituting $U(\xi) = S(Y)$, then the governing Eq. (2.59) becomes

$$-vS(Y) + \frac{1}{2}S^2(Y) - c(1 - Y^2)\frac{dS(Y)}{dY} = 0 \tag{2.60}$$

Step 4: Assuming the general solution of Eq. (2.60) is of the form $S(Y) = \sum_{m=0}^{M} a_m Y^m$, the parameter M will be found by balancing the order of the linear term(s) of the highest order with the nonlinear term(s). Here, after substituting the above expression into Eq. (2.60) and balancing the order of the second term against the order of the Y^2 component of the third one, we get the parameter M as

$$2M = 2 + M - 1 \Rightarrow M = 1$$

Step 5: Therefore, the solution has the form

$$S(Y) = a_0 + a_1 Y \tag{2.61}$$

Substituting (2.61) in (2.60)

$$-v(a_0 + a_1 Y) + \frac{1}{2}(a_0 + a_1 Y)^2 - c(1 - Y^2)a_1 Y = 0 \tag{2.62}$$

Each coefficient of the power of Y has to vanish. So, we arrive at

Y^2 coefficient:

$$\frac{1}{2}a_1^2 + a_1 c = 0 \Rightarrow a_1 = -2c \tag{2.63}$$

Y^1 coefficient:

$$-va_1 + a_0 a_1 = 0 \Rightarrow a_0 = v \tag{2.64}$$

Y^0 coefficient:

$$-va_0 + \frac{1}{2}a_0^2 - a_1 c = 0, \Rightarrow v^2 = 4c^2 \tag{2.65}$$

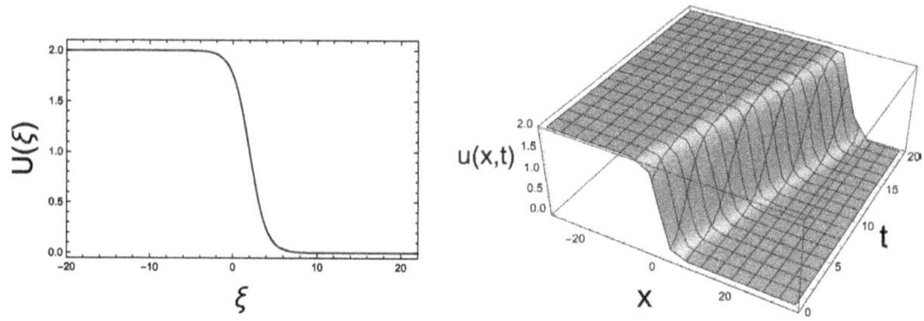

Figure 2.19. Shock solution of the Burgers' Eq. (2.57) (given in Eq. (2.66)).

These are three equations (2.63) – (2.65) have four unknowns $(a_0,\ a_1,\ c,\ v)$. We choose v as a free parameter. So, $a_0 = v,\ a_1 = -v,\ c = \frac{v}{2}$.

Therefore,

$$S(Y) = v - vY = v\left[1 - \tanh(\xi)\right] = v\left[1 - \tanh\{c(x - vt)\}\right].$$

Hence, the solution is

$$u(x,t) = U(\xi) = v\left[1 - \tanh\left\{\frac{v}{2}\left(x - vt\right)\right\}\right], \tag{2.66}$$

which is the well-known *shock-wave structure*, shown in Fig. 2.19.

KdV equation

Let us consider the KdV equation

$$\frac{\partial u}{\partial t} + 6u\frac{\partial u}{\partial x} + \frac{\partial^3 u}{\partial x^3} = 0. \tag{2.67}$$

To find the solution of the KdV Eq. (2.67) by the tanh method, first we follow the same steps **step 1** and **step 2** as before, and we get

$$-cvu(\xi) + 3cu^2(\xi) + c^3\frac{d^2u(\xi)}{d\xi^2} = 0. \tag{2.68}$$

Step 3: Taking into account the new variable $Y = \tanh(\xi)$ and substituting $u(\xi) = S(Y)$, the governing equation becomes

$$-vS(Y) + 3S^2(Y) + c^2(1 - Y^2)\left[-2Y\frac{dS(Y)}{dY} + (1 - Y^2)\frac{d^2S(Y)}{dY^2}\right] = 0 \tag{2.69}$$

Step 4: Substituting $S(Y) = \sum_{m=0}^{M} a_m Y^m$ into Eq. (2.69) and balacing the order of the second term (nonlinear) against the order of the Y^2 component of the 3rd one (or the Y^4 component of the last one) to get the parameter M. Hence,

$$2M = 3 + M - 1 = 4 + M - 2$$
$$\Rightarrow\ M = 2$$

Step 5: Therefore, the solution has the form

$$S(Y) = a_0 + a_1 Y + a_2 Y^2 \tag{2.70}$$

Using (2.70), we get from Eq. (2.69) the following system of equations:

$$-va_0 + 3a_0^2 + 2a_2 = 0 \tag{2.71}$$
$$-a_1 + 3a_0 a_1 - 2a_1 = 0 \tag{2.72}$$
$$-va_2 + 3a_1^2 + 6a_0 a_2 - 8a_2 = 0 \tag{2.73}$$
$$6a_1 a_2 + 2a_1 = 0 \tag{2.74}$$
$$3a_2^2 + 6a_2 = 0 \tag{2.75}$$

We are now able to proceed as before. However, from the fact that $M = 2$, we observe from the structure of Eq. (2.68) that $S(Y)$ should be proportional to $(1-Y^2)$. So, we are able to introduce $S(Y) = \nu(1-Y^2)$. Dividing Equation (2.69) by $\nu(1 - Y^2)$, we obtain

$$-v + 3\nu(1 - Y^2) + C^2\left[-2Y\frac{d(1-Y^2)}{dY} + (1-Y^2)\frac{d^2(1-Y^2)}{dY^2}\right] = 0$$

Only terms proportional to Y^2 and Y^0 are left.

Thus, Y^0 coefficient:

$$-v + 3\nu - 2c^2 = 0$$

Y^2 coefficient:

$$-3\nu + 4c^2 + 2c^2 = 0$$

We have three unknowns (v, ν, c) and two equations, and we may choose c as a free parameter. The other variables are then found to be

$$\nu = 2c^2, \quad v = 4c^2$$

Finally, the solitary wave solution of the KdV equation reads

$$\begin{aligned} u(x,t) &= 2c^2(1 - \tanh^2[c(x - 4c^2 t)]) \\ &= 2c^2 \operatorname{sech}^2[c(x - 4c^2 t)] \end{aligned} \tag{2.76}$$

which is the well-known bell-shaped form shown in Fig. 2.20.

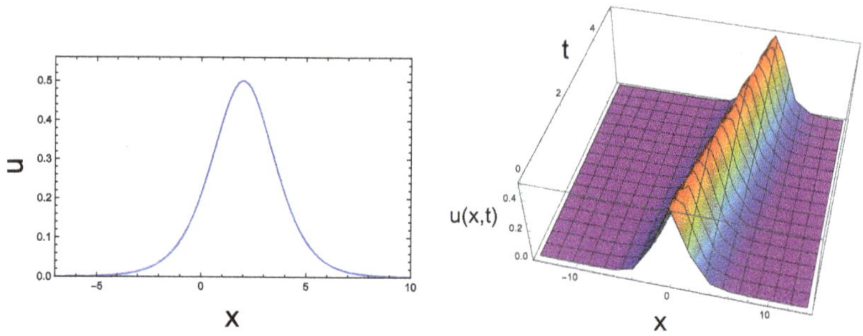

Figure 2.20. Soliton solution of the KdV equation (2.67) (given in Eq. (2.76)).

Modified KdV equation

If one changes the quadratic nonlinearity of the KdV equation to a cubic one, one gets the modified KdV (mKdV) equation as

$$\frac{\partial u}{\partial t} + 6u^2\frac{\partial u}{\partial x} + \frac{\partial^3 u}{\partial x^3} = 0 \tag{2.77}$$

For instance, this equation appears in electric theory, double layer theory, and serves as a model of solutions in a multicomponent plasma in a harmonic lattice. Repeating the same steps **step 1** and **step 2** as before, we get

$$-cv\frac{du(\xi)}{d\xi} + 6u^3c\frac{du(\xi)}{d\xi} + c^3\frac{d^3u(\xi)}{d\xi^3} = 0 \tag{2.78}$$

Step 3: After integration

$$-cvu(\xi) + 2cu^2(\xi) + c^3\frac{du(\xi)}{d\xi^2} = 0 \tag{2.79}$$

Step 4: Taking into account the new variable Y, the governing equation becomes

$$-vS(Y) + 2s^2(Y) + c^2(1-Y^2)\left[-2Y\frac{dS(Y)}{dY} + (1-Y^2)\frac{d^2S(Y)}{dY^2}\right] = 0 \tag{2.80}$$

The parameter M is determined by balancing the order of the second term (nonlinear) against the order of the Y^2 component of the third (or the Y^4 component of the last one). Hence, $2M = 3 + M - 1 = 4 + M - 2$, so that $M = 1$.

It turns out, however, that the series expansion (2.55) (with $M = 1$) does not lead to a solution unless a trivial one ($u =$ constant).

Two alternatives are: We may try a series expansion with $M = \infty$. Performing such an expansion, we are able to arrive at a recursion relation between c, v and $a_i (i \geq 1)$. However, it requires a lot of insight to define the unknowns, and it involves tedious algebra, so it falls outside the scope of the article. The other choice originates from the structure of Eq. (2.80), which suggests the ansatz $S(Y) = \nu(1 - Y^2)^{1/2}$. Then, after substituting this into Eq. (2.80), one can obtain

$$
\begin{aligned}
&- v\nu + 2\nu^2(1 - Y^2) + c^2(1 - Y^2)^{1/2}\left[-2Y\frac{dS(1 - Y^2)^{1/2}}{dY}\right.\\
&\quad + \left.(1 - Y^2)\frac{d^2(1 - Y^2)^{1/2}}{dY^2}\right] = 0
\end{aligned}
$$

Eventually, we arrive at the following relations:

$$
\begin{aligned}
-v + 2\nu^2 - 2c^2 &= 0\\
-2\nu^2 + 2c^2 &= 0
\end{aligned}
$$

We have three unknowns (v, ν, c) and two equations, and we may choose c as a free parameter. The other variables are then found to be $\nu = c$, $v = c^2$. Therefore, the solitary wave solution of the mKdV equation reads

$$
\begin{aligned}
u(x,t) &= c(1 - \tanh^2[c(x - c^2 t)])^{1/2}\\
&= csech[c(x - c^2 t)] \qquad (2.81)
\end{aligned}
$$

and shown in Fig. (2.21).

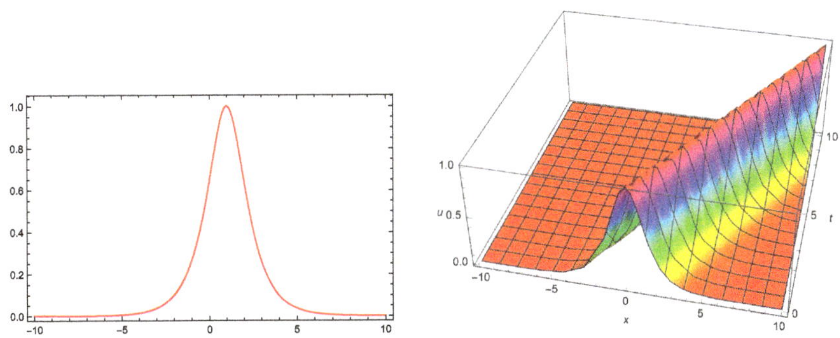

Figure 2.21. Soliton solution of the mKdV equation (2.77) (given in Eq. (2.81)).

KdVB equation:

We consider the KdVB equation

$$\frac{\partial\phi}{\partial t} - b\frac{\partial^2\phi}{\partial x^2} + c\phi\frac{\partial\phi}{\partial x} + d\frac{\partial^3\phi}{\partial x^3} = 0 \tag{2.82}$$

where $a,\ b,\ c,\ d$ are real constants.

Take the transformation $\xi = l(x - vt)$, so $u(x,t) = u(\xi)$, and after following **step 1** and **step 2** of the hyperbolic tangent method, we obtain

$$d\frac{d^2u}{d\xi^2} - b\frac{du}{d\xi} + c\frac{u^2}{2} - Vu = 0. \tag{2.83}$$

Step 3: Introduce the new variable $Y = \tanh(\xi)$ and substitute $S(Y) = u(\xi)$, then Eq. (2.83) becomes

$$d(1-Y^2)^2\frac{d^2S}{dY^2} - \{2dY(1-Y^2) + b(1-Y^2)\}\frac{dS}{dY} + \frac{c}{2}S^2 - VS = 0 \tag{2.84}$$

Step 4: Substituting $S(Y) = \sum_{m=0}^{M} a_m Y^m$ into Eq. (2.84) and balancing the order of the leading order of nonlinear terms against the highest order derivative term to get the parameter M. Hence,

$$2M = 3 + M - 1 \Rightarrow M = 2$$

Step 5: Therefore, the solution has the form

$$S(Y) = a_0 + a_1Y + a_2Y^2 \tag{2.85}$$

Using (2.85), we get from Eq. (2.84) the following system of equations:

$$2a_2d - ba_1 + \frac{1}{2}ca_0^2 - Va_0 - c_1 = 0 \tag{2.86}$$

$$2a_1d + 2ba_2 - ca_0a_1 + Va_1 = 0 \tag{2.87}$$

$$ba_1 - 8a_2d + \frac{1}{2}ca_1^2 + ca_0a_2 - Va_2 = 0 \tag{2.88}$$

$$2a_1d + 2ba_2 + ca_1a_2 = 0 \tag{2.89}$$

$$6a_2d + \frac{1}{2}ca_2^2 = 0 \tag{2.90}$$

From these equations, the unknowns are determined as follows:

$$a_0 = \frac{1}{c}(V + 12d), \quad a_1 = -\frac{12b}{5c}, \quad a_2 - \frac{12d}{c} \tag{2.91}$$

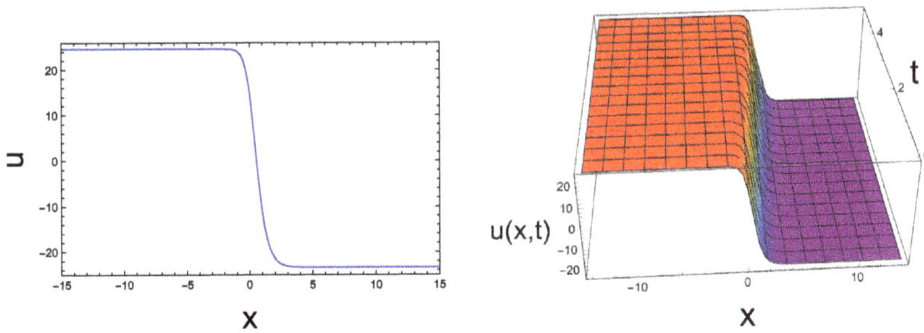

Figure 2.22. Sock solution of KdVB equation (2.82) (given in Eq. (2.92)).

together with a relation between b and d as $b^2 = 100d^2$. Equation (2.87) determines V in terms of a_0, a_1, a_2 and c_1. Finally, the solution of the KdV–Burgers equation (2.82) is

$$\phi(x,t) = a_0 + a_1 \tanh(x - Vt) + a_2 \tanh^2(x - Vt) \tag{2.92}$$

shown in Fig. 2.22 where a_0, a_1 and a_2 are given above together with the relation $b^2 = 100d^2$.

2.6 Tanh–Coth Method

An extended tanh method for determining precise solutions to NLPDEs, especially ones that represent wave phenomena like solitons, is the tanh-coth method. It combines the tanh and coth functions, which are frequently seen in NLPDE solutions, particularly when nonlinear wave equations and soliton theory are involved. This technique is a straightforward way to get solitary wave solutions to several NLPDEs.

Let us consider an NPDE

$$P(u, u_t, u_x, u_{xx}, u_{xxx}, \ldots) = 0. \tag{2.93}$$

To find the traveling wave solution of Eq. (2.93) by the Tanh–Coth Method, we follow the following steps:

Step 1: Combining the two independent variables t and x into the new wave variable $\xi = x - ct$, the NPDE (2.93) is converted to an ODE

$$Q(u, u', u'', u''', \ldots) = 0 \tag{2.94}$$

Equation (2.94) is then integrated as long as all terms contain derivatives, provided the integration constants are considered to be zero.

Step 2: Introduce a new independent variable $Y = \tanh(\mu\xi)$ (where μ is the wave number) that leads to the change of derivatives:

$$\left.\begin{aligned}
\tfrac{d}{d\xi} &= \mu(1-Y^2)\tfrac{d}{dY}\\
\tfrac{d^2}{d\xi^2} &= 2\mu^2 Y(1-Y^2)\tfrac{d}{dY} + \mu^2(1-Y^2)\tfrac{d^2}{dY^2}\\
\tfrac{d^3}{d\xi^3} &= 2\mu^3(1-Y^2)(3Y^2-1)\tfrac{d}{dY} - 6\mu^3 Y(1-Y^2)^2\tfrac{d^2}{dY^2}\\
&\quad + \mu^3(1-Y^2)^3\tfrac{d^3}{dY^3}\\
\tfrac{d^4}{d\xi^4} &= -8\mu^4 Y(1-Y^2)(3Y^2-1)\tfrac{d}{dY} + 4\mu^4(1-Y^2)^2(9Y^2-2)\tfrac{d^2}{dY^2}\\
&\quad -12\mu^4 Y(1-Y^2)^3\tfrac{d^3}{dY^3} + \mu^4(1-Y^2)^4\tfrac{d^4}{dY^4}
\end{aligned}\right\} \tag{2.95}$$

Step 3: In the tanh method, only the positive integer value of M is considered. But in the tanh–coth method, we consider both positive and negative integer values of M. Hence, the finite expansion can be written as

$$u(\mu,\xi) = S(Y) = \sum_{m=0}^{M} a_K Y^K + \sum_{m=0}^{M} b_K Y^{-K} \tag{2.96}$$

where M is a positive integer, that will be determined. For noninteger M, a transformation formula is used to overcome this difficulty, and it reduces to the standard tanh method $b_k = 0, 1 \leq k \leq M$. Substituting (2.96) into the reduced ODE results in an algebraic equation in power of Y. By using the homogeneous balancing method (same as step 4 in the tanh method), we can obtain the value of M.

Step 4: After determining the value of M, collect all coefficients of power of Y in the resulting equation where these coefficients have to be zero. This will give a system of algebraic equations involving the parameters a_k, b_k, μ, and c. Having determined these parameters, we may obtain solitons in terms of sech^2 or kinks in terms of tanh. However, this method may give periodic solutions.

KdV equation

The KdV equation is

$$u_t + auu_x + u_{xxx} = 0 \tag{2.97}$$

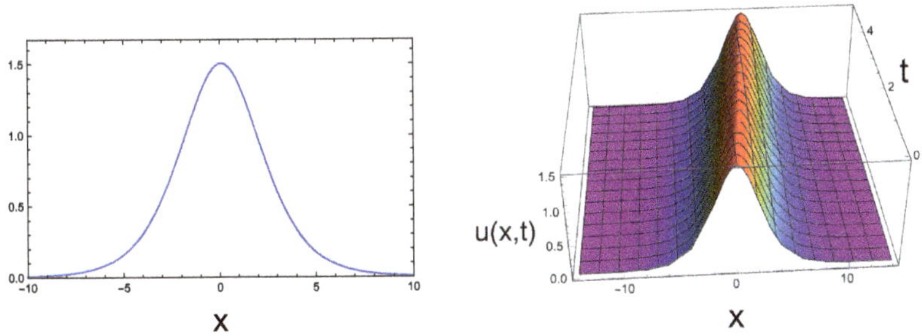

Figure 2.23. Regular soliton solution of the KdV equation.

Step 1: Substitute the wave variable $\xi = x - ct$ in Eq. (2.97), where c is the wave speed, and integrate once to obtain

$$-cu + \frac{a}{2}u^2 + u'' = 0 \tag{2.98}$$

Step 2: Taking into account the new variable $Y = \tanh(\xi)$ and substituting $u(\xi) = S(Y)$, (putting Eq. (2.95) in Eq. (2.98)) the governing equation becomes

$$-vS(Y) + 3s^2(Y) + c^2(1 - Y^2)\left[-2Y\frac{dS(Y)}{dY} + (1 - Y^2)\frac{d^2S(Y)}{dY^2}\right] = 0 \tag{2.99}$$

Step 3: Substituting $S(Y) = \sum_{m=0}^{M} a_m Y^m$ into Eq. (2.99) and balancing the order of the second term (nonlinear) against the order of the Y^2 component of the third one (or the Y^4 component of the last one) to get the parameter M. Hence,

$$2M = 3 + M - 1 = 4 + M - 2$$
$$\Rightarrow \ M = 2.$$

Step 4: The tanh–coth method allows us to use the substitution

$$u(x,t) = S(Y) = \sum_{j=0}^{2} a_j Y^j + \sum_{i=1}^{2} b_j Y^{-1} \tag{2.100}$$

substituting (2.100) into (2.98), collecting the coefficient of each power of Y^r, $0 \leq r \leq 8$, setting each coefficient to zero, and solving the resulting

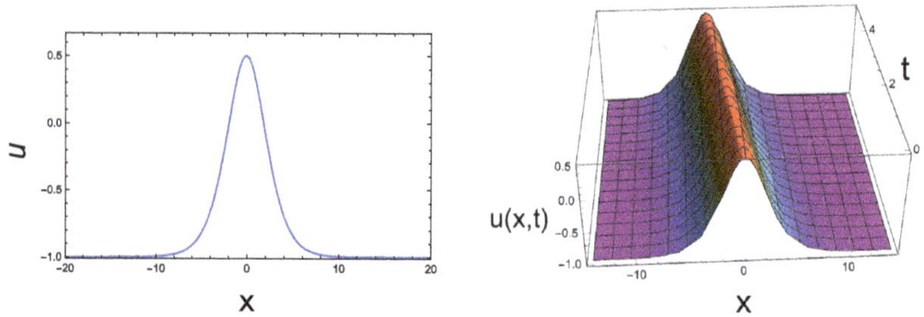

Figure 2.24. Singular soliton solution of the KdV equation.

system of algebraic equations, we find the following sets of solutions:

$$(i)\quad a_0=\frac{3c}{a},\quad a_1=a_2=b_1=0,\quad b_2=-\frac{3c}{a},\quad \mu=\frac{1}{2}\sqrt{c},\quad c>0 \tag{2.101}$$

$$(ii)\quad a_0=-\frac{c}{a},\quad a_1=a_2=b_1=0,\quad b_2=\frac{3c}{a},\quad \mu=\frac{1}{2}\sqrt{-c},\quad c<0 \tag{2.102}$$

$$(iii)\quad a_0=\frac{3c}{a},\quad a_1=a_2=b_1=0,\quad b_2=\frac{3c}{a},\quad \mu=\frac{1}{2}\sqrt{c},\quad c>0 \tag{2.103}$$

$$(iv)\quad a_0=-\frac{c}{a},\quad a_1=a_2=b_1=0,\quad b_2=\frac{3c}{a},\quad \mu=\frac{1}{2}\sqrt{-c},\quad c<0 \tag{2.104}$$

consequently, we obtain the following soliton solutions:

$$\left.\begin{aligned} u_1(x,t) &= \tfrac{3c}{a}\text{sech}^2\left[\tfrac{1}{2}\sqrt{c}(x-ct)\right],\quad c>0\\ u_2(x,t) &= -\tfrac{c}{a}\left(1-3\tanh^2\left[\tfrac{1}{2}\sqrt{-c}(x-ct)\right]\right),\quad c<0 \end{aligned}\right\} \tag{2.105}$$

Moreover, the traveling wave solutions

$$\left.\begin{aligned} u_3(x,t) &= -\tfrac{3c}{a}\text{cosech}^2\left[\tfrac{1}{2}\sqrt{c}(x-ct)\right],\quad c>0\\ u_4(x,t) &= -\tfrac{c}{a}\left(1-3coth^2\left[\tfrac{1}{2}\sqrt{-c}(x-ct)\right]\right),\quad c<0 \end{aligned}\right\} \tag{2.106}$$

Note that the tanh–coth method provides a wide variety of solutions over the tanh method.

mKdV equation:

The modified KdV equation is

$$u_t + au^2u_x + u_{xxx} = 0 \tag{2.107}$$

We first substitute the wave variable $\xi = x - ct$ (where c is the wave speed) into the mKdV Eq. (2.107) and following the procedure **steps 1–4** as above, one can easily obtain the solutions of the mKdV equation as

$$\left.\begin{aligned} u_1(x,t) &= \sqrt{\tfrac{3c}{a}}\, sech\left[\sqrt{c}(x-ct)\right], \quad c>0, \quad a>0 \\ u_2(x,t) &= \sqrt{\tfrac{3c}{a}}\left(\tanh\left[\sqrt{-\tfrac{1}{2}c}(x-ct)\right]\right), \quad c<0, \quad a<0 \end{aligned}\right\} \tag{2.108}$$

respectively, are readily obtained. Moreover, the following traveling wave periodic solutions

$$\left.\begin{aligned} u_3(x,t) &= \sqrt{\tfrac{3c}{a}}\left(coth\left[\sqrt{-\tfrac{1}{2}c}(x-ct)\right]\right), \quad c<0, \quad a<0 \\ u_4(x,t) &= \sqrt{\tfrac{6c}{a}}\, sech\left[\sqrt{-c}(x-ct)\right], \quad c<0, \quad a<0 \\ u_5(x,t) &= \sqrt{\tfrac{-6c}{a}}\, cosech\left[\sqrt{c}(x-ct)\right], \quad c>0, \quad a<0 \end{aligned}\right\} \tag{2.109}$$

can also be derived by using the sign of the wave speed c and the parameter a.

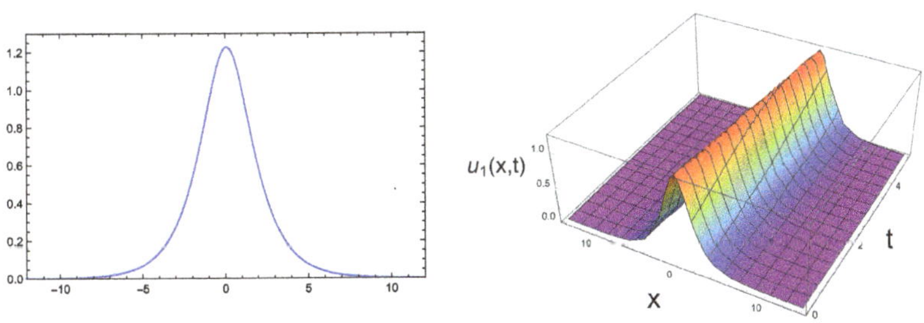

Figure 2.25. Soliton solution of the mKdV equation (2.107) given in Eq. (2.108).

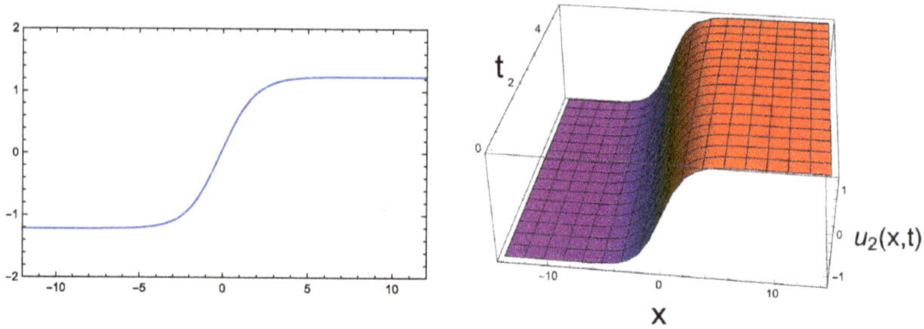

Figure 2.26. Shock solution of the mKdV equation (2.107) given in Eq. (2.108).

Burgers' Equation:

The standard Burgers' equation is given by

$$\frac{\partial u}{\partial t} + u\frac{\partial u}{\partial x} - \frac{\partial^2 u}{\partial x^2} = 0. \tag{2.110}$$

Step 1: Substitute the wave variable $\xi = x - ct$ in Eq. (2.110), where c is the wave speed, and integrating the transformed equation with the help of the initial condition $u(\xi) \to 0$, $\frac{du(\xi)}{d\xi} \to 0$ at $\xi \to \pm\infty$, we get

$$-cu(\xi) + \frac{1}{2}u^2(\xi) - \frac{du(\xi)}{d\xi} = 0 \tag{2.111}$$

Step 2: Taking into account the new variable $Y = \tanh(\mu\xi)$ and substituting $u(\xi) = S(Y)$, the governing equation (2.111) becomes

$$-cS(Y) + \frac{1}{2}S^2(Y) - \mu(1 - Y^2)\frac{dS(Y)}{dY} = 0 \tag{2.112}$$

Step 3: Substituting $U(\xi) = S(Y) = \Sigma_{m=0}^{M} a_m Y^M + \Sigma_{m=1}^{M} b_m Y^{-M}$ into Eq. (2.60) and balancing the order of the second term against the order of the Y^2 component of the 3rd one to get the parameter M. Hence,

$$2M = 2 + M - 1 \Rightarrow M = 1$$

Step 4: Substituting the value of M, we get the general solution as

$$S(Y) = a_0 + a_1 Y + b_1 Y^{-1} \tag{2.113}$$

into Eq. (2.112) and collecting the coefficient of each power of Y^r, $0 \leq r \leq 2$, setting each coefficient to zero, and solving the resulting system of

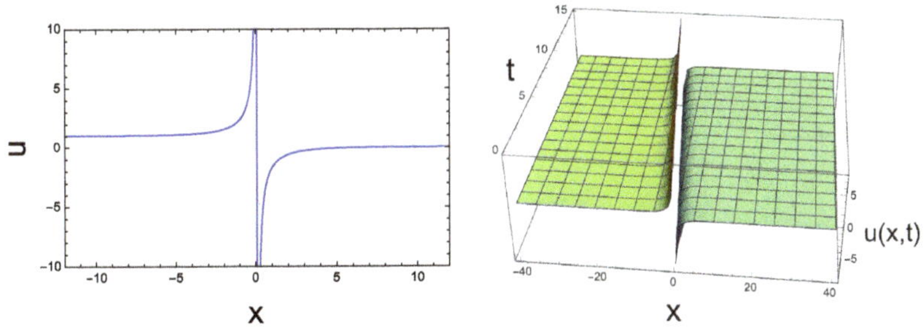

Figure 2.27. Shock solution of the Burger's equation given in Eq. (2.114).

algebraic equations, we find the solution of the Burger equation as

$$\left.\begin{aligned} u(x,t) &= c - \tfrac{c}{2}[\tanh[\tfrac{c}{4}(x-ct)] + coth[\tfrac{c}{4}(x-ct)]], \quad c > 0, \\ u(x,t) &= c + \tfrac{c}{2}[\tanh[-\tfrac{c}{4}(x-ct)] + coth[-\tfrac{c}{4}(x-ct)]], \quad c < 0. \end{aligned}\right\} \tag{2.114}$$

2.7 (G'/G) Expansion Method

The $\left(\frac{G'}{G}\right)$ expansion method is a technique developed specifically to find new traveling wave solutions of nonlinear evolution equations (NLEEs) that cannot be found using the tanh method or tanh–coth method. In this method, the traveling wave solutions of an NLEEs can be expressed by a polynomial in $\left(\frac{G'}{G}\right)$, where $G = G(\xi)$ satisfies a second order linear ODE $G'' + \lambda G' + \mu G = 0$, $G' = \frac{d}{d\xi}G(\xi)$, $\xi = x - Vt$. The homogeneous balance between the highest-order derivatives and nonlinear terms that appear in a particular NLEE can be used to calculate the polynomial's degree. A series of algebraic equations that come from the method can be solved to determine the coefficients of the polynomial. The $\left(\frac{G'}{G}\right)$ expansion method will be shown to yield some new traveling wave solutions of various NLEEs. Let's discuss the $\left(\frac{G'}{G}\right)$ expansion method.

Suppose an NLEE, say in two independent variables x and t, is given by

$$P(u, u_t, u_x, u_{tt}, u_{xx}, \ldots) = 0 \tag{2.115}$$

where $u = u(x,t)$ is an unknown function, and P is a polynomial in $u = u(x,t)$ and its various partial derivatives, in which the highest-order derivatives and nonlinear terms are involved. In the following, we give the main step of the $\left(\frac{G'}{G}\right)$ expansion method.

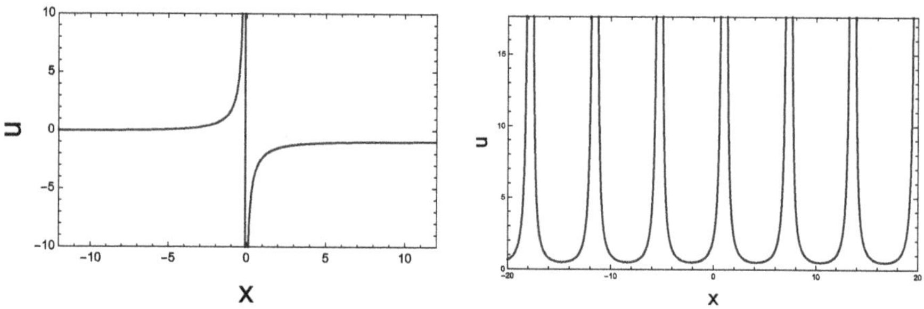

Figure 2.28. Regular shock solution of the Burgers' equation.

Step 1: Combining the independent variables x and t into one variable $\xi = x - Vt$. So,

$$u(x,t) = u(\xi), \quad \xi = x - Vt \tag{2.116}$$

where V is the speed of the traveling wave. Using transformation (2.116) converts Eq. (2.115) into an ODE for $u(x,t) = u(\xi)$

$$P(u, -Vu', u', V^2u'', -Vu'', u'', \ldots) = 0 \tag{2.117}$$

Step 2: Suppose that the solution of the ODE (2.117) can be expressed by a polynomial in $(\frac{G'}{G})$ as

$$u(\xi) = \sum_{i=0}^{m} \alpha_i \left(\frac{G'}{G}\right)^i \tag{2.118}$$

where $G = G(\xi)$ satisfies the second-order linear ODE in the form

$$G'' + \lambda G' + \mu G = 0 \tag{2.119}$$

The constants $\alpha_m, \ldots, \lambda$ and μ will be determined later, with $\alpha_m \neq 0$. The unknown part in (2.118) is also a polynomial in $(\frac{G'}{G})$, but the degree of which is generally equal to or less than $m - 1$. The positive integer m can be determined by considering the homogeneous balance between the highest-order derivatives and nonlinear terms appearing in the ODE (2.117).

Step 3: By substituting (2.118) into Eq. (2.117), using ODE (2.119), and collecting all terms with the same order of $(\frac{G'}{G})$ together, the left-hand side of Eq. (2.117) is converted into another polynomial in $(\frac{G'}{G})$. Equating each coefficient of this polynomial to zero yields a set of algebraic equations for $\alpha_m, \ldots, \alpha_0, V, \lambda$ and μ.

Step 4: Assuming that the constants $\alpha_m, \ldots, V, \lambda$ and μ can be obtained by solving the algebraic equation in *Step 3*, since the general solutions of the ODE (2.119) have been well known to us, then substituting $\alpha_m, \ldots, V$ and the general solutions of Eq. (2.119) into (2.118), we have more traveling wave solutions of the nonlinear evolution equation (2.115).

KdV Equation:

Let us consider the KdV equation as

$$u_t + uu_x + au_{xxx} = 0. \tag{2.120}$$

To solve Eq. (2.120), we follow the following steps:

Step 1: Taking the transformation $\xi = x - Vt$ on Eq. (2.120) and after integration with respect to ξ, we get

$$C - Vu + \frac{u^2}{2} + au'' = 0 \tag{2.121}$$

where C is an integration constant.

Step 2: We assume that the solution of the ODE (2.121) may be written as a polynomial in (G/G') as

$$u(\xi) = \sum_{i=0}^{m} \alpha_i \left(\frac{G'}{G}\right)^i \tag{2.122}$$

where $G = G(\xi)$ satisfies Eq. (2.119). The homogeneous balancing method allows us to determine that $m = 2$. Consequently, Eq. (2.122) can be expressed as

$$u(\xi) = \alpha_2 \left(\frac{G'}{G}\right)^2 + \alpha_1 \left(\frac{G'}{G}\right) + \alpha_0 \tag{2.123}$$

where $\alpha_2 \neq 0$.

Step 3: Substituting (2.123) into (2.121), along with $\frac{G''}{G} = -(\mu + \lambda\frac{G'}{G})$, one can derive

$$\left(\frac{\alpha_2^2}{2} + 6a\alpha_2\right)\left(\frac{G'}{G}\right)^4 + \left(\alpha_2\alpha_1 + 2a\alpha_1 + 10a\lambda\alpha_2\right)\left(\frac{G'}{G}\right)^3$$
$$+\left[\frac{1}{2}\left(\alpha_1^2 + 2\alpha_2\alpha_0\right) - V\alpha_2 + a\left(8\alpha_2\mu + 3\alpha_1\lambda + 4\alpha_2\lambda^2\right)\right]\left(\frac{G'}{G}\right)^2$$
$$+\left(-V\alpha_1 + \alpha_1\alpha_0 + 6a\alpha_2\lambda\mu + 2a\alpha_1\mu + a\alpha_1\lambda^2\right)\left(\frac{G'}{G}\right)$$
$$+\left(C - V\alpha_0 + \frac{1}{2}\alpha_0^2 + 2a\alpha_2\mu^2 + a\alpha_1\lambda\mu\right) = 0 \tag{2.124}$$

Thus, Eq. (2.121) is transformed into a new polynomial in G'/G. Equating each coefficient of this polynomial to zero results in a set of simultaneous algebraic equations for α_2, α_1, α_0, V, λ, μ, and C.

$$\left.\begin{array}{c} c - V\alpha_0 + \frac{1}{2}\alpha_0^2 + 2a\alpha_2\mu^2 + a\alpha_1\lambda\mu = 0 \\ -V\alpha_1 + \alpha_1\alpha_0 + 6a\alpha_2\lambda\mu + 2a\alpha_1\mu + a\alpha_1\lambda^2 = 0 \\ \frac{(\alpha_1^2+2\alpha_2\alpha_0)}{2} - V\alpha_2 + a(8\alpha_2\mu + 3\alpha_1\lambda + 4\alpha_2\lambda^2) = 0 \\ \alpha_2\alpha_1 + 2a\alpha_1 + 10a\lambda\alpha_2 = 0 \\ \frac{\alpha_2^2}{2} + 6a\alpha_2 = 0 \end{array}\right\} \tag{2.125}$$

Step 4: From the above system of Eq. (2.125), we get the value of the parameters as

$$\alpha_2 = -12a, \quad \alpha_1 = -12a\lambda, \quad V = \alpha_0 + 8a\mu + a\lambda^2,$$
$$C = \frac{\alpha_0^2}{2} + (8a\mu + a\lambda^2)\alpha_0 + (2+\lambda)12a^2\mu^2$$

where λ, μ, α_0 are constant. By putting the values of α_1 and α_2 in Eq. (2.123), we get

$$u(\xi) = -12a\left(\frac{G'}{G}\right)^2 - 12a\lambda\left(\frac{G'}{G}\right) + \alpha_0 \tag{2.126}$$

where $\xi = x - Vt = x - (\alpha_0 + a\lambda^2 + 8a\mu)t$.

From Eq. (2.119), we get

$$\left(\frac{d^2}{d\xi^2} + \lambda\frac{d}{d\xi} + \mu\right)G = 0 \tag{2.127}$$

The auxiliary equation of Eq. (2.127) is

$$m^2 + \lambda m + \mu = 0 \implies m = \frac{-\lambda \pm \sqrt{\lambda^2 - 4\mu}}{2} \tag{2.128}$$

So, here three cases arise.

Case-I

For $\lambda^2 - 4\mu > 0$,

The solution of Eq. (2.127) is

$$G = e^{-\frac{\lambda}{2}\xi}\left[C_1 \cosh\left(\frac{\sqrt{\lambda^2 - 4\mu}\xi}{2}\right) + C_2 \sinh\left(\frac{\sqrt{\lambda^2 - 4\mu}\xi}{2}\right)\right]$$

Substituting the value of G'/G in Eq. (2.126), we get

$$u_1(\xi) = -3a(\lambda^2 - 4\mu)\left[\frac{C_1 \sinh\left(\frac{\sqrt{\lambda^2-4\mu}\xi}{2}\right) + C_2 \cosh\left(\frac{\sqrt{\lambda^2-4\mu}\xi}{2}\right)}{C_1 \cosh\left(\frac{\sqrt{\lambda^2-4\mu}\xi}{2}\right) + C_2 \sinh\left(\frac{\sqrt{\lambda^2-4\mu}\xi}{2}\right)}\right]^2 + 3a\lambda^2 + \alpha_0 \tag{2.129}$$

where $\xi = x - (\alpha_0 + 8a\mu + a\lambda^2)t$ and C_1 and C_2 are arbitrary constants.

If C_1 and C_2 are taken as special values, a variety of results can be found. If $C_1 > 0,\ C_1^2 > C_2^2$, then $u_1(\xi)$ can be written as

$$u_1(\xi) = 3a(\lambda^2 - 4\mu)\text{sech}^2\left(\frac{\sqrt{\lambda^2 - 4\mu}\xi}{2}\right) + 12a\mu + \alpha_0 \tag{2.130}$$

Shown in Fig. (2.29)

Case-II

For $\lambda^2 - 4\mu < 0$,

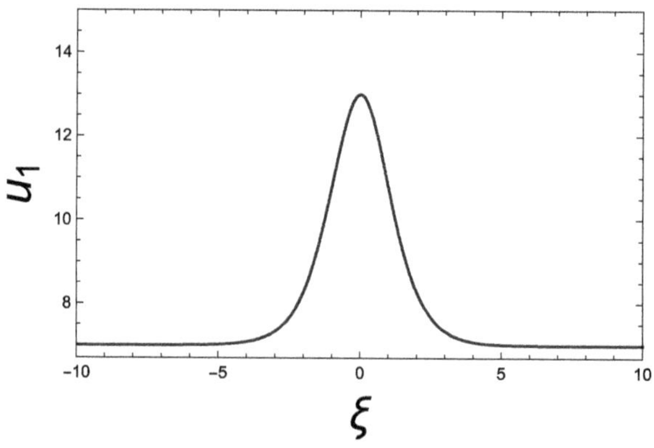

Figure 2.29. Regular soliton solution of Eq. (2.129) given in Eq. (2.130).

The solution of Eq. (2.127) is

$$G = e^{-\frac{\lambda}{2}\xi}\left[C_1 \cos\left(\frac{\sqrt{4\mu - \lambda^2}\xi}{2}\right) + C_2 \sin\left(\frac{\sqrt{4\mu - \lambda^2}\xi}{2}\right)\right]$$

Substituting the value of G'/G in Eq. (2.126), we get

$$u_2(\xi) = -3a(4\mu - \lambda^2)\left[\frac{-C_1 \sin\left(\frac{\sqrt{4\mu-\lambda^2}\xi}{2}\right) + C_2 \cos\left(\frac{\sqrt{4\mu-\lambda^2}\xi}{2}\right)}{C_1 \cos\left(\frac{\sqrt{4\mu-\lambda^2}\xi}{2}\right) + C_2 \sin\left(\frac{\sqrt{4\mu-\lambda^2}\xi}{2}\right)}\right]^2 + 3a\lambda^2 + \alpha_0 \tag{2.131}$$

where $\xi = x - (\alpha_0 + 8a\mu + a\lambda^2)t$ and C_1 and C_2 are arbitrary constants.

Case-III

For $\lambda^2 - 4\mu = 0$,

The solution of Eq. (2.127) is

$$G = (C_1 + C_2\xi)e^{-\frac{\lambda}{2}\xi}$$

Substituting the value of G'/G in Eq. (2.126), we get

$$u_3(\xi) = \alpha_0 + 3a\lambda^2 - 12a\left(\frac{C_2}{C_1 + C_2\xi}\right)^2 \tag{2.132}$$

where $\xi = x - (\alpha_0 = 36a\lambda^2)t$ and C_1 and C_2 are arbitrary constants.

mKdV Equation

The mKdV equation is

$$u_t - u^2 u_x + a u_{xxx} = 0 \tag{2.133}$$

where $a > 0$.

Step 1: Taking the traveling wave transformation $\xi = x - Vt$ on Eq. (2.133), and after integrating both sides with respect to ξ, we get

$$C - Vu - \frac{u^3}{3} + au'' = 0 \tag{2.134}$$

where C is an integration constant.

Step 2: Suppose that the solution of ODE (2.134) can be expressed by a polynomial in (G'/G) as mentioned in Eq. (2.118), where $G = G(\xi)$ satisfies Eq. (2.119). By the homogeneous balancing method, we can calculate that $m = 1$. Then Eq. (2.118) can be written as

$$u(\xi) = \alpha_1\left(\frac{G'}{G}\right) + \alpha_0 \tag{2.135}$$

where $\alpha_1 \neq 0$.

Step 3: Substituting the value of u from Eq. (2.135) to Eq. (2.134), along with $\frac{G''}{G} = -\left(\mu + \lambda\frac{G'}{G}\right)$, we obtained

$$\begin{aligned}&\left(2a\alpha_1 - \frac{\alpha_1}{3}\right)\left(\frac{G'}{G}\right)^3 + \left(3a\alpha_1\lambda - \alpha_1^2\alpha_0\right)\left(\frac{G'}{G}\right)^2\\ &+\left(2a\alpha_1\mu + a\lambda^2\alpha_1 - \alpha_1\alpha_0^2 - V\alpha_1\right)\left(\frac{G'}{G}\right)\\ &+\left(C - V\alpha_0 + \frac{1}{3}\alpha_0^3 + a\alpha_1\mu\lambda\right) = 0\end{aligned} \tag{2.136}$$

Thus, we converted Eq. (2.134) into another polynomial in G'/G. Equating each coefficient of this polynomial to zero yields a set of simultaneous algebraic equations for α_1, α_0, V, λ, μ and C.

$$\left.\begin{aligned}C - V\alpha_0 - \tfrac{1}{3}\alpha_0^3 + a\alpha_1\mu\lambda = 0,\\ 2a\alpha_1\mu + a\alpha_1\lambda^2 - \alpha_1\alpha_0^2 - V\alpha_1 = 0,\\ 3a\alpha_1\lambda - \alpha_1^2\alpha_0 = 0,\\ 2a\alpha_1 - \tfrac{\alpha_1}{3} = 0.\end{aligned}\right\} \tag{2.137}$$

Step 4: Solving the above set of Eq. (2.137), we obtained $\alpha_0 = \pm\frac{1}{2}\lambda\sqrt{6a}$, $\alpha_1 = \pm\sqrt{6a}$, $V = 2a\mu - \frac{1}{2}a\lambda^2$ and $C = 0$, where λ, μ are constant. By putting the values of α_1 and α_0 in Eq. (2.135), we get

$$u(\xi) = \pm\sqrt{6a}\left(\frac{G'}{G}\right) \pm \frac{1}{2}\lambda\sqrt{6a}, \tag{2.138}$$

where $\xi = (x - Vt) = x - (2a\mu - \frac{1}{2}a\lambda^2)t$.

From Eq. (2.119), we get

$$G'' + \lambda G' + \mu G = 0 \implies (D^2 + \lambda D + \mu)G = 0 \quad \left[D \equiv \frac{d}{d\xi}\right]. \tag{2.139}$$

The auxiliary equation of Eq. (2.139) is

$$m^2 + \lambda m + \mu = 0 \implies m = \frac{-\lambda \pm \sqrt{\lambda^2 - 4\mu}}{2}. \tag{2.140}$$

So, depending upon the values of λ and μ, there arise three conditions

$$\lambda^2 - 4\mu > 0, \quad \lambda^2 - 4\mu < 0, \text{ and } \lambda^2 - 4\mu = 0.$$

Substituting the general solutions of this equation into Eq. (2.138), we get three types of traveling wave solutions of the mKdV equation for these three conditions.

For $\lambda^2 - 4\mu > 0$,

The general solution of Eq. (2.139) is

$$G = e^{-\frac{\lambda}{2}\xi}\left[C_1\cosh\left(\frac{\sqrt{\lambda^2 - 4\mu}\xi}{2}\right) + C_2\sinh\left(\frac{\sqrt{\lambda^2 - 4\mu}\xi}{2}\right)\right]$$

So, after substituting the value of G'/G in Eq. (2.138), we get

$$u_{1,2}(\xi) = \pm\frac{1}{2}\sqrt{6a(\lambda^2 - 4\mu)}\left[\frac{C_1\sinh\left(\frac{\sqrt{\lambda^2-4\mu}\xi}{2}\right) + C_2\cosh\left(\frac{\sqrt{\lambda^2-4\mu}\xi}{2}\right)}{C_1\cosh\left(\frac{\sqrt{\lambda^2-4\mu}\xi}{2}\right) + C_2\sinh\left(\frac{\sqrt{\lambda^2-4\mu}\xi}{2}\right)}\right]^2 + 3a\lambda^2 + \alpha_0 \tag{2.141}$$

where $\xi = x - (2a\mu - \frac{1}{2}a\lambda^2)t$ and C_1 and C_2 are arbitrary constants. If C_1 and C_2 are taken as special values, the various known results in the literature can be found. If $C_1 \neq 0$, $C_2 = 0$, $\lambda > 0$, $\mu < 0$ then $u_{1,2}$ can be

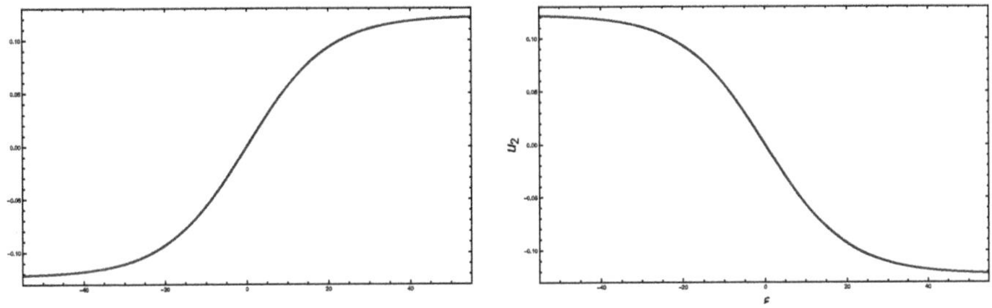

Figure 2.30. Shock wave solution of Eq. (2.133) given in Eq. (2.141).

written as

$$u_{1,2}(\xi) = \pm\frac{1}{2}\lambda\sqrt{6a}\tanh\left(\frac{\lambda\xi}{2}\right), \quad \xi = x + \frac{1}{2}a\lambda^2 t \tag{2.142}$$

which are the shock wave solution of the mKdV equation (2.133).

For $\lambda^2 - 4\mu < 0$,

The general solution of Eq. (2.139) is

$$G = e^{-\frac{\lambda}{2}\xi}\left[C_1\cos\left(\frac{\sqrt{4\mu-\lambda^2}\xi}{2}\right) + C_2\sin\left(\frac{\sqrt{4\mu-\lambda^2}\xi}{2}\right)\right].$$

Substituting the value of G'/G in Eq. (2.138), we get

$$u_{3,4}(\xi) = \pm\frac{1}{2}\sqrt{6a(4\mu-\lambda^2)}\left[\frac{-C_1\sin\left(\frac{\sqrt{4\mu-\lambda^2}\xi}{2}\right) + C_2\cos\left(\frac{\sqrt{4\mu-\lambda^2}\xi}{2}\right)}{C_1\cos\left(\frac{\sqrt{4\mu-\lambda^2}\xi}{2}\right) + C_2\sin\left(\frac{\sqrt{4\mu-\lambda^2}\xi}{2}\right)}\right]^2 \tag{2.143}$$

where $\xi = x - (2a\mu - \frac{1}{2}a\lambda^2)t$ and C_1 and C_2 are arbitrary constants.

For $\lambda^2 - 4\mu = 0$,

The general solution of Eq. (2.139) is

$$G = (C_1 + C_2\xi)e^{-\frac{\lambda}{2}\xi}.$$

Substituting the value of (G'/G) in Eq. (2.138), we get

$$u_{5,6}(\xi) = \pm\sqrt{6a}\left(\frac{C_2}{C_1 + C_2\xi}\right)^2 \tag{2.144}$$

where C_1 and C_2 are arbitrary constants.

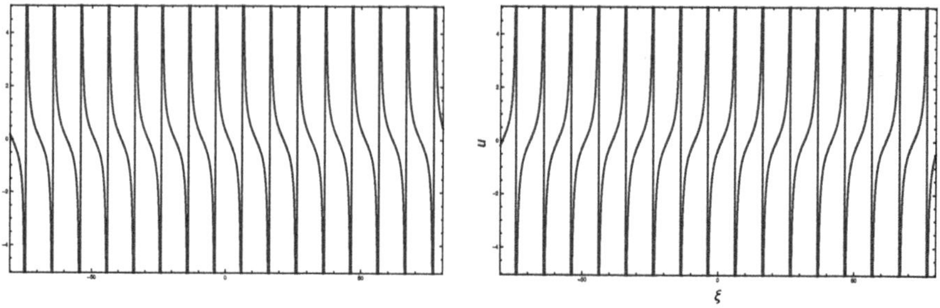

Figure 2.31. Multiple shock wave solution of Eq. (2.133).

We can also obtain the solution of the Burgers equation $u_t + uu_x = u_{xx}$ by the G'/G expansion method as

1. For $\lambda^2 - 4\mu > 0$,

$$u_1(\xi) = -2\sqrt{\lambda^2 - 4\mu}\left(\frac{C_1 \sinh \frac{1}{2}\sqrt{\lambda^2 - 4\mu}\xi + C_2 \cosh \frac{1}{2}\sqrt{\lambda^2 - 4\mu}\xi}{C_1 \cosh \frac{1}{2}\sqrt{\lambda^2 - 4\mu}\xi + C_2 \sinh \frac{1}{2}\sqrt{\lambda^2 - 4\mu}\xi}\right)$$

2. For $\lambda^2 - 4\mu < 0$,

$$u_2(\xi) = -2\sqrt{\lambda^2 - 4\mu}\left(\frac{-C_1 \sin \frac{1}{2}\sqrt{4\mu - \lambda^2}\xi + C_2 \cos \frac{1}{2}\sqrt{4\mu - \lambda^2}\xi}{C_1 \cos \frac{1}{2}\sqrt{4\mu - \lambda^2}\xi + C_2 \sin \frac{1}{2}\sqrt{4\mu - \lambda^2}\xi}\right)$$

3. For $\lambda^2 - 4\mu = 0$,

$$u_3(\xi) = -\frac{2C_2}{C_1 + C_2\xi}$$

where $\xi = x \pm \sqrt{\lambda^2 - 4\mu}t$.

And in a similar fashion, one can easily obtain the solution for the modified Burgers–KdV equation $u_t + pu^2u_x + qu_{xx} - ru_{xxx} = 0$ as

1. For $\lambda^2 - 4\mu > 0$,

$$u_1^{\pm}(\xi) = \pm\sqrt{\frac{r}{6q}}\left(\frac{q}{r} + \left\|\frac{q}{r}\right\|\left(\frac{C_1 \sinh \frac{1}{6}|\frac{q}{r}|\xi + C_2 \cosh \frac{1}{6}|\frac{q}{r}|\xi}{C_1 \cosh \frac{1}{6}|\frac{q}{r}|\xi + C_2 \sinh \frac{1}{6}|\frac{q}{r}|\xi}\right)\right)$$

where $\xi = x - \frac{2q^2}{9r}t$, the solutions $u_1^+(\xi)$ (respectively $u_1^-(\xi)$) correspond to $rq > 0$ (respectively $rq < 0$).

2. For $\lambda^2 - 4\mu = 0$,

$$u_3(\xi) = \pm\sqrt{\frac{6r}{p}}\frac{C_2}{C_1 + C_2\xi}$$

where $\xi = x$.

2.8 Modified $\left(\frac{G'}{G}\right)$ Expansion Method

The modified $\left(\frac{G'}{G}\right)$ expansion method extends the standard $\left(\frac{G'}{G}\right)$ method, which is extensively utilized in the analysis of NLPDE, particularly in the context of finding exact solutions to these equations. This approach is an effective tool in applied mathematics and theoretical physics for generating exact solutions to nonlinear equations. The modified $\left(\frac{G'}{G}\right)$ expansion technique is a revision of this method, aiming to increase its efficiency and adaptability by introducing more flexibility in the solution form.

Consider an NLPDE provided by

$$F(\phi, \phi_x, \phi_t, \phi_{xx}, \phi_{tt}, \phi_{xt}, \ldots) = 0 \tag{2.145}$$

where $\phi = \phi(x,t)$ is an unknown function and F is a polynomial of ϕ and its partial derivatives, in which the highest-order derivatives and nonlinear terms are involved. To find the solution of the above NPDE by the modified $\frac{G'}{G}$ expansion method, we go through the following steps:

Step 1: The traveling wave variable

$$\phi(\xi, \tau) = \phi(\chi) = l\xi - \rho\tau \tag{2.146}$$

where ρ and l are constants, permits us to reduce into an ODE:

$$F(\phi, \phi', \phi'', \phi''', \ldots) = 0 \tag{2.147}$$

Step 2: Suppose that the solution of (2.145) can be expressed by a polynomial $\frac{G'}{G}$ as follows:

$$\phi(\chi) = \alpha_0 + \sum_{i=1}^{m}\left[\alpha_i\left(\frac{G'}{G}\right)^i + \alpha_{-i}\left(\frac{G'}{G}\right)^{-i}\right] \tag{2.148}$$

where $G = G(\xi)$ satisfies the second-order linear ODE

$$G'' + \gamma G = 0 \tag{2.149}$$

where α_0, α_i, α_{-i} and γ are the constants to be determined.

Step 3: The parameter m in (2.148) can be found by balancing the highest-order derivatives term and the highest nonlinear term in (2.145).

(i) If m is a positive integer, then go to Step 2.
(ii) If m is not a positive integer, we put $\phi = v^m$ and then return to Step 1.

Step 4: Substituting (2.148) into (2.147) and using (2.149), collecting all terms with the same power of $\frac{G'}{G}$ together, and then equating each coefficient of the resulted polynomial to zero, yields a system of algebraic equations for α_0, α_1, α_{-1}, and γ.

Step 5: Since the general solution (2.148) is well known, we can substitute α_0, α_i, α_{-i}, and γ and get the traveling wave solutions of the given NLPDE.

KdV Equation:

Here, we studied the solution of the KdV equation by following the steps of the modified $\left(\frac{G'}{G}\right)$ expansion method. We consider a KdV equation as

$$\frac{\partial \phi_1}{\partial \tau} + C\phi_1 \frac{\partial \phi_1}{\partial \xi} + D\frac{\partial^3 \phi_1}{\partial \xi^3} = 0 \tag{2.150}$$

Take the traveling wave variable $\chi = l\xi - \rho\tau$ and follow Step 1 to obtain the ODE as

$$-\rho\phi_1 + \frac{lC}{2}\phi_1^2 + l^3 D\phi'' = 0 \tag{2.151}$$

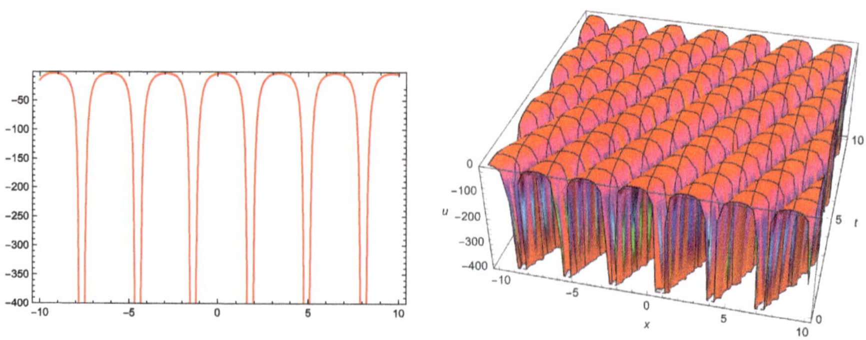

Figure 2.32. Solution of Eq. (2.150) given in (2.153).

Again, after proceeding as Step 2 and Step 3, we get the general solution as

$$\phi_1 = \alpha_0 + \alpha_1\left(\frac{G'}{G}\right) + \alpha_2\left(\frac{G'}{G}\right)^2 + \alpha_{-1}\left(\frac{G'}{G}\right)^{-1} + \alpha_{-2}\left(\frac{G'}{G}\right)^{-2} \tag{2.152}$$

where G is the solution of $G'' + \gamma G = 0$. Substituting Eq. (2.152) into Eq. (2.151) and setting the coefficients of $\frac{G'}{G}$ together to zero, we have the following set of overdetermined equations in the unknowns α_0, α_1, α_{-1}, α_{-2} and l. Solve the set of equations of coefficients of $\frac{G'}{G}$ together. By using Mathematica, we obtain the following solution:

(i) For $\alpha_0 = -\frac{4D\gamma l^2}{C}$, $\alpha_1 = \alpha_{-1} = 0$, $\alpha_{-2} = -\frac{12Dl^2\gamma^2}{C}$, and $\rho = \pm 4Dl^3\gamma$

$$\phi_1 = -\frac{4D\gamma l^2}{C} - \frac{12Dl^2\gamma^2}{C}\left(\frac{c_1\cos\sqrt{\gamma}\chi + c_2\sin\sqrt{\gamma}\chi}{c_2\sqrt{\gamma}\cos\sqrt{\gamma}\chi - c_1\sqrt{\gamma}\sin\sqrt{\gamma}\chi}\right)^2 \tag{2.153}$$

(ii) For $\alpha_2 = \frac{4\gamma\alpha_0}{3+\gamma^2}$ and $k = 8Dl^3\gamma$

$$\phi_1 = \alpha_0 + \frac{4\gamma\alpha_0}{3+\gamma^2}\left(\frac{\sqrt{\gamma}c_2\cos\sqrt{\gamma}\chi - c_1\sqrt{\gamma}\sin\sqrt{\gamma}\chi}{c_1\cos\sqrt{\gamma}\chi + c_2\sin\sqrt{\gamma}\chi}\right)^2 \tag{2.154}$$

2.9 Cosine Expansion Method

When solving some kinds of differential equations, especially when one is interested to obtain the solution of the form of periodic or oscillatory functions, the cosine expansion approach is frequently employed. It's a method for representing a function as the sum of cosine functions. This approach is useful in many areas of mathematics, including signal processing, PDE solution, and even Fourier series expansions.

Suppose a NLPDE is given by

$$F(\phi, \phi_\xi, \phi_\tau, \phi_{\xi\xi}, \phi_{\tau\tau}, \phi_{\xi\tau}, \ldots) = 0 \tag{2.155}$$

where $\phi = \phi(\xi, \tau)$ is an unknown function and F is a polynomial in and its partial derivatives, in which the highest order derivatives and nonlinear terms are involved. Let us give the main steps of the cosine expansion method as follows:

Step 1:

Take the traveling wave variable

$$\chi = \xi - c\tau \tag{2.156}$$

where c is constant to convert the NLPDE (2.155) into the following ODE

$$F(\phi, \phi', \phi'', \phi''', \ldots) = 0 \tag{2.157}$$

Step 2:

Suppose the solution of Eq. (2.157) can be expressed as

$$\phi(\chi) = h \cos^{\beta}(\mu\chi) \tag{2.158}$$

where h and μ are constants to be determined.

Step 3:

The parameter β in Eq. (2.158) can be found by homogeneous balancing the highest order derivatives term and highest nonlinear term in Eq. (2.157).

Step 4:

Substituting the value of ϕ from Eq. (2.158) with the fixed parameter β into the obtained ODE (2.157). Collecting all terms with the same power of $\cos(\mu\chi)$ together and set the coefficient to zero, to get a set of algebraic equations with respect to the unknowns μ and h. Solve the set of algebraic equations to get μ and h then we have the traveling wave solutions of the given nonlinear equation.

KdV equation:

Here, consider a general KdV equation as

$$\frac{\partial \phi}{\partial \tau} + a\phi \frac{\partial \phi}{\partial \xi} + b\frac{\partial^3 \phi}{\partial \xi^3} = 0 \tag{2.159}$$

To find the solution of Eq. (2.159) by *cosione expantion method*, we proceed through the following steps:

Step 1: By using the wave variable $\chi = \xi - c\tau$ and after integration, transform the given KdV equation (2.159) to

$$-c\phi + \frac{a}{2}\phi^2 + b\phi'' = 0 \tag{2.160}$$

With the help of *step 2,* we can calculate $\beta = -2$ and the general solution of Eq. (2.160) takes the form

$$\phi = h \cos^{-2}(\mu\chi) \tag{2.161}$$

substituting Eq. (2.161) into Eq. (2.160) collecting all terms with the same powers of $\cos(\mu\chi)$ to zero together, we obtain

$$b\mu^2(-2)^2 = -c,$$
$$\frac{a}{2}h = -b\mu^2(-2)(-2-1).$$

Solve the set of above equations, we obtain the following solutions

$$\phi(\xi,\tau) = \frac{3c}{a}\sec^2\left[\sqrt{\frac{-c}{4b}}(x-ct)\right], \quad c<0$$

$$\text{and} \quad \phi(\xi,\tau) = \frac{3c}{a}\sec h^2\left[\sqrt{\frac{c}{4b}}(x-ct)\right], \quad c>0$$

mKdV Equation:

Here, consider a general modified KdV equation as

$$\frac{\partial\phi}{\partial\tau} + a\phi^2\frac{\partial\phi}{\partial\xi} + b\frac{\partial^3\phi}{\partial\xi^3} = 0 \tag{2.162}$$

To find the solution of Eq.(2.162) by *cosine expansion method*, we proceed through the following steps:

Step 1: By using the wave variable $\chi = \xi - c\tau$ and after integration, transform the given KdV equation (2.162) to

$$-c\phi + \frac{a}{3}\phi^3 + b\phi'' = 0 \tag{2.163}$$

With the help of *step 2*, we can calculate $\beta = -1$ and the general solution of Eq. (2.163) takes the form

$$\phi = h\cos^{-1}(\mu\chi) \tag{2.164}$$

substituting Eq. (2.164) into Eq. (2.163) collecting all terms with the same powers of $\cos(\mu\chi)$ to zero together, we obtain

$$\mu^2 = \frac{1}{a-c^2},$$
$$h^2 = 2$$

Solve the set of above equations, we obtain the following solutions

$$\phi(\xi,\tau) = \sqrt{2}\sec\left[\sqrt{\frac{1}{a-c^2}}(x-ct)\right], \quad a>c^2$$

$$\text{and} \quad \phi(\xi,\tau) = \sqrt{2}\sec h\left[\sqrt{\frac{1}{c^2-a}}(x-ct)\right], \quad a<c^2.$$

2.10 Sine Expansion Method

The sine expansion method is a technique for expressing a function as a series of sine functions. It is also frequently used to solve differential equations, particularly those with precise boundary conditions. This method, like the cosine expression method, is commonly used to solve PDEs, particularly when dealing with boundary value problems when the function is zero at the boundaries (Dirichlet boundary conditions).

Suppose an NLPDE is given by

$$F(\phi, \phi_\xi, \phi_\tau, \phi_{\xi\xi}, \phi_{\tau\tau}, \phi_{\xi\tau}, \ldots) = 0 \tag{2.165}$$

where $\phi = \phi(\xi, \tau)$ is an unknown function and F is a polynomial in and its partial derivatives, in which the highest order derivatives and nonlinear terms are involved. Let us give the main steps of the cosine expansion method as follows:

Step 1:

Take the traveling wave variable

$$\chi = \xi - c\tau \tag{2.166}$$

where c is constant, to convert the NLPDE (2.165) into the following ODE

$$F(\phi, \phi', \phi'', \phi''', \ldots) = 0 \tag{2.167}$$

Step 2:

Suppose the solution of Eq. (2.167) can be expressed as

$$\phi(\chi) = h \sin^{\beta}(\mu\chi) \tag{2.168}$$

where h and μ are constants to be determined.

Step 3:

The parameter β in Eq. (2.168) can be found by homogeneous balancing of the highest order derivatives term and highest nonlinear term in Eq. (2.167).

Step 4:

Substituting the value of ϕ from Eq. (2.168) with the fixed parameter β into the obtained ODE (2.167). Collecting all terms with the same power of $\sin(\mu\chi)$ together and setting the coefficient to zero, to get a set of algebraic equations with respect to the unknowns μ and h. Solve the set of algebraic equations to get μ and h then we have the traveling wave solutions of the given nonlinear equation.

By following the above method, one can easily obtain the solution to the following equation

$$\frac{\partial \phi}{\partial \tau} + a\phi \frac{\partial \phi}{\partial \xi} + b\frac{\partial^3 \phi}{\partial \xi^3} = 0 \tag{2.169}$$

as

$$\phi(\xi, \tau) = \frac{3c}{a} \cos ec^2 \left[\sqrt{\frac{-c}{4b}}(x - ct) \right], \quad c < 0$$

and

$$\phi(\xi\tau) = \frac{-3c}{a} \cos ech^2 \left[\sqrt{\frac{c}{4b}}(x - ct) \right], \quad c > 0.$$

2.11 Sine–Cosine Expansion Method

The technique known as the sine–cosine expansion method is used to solve specific kinds of PDEs by expanding the solution in terms of both sine and cosine functions. This can be considered as a combination of the sine and cosine methods and offers new solutions to the cosine and sine methods. When the problem has periodic boundary conditions then we call for a combination of sine and cosine terms to adequately represent the solution. When a more comprehensive solution that is not solely based on sine or cosine is required, or when problems involving boundary conditions on intervals where both sine and cosine functions naturally appear, this approach is usually used.

Suppose an NLPDE is given by

$$F(\phi, \phi_\xi, \phi_\tau, \phi_{\xi\xi}, \phi_{\tau\tau}, \phi_{\xi\tau}, \ldots) = 0 \tag{2.170}$$

where $\phi = \phi(\xi, \tau)$ is an unknown function and F is a polynomial in and its partial derivatives, in which the highest-order derivatives and nonlinear terms are involved. Let us give the main steps of the sine-cosine expansion method as follows:

Step 1: Take the traveling wave variable

$$\chi = l\xi - \rho\tau \tag{2.171}$$

where ρ and l are constants, to convert the NLPDE (2.170) into the following ODE

$$F(\phi, \phi', \phi'', \phi''', \ldots) = 0. \tag{2.172}$$

Step 2: Suppose the solution of Eq. (2.172) can be expressed as

$$\phi(\chi) = A_0 + \sum_{i=1}^{n} \cos^{i-1}(\omega(\chi))\left[A_i \sin(\omega(\chi)) + B_i \cos(\omega(\chi))\right] \tag{2.173}$$

where $A_i(i = 0, 1, 2, \ldots, n)$ and $B_i(i = 0, 1, 2, \ldots, n)$ are constants to be determined.

Step 3: The parameter n in Eq. (2.173) can be found by homogeneous balancing the highest order derivatives term and the highest nonlinear term in Eq. (2.172).

(i) If m is a positive integer, then go to Step 4.
(ii) If m is not a positive integer, we put $\phi = v^m$ and then return to Step 1.

Step 4: Substituting the value of ϕ from Eq. (2.173) with the fixed parameter n into the obtained ODE (2.172), where $\omega(\chi)$ satisfies

$$\frac{d}{d\chi} = \gamma\sqrt{a + b\sin^2(\omega(\chi))}, \quad \gamma = \pm 1 \tag{2.174}$$

where a and b are constants. Collecting all terms with the same power of $\omega'^{s} \sin^i \omega \cos^j \omega$ together. Set to zero the coefficient of $(i = 0, 1, 2; s = 0, 1; j = 0, 1, 2, \ldots, n)$ to get a set of algebraic equations $c_{ijs}(A_s; B_s) = 0$, with respect to the unknowns $A_0, A_i(i = 1, \ldots, n)$ and $B_i(i = 1, \ldots, n)$.

Step 5: Solve the set of algebraic equations to get A_0, A_i, and B_i, and then we have the traveling wave solutions of the given nonlinear equation. In the sine–cosine expansion method, there are three cases of (2.174).

Case I: In these cases, $a = 0,\ b = 1$, then the equation reduces to a first-order ODE

$$\frac{d(\omega(\chi))}{d\chi} = \sin(\omega(\chi)) \tag{2.175}$$

which has the solutions:

$\sin(\omega(\chi)) = \operatorname{sech}(\chi)$, or $\cos(\omega(\chi)) = -\tanh(\chi)$ and $\sin(\omega(\chi)) = i\,cosech(\chi)$, or $\cos(\omega(\chi)) = -\coth(\chi)$, where $i = \sqrt{-1}$.

Case II: In these cases, $a = 1,\ b = -m^2$, then the equation reduces to the first-order ODE

$$\frac{d(\omega(\chi))}{d\chi} = \pm\sqrt{1 - m^2 \sin^2(\omega(\chi))}, \tag{2.176}$$

where m is the modulus of Jacobi elliptic functions, which has the solutions

$\sin(\omega(\chi)) = msn(\chi;m)$, or $\cos(\omega(\chi)) = dn(\chi;m)$ and $\sin(\omega(\chi)) = ns(\chi;m)$, or $\cos(\omega(\chi)) = ics(\chi,m)$, where $i = \sqrt{-1}$.

KdV equation:

Here, consider a general KdV equation as

$$\frac{\partial \phi_1}{\partial \tau} + C\phi_1 \frac{\partial \phi_1}{\partial \xi} + D\frac{\partial^3 \phi_1}{\partial \xi^3} = 0. \tag{2.177}$$

To find the solution of Eq. (2.177) by *sine–cosine expansion method*, we proceeded through the following steps:

Step 1: By using the wave variable $\chi = l\xi - \rho\tau$ and after integration, transform the given KdV equation (2.177) to

$$-\rho\phi + \frac{lc}{2}\phi_1^2 + l^3 D\phi'' = 0. \tag{2.178}$$

With the help of *step 2* and *step 3*, we can calculate $n = 2$ and the general solution of Eq. (2.178) take the form

$$\begin{aligned}\phi_1 &= A_0 + A_1 \sin(\omega(\chi)) + B_1 \cos(\omega(\chi)) \\ &\quad + A_2 \sin(\omega(\chi))\cos(\omega(\chi)) + B_2 \cos^2(\omega(\chi))\end{aligned} \tag{2.179}$$

substituting Eq. (2.179) into Eq. (2.178) and collecting all terms with the same powers of $\sin^i \omega \cos^j \omega (i = 0, 1; j = 0, 1, \ldots, n)$ to zero together. Set to zero the coefficients of the unknowns A_0, A_1, A_2, B_1, B_2, and l. Solve the set of equations of $\sin^i \omega \cos^j \omega$, By using Mathematica, we obtain the following solutions:

$$\begin{aligned}A_0 &= \frac{a^2\rho}{c(a^2+ab+b^2)l} + \frac{ab\rho}{c(a^2+ab+b^2)l} + \frac{b^2\rho}{c(a^2+ab+b^2)l} \\ &\quad + \frac{aDl^2\rho}{c\sqrt{(a^2+ab+b^2)D^2l^6}} + \frac{2bDl^2\rho p}{c\sqrt{(a^2+ab+b^2)D^2l^6}}\end{aligned}$$

$$A_1 = 0, \quad A_2 = 0, \quad B_1 = 0, \quad B_2 = \frac{3bDl^2\rho}{c\sqrt{(a^2+ab+b^2)D^2l^6}},$$

$$\gamma = -\frac{\sqrt{\rho}}{(a^2D^2l^6 + abD^2l^6 + b^2D^2l^6)^{1/4}}$$

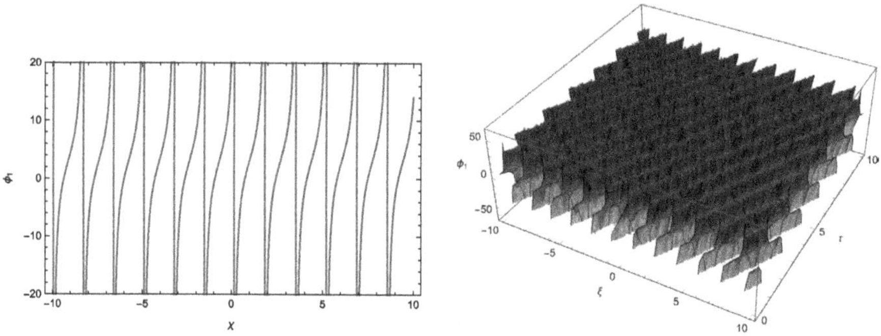

Figure 2.33. Solution of Eq. (2.177) given in Eq. (2.180).

So, the generalized solution can be written as

$$\begin{aligned}\phi_1 &= \frac{a^2\rho}{c(a^2+ab+b^2)l} + \frac{ab\rho}{c(a^2+ab+b^2)l} + \frac{b^2\rho}{c(a^2+ab+b^2)l} \\ &\quad + \frac{aDl^2\rho}{c\sqrt{(a^2+ab+b^2)D^2l^6}} + \frac{2bDl^2\rho}{c\sqrt{(a^2+ab+b^2)D^2l^6}} \\ &\quad - \frac{\sqrt{\rho}}{(a^2D^2l^6+abD^2l^6+b^2D^2l^6)^{1/4}}\cos^2(\omega(\chi))\end{aligned} \tag{2.180}$$

For three cases

Case I:

$$\begin{aligned}\phi_1 &= \frac{a^2\rho}{c(a^2+ab+b^2)l} + \frac{ab\rho}{c(a^2+ab+b^2)l} + \frac{b^2\rho}{c(a^2+ab+b^2)l} \\ &\quad + \frac{aDl^2\rho}{c\sqrt{(a^2+ab+b^2)D^2l^6}} + \frac{2bDl^2\rho}{c\sqrt{(a^2+ab+b^2)D^2l^6}} \\ &\quad + \frac{3bDl^2\rho}{c\sqrt{(a^2+ab+b^2)D^2l^6}}\tanh^2(\omega(\chi))\end{aligned} \tag{2.181}$$

or

$$\begin{aligned}\phi_1 &= \frac{a^2\rho}{c(a^2+ab+b^2)l} + \frac{ab\rho}{c(a^2+ab+b^2)l} + \frac{b^2\rho}{c(a^2+ab+b^2)l} \\ &\quad + \frac{aDl^2\rho}{c\sqrt{(a^2+ab+b^2)D^2l^6}} + \frac{2bDl^2\rho}{c\sqrt{(a^2+ab+b^2)D^2l^6}} \\ &\quad + \frac{3bDl^2\rho}{c\sqrt{(a^2+ab+b^2)D^2l^6}}\coth^2(\omega(\chi))\end{aligned} \tag{2.182}$$

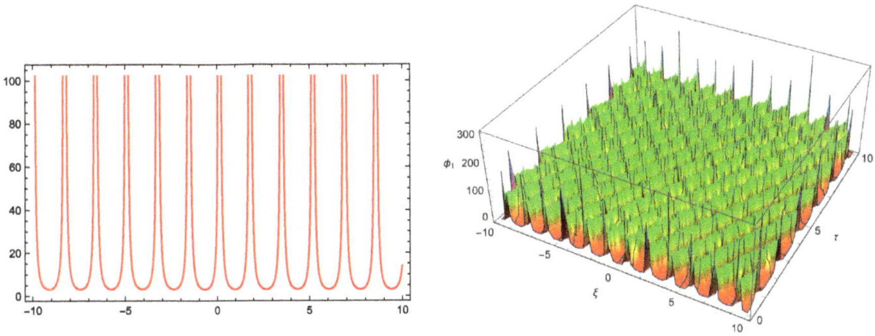

Figure 2.34. Solution of Eq. (2.177) given in Eq. (2.182).

Case II:

$$\phi_1 = \frac{a^2\rho}{c(a^2+ab+b^2)l} + \frac{ab\rho}{c(a^2+ab+b^2)l} + \frac{b^2\rho}{c(a^2+ab+b^2)l} + \frac{aDl^2\rho}{c\sqrt{(a^2+ab+b^2)D^2l^6}} + \frac{2bDl^2\rho}{c\sqrt{(a^2+ab+b^2)D^2l^6}} - \frac{3bDl^2\rho}{c\sqrt{(a^2+ab+b^2)D^2l^6}}cn^2(\chi;m) \tag{2.183}$$

$$\phi_1 = \frac{a^2\rho}{c(a^2+ab+b^2)l} + \frac{ab\rho}{c(a^2+ab+b^2)l} + \frac{b^2\rho}{c(a^2+ab+b^2)l} + \frac{aDl^2\rho}{c\sqrt{(a^2+ab+b^2)D^2l^6}} + \frac{2bDl^2\rho}{c\sqrt{(a^2+ab+b^2)D^2l^6}} - \frac{3bDl^2\rho}{c\sqrt{(a^2+ab+b^2)D^2l^6}}\frac{ids^2(\chi;m)}{m} \tag{2.184}$$

Case III:

$$\phi_1 = \frac{a^2\rho}{c(a^2+ab+b^2)l} + \frac{ab\rho}{c(a^2+ab+b^2)l} + \frac{b^2\rho}{c(a^2+ab+b^2)l} + \frac{aDl^2\rho}{c\sqrt{(a^2+ab+b^2)D^2l^6}} + \frac{2bDl^2\rho}{c\sqrt{(a^2+ab+b^2)D^2l^6}} - \frac{3bDl^2\rho}{c\sqrt{(a^2+ab+b^2)D^2l^6}}dn^2(\chi;m) \tag{2.185}$$

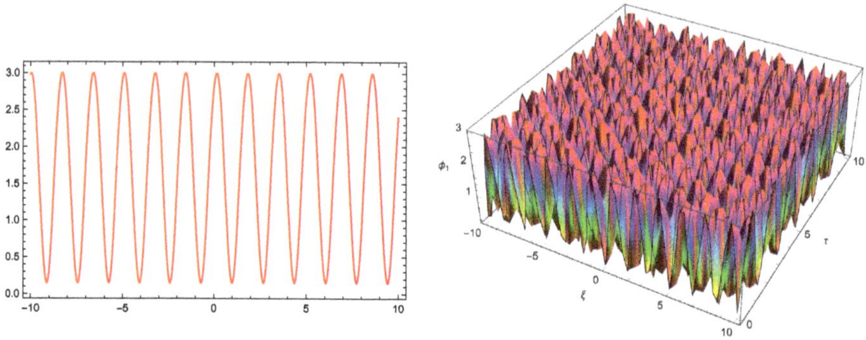

Figure 2.35. Solution of Eq. (2.177) given in Eq. (2.185).

or

$$\begin{aligned}\phi_1 = &\frac{a^2\rho}{c(a^2+ab+b^2)l} + \frac{ab\rho}{c(a^2+ab+b^2)l} + \frac{b^2\rho}{c(a^2+ab+b^2)l}\\ &+ \frac{aDl^2\rho}{c\sqrt{(a^2+ab+b^2)D^2l^6}} + \frac{2bDl^2\rho}{c\sqrt{(a^2+ab+b^2)D^2l^6}}\\ &- \frac{3bDl^2\rho}{c\sqrt{(a^2+ab+b^2)D^2l^6}} ics^2(\chi;m)\end{aligned} \tag{2.186}$$

mKdV Equation:

We consider the mKdV equation

$$u_t + au^2u_x + bu_{xxx} = 0, \quad a, b > 0 \tag{2.187}$$

where a and b are constants, and $u(x,t)$ is an unknown function. Following the analysis presented above in *set 1*, Eq. (2.187) will be transformed to

$$-cu_\xi + \frac{a}{3}u_\xi^3 + bu_{\xi\xi\xi} = 0 \tag{2.188}$$

Integrating (2.188) once and using the constant of integration to be zero, we find

$$-cu + \frac{a}{3}u^3 + bu_{\xi\xi} = 0 \tag{2.189}$$

Substituting Eq. 2.179 into (2.189) yields

$$\begin{aligned}&-c\lambda\cos^\beta(\mu\xi) + \frac{a}{3}\lambda^3\cos^{3\beta}(\mu\xi) - b\mu^2\beta^2\lambda\cos^\beta(\mu\xi) + b\lambda\mu^\beta(\beta-1)\\ &\cos^{\beta-2}(\mu\xi) = 0\end{aligned} \tag{2.190}$$

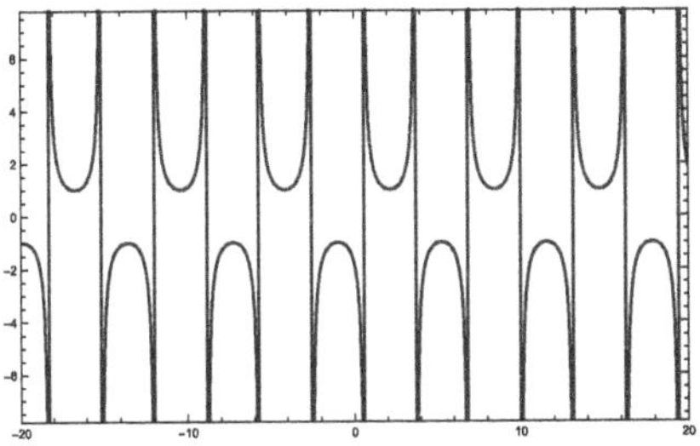

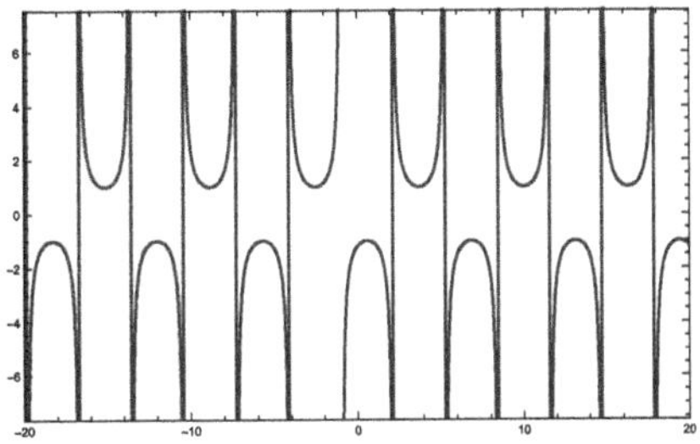

Figure 2.36. (a) Regular soliton solution of 2D.

It is obvious that Eq. (2.190) is satisfied if the following system of algebraic equations holds

$$\beta - 1 \neq 0, \quad \beta - 2 = 3\beta, \quad \frac{a}{3}\lambda^2 = -b\mu^2\beta(\beta-1), \quad -c = b\mu^2\beta^2 \tag{2.191}$$

Solving this system leads to

$$\beta = -1, \quad \mu = \sqrt{\frac{-c}{b}}, \quad \lambda = \frac{6c}{a} \tag{2.192}$$

The same results are obtained if we use the sine method. In view of (2.192), we obtain the solutions

$$u(x,t) = \sqrt{\frac{6c}{a}} \sec\left[\sqrt{\frac{-c}{b}}(x-ct)\right], \quad c < 0 \tag{2.193}$$

and

$$u(x,t) = \sqrt{\frac{6c}{a}} csc\left[\sqrt{\frac{-c}{b}}(x-ct)\right], \quad c < 0 \tag{2.194}$$

For $c > 0$, the following solutions

$$u(x,t) = \sqrt{\frac{6c}{a}} sech\left[\sqrt{\frac{c}{b}}(x-ct)\right], \quad c > 0 \tag{2.195}$$

and

$$u(x,t) = \sqrt{\frac{6c}{a}} csch\left[\sqrt{\frac{c}{b}}(x-ct)\right], \quad c > 0, \quad i = \sqrt{-1} \tag{2.196}$$

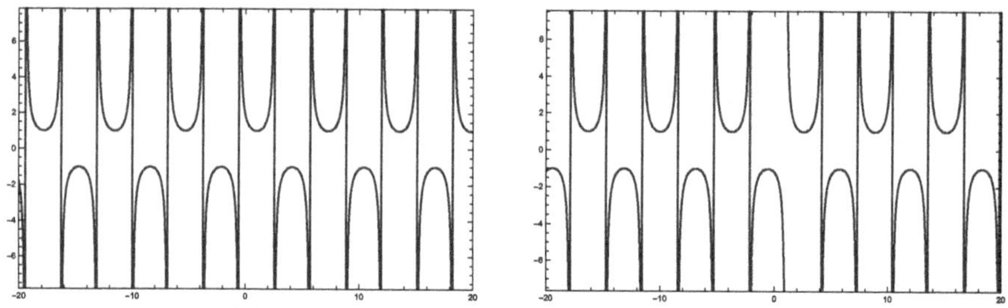

(a) (b)

Figure 2.37. (a) Regular soliton solution of 2D.

It is worth noting that in (2.190), if $\beta = 1$, this holds only if $a = 0$.

2.12 Sech Expansion Method

In mathematical physics, the sech expansion method is a methodology used to discover exact solutions to nonlinear differential equations, especially when dealing with soliton equations like the KdV and mKdV equations. When looking for solutions that have sech functional forms, which are typical of solitary wave solutions (solitons), this approach is frequently employed.

Suppose that a NLPDE is given by

$$F(\phi, \phi_\xi, \phi_\tau, \phi_{\xi\xi}, \phi_{\tau\tau}, \phi_{\xi\tau}, \ldots) = 0 \tag{2.197}$$

where $\phi = \phi(\xi, \tau)$ is an unknown function and F is a polynomial in ϕ and its partial derivatives. In the following, we give the main steps of the Sech expansion method:

Step 1: Take the traveling wave variable

$$\chi = l\xi - \rho\tau \tag{2.198}$$

where ρ and l are constants, to convert the NLPDE (2.197) into the following ODE

$$F(\phi, \phi', \phi'', \phi''', \ldots) = 0 \tag{2.199}$$

Step 2: Suppose that Eq. (2.199) has a solution of the form

$$\phi(\chi) = h\text{sech}^p(\chi) \tag{2.200}$$

where h and p are unknown parameters.

Step 3: The parameter p in Eq. (2.200) can be found by homogeneous balancing the highest order derivatives term and highest nonlinear term in Eq. (2.199).

Step 4: Substituting the value of ϕ from Eq. (2.200) with the fixed parameter p into the obtained ODE (2.199). Collecting all terms with the same power of sech(χ) together and setting the coefficient to zero, to get a set of algebraic equations with respect to the unknowns p and h. Solve the set of algebraic equations to get p and h then we have the traveling wave solutions of the given nonlinear equation.

KdV equation:

Here, consider a general KdV equation as

$$\frac{\partial \phi}{\partial \tau} + a\phi \frac{\partial \phi}{\partial \xi} + b\frac{\partial^3 \phi}{\partial \xi^3} = 0 \tag{2.201}$$

The solution of Eq. (2.201) by Sech Method is obtained as

$$\phi = \frac{12bl^2}{a}\text{sech}^2(l\xi - \rho\tau) \tag{2.202}$$

mKdV equation:

Here, consider a general modified KdV equation as

$$\frac{\partial \phi}{\partial \tau} + a\phi^2 \frac{\partial \phi}{\partial \xi} + b\frac{\partial^3 \phi}{\partial \xi^3} = 0 \tag{2.203}$$

The solution of Eq. (2.203) can be easily obtained by the *sech expansion method* as

$$\phi(\chi) = \pm\sqrt{\frac{6bl^2}{a}}\text{sech}(\chi) \tag{2.204}$$

2.13 Sech–Tanh Expansion Method

To obtain precise solutions to NLPDEs, the sech–tanh approach is frequently employed. If the solutions of a PDE is expected involving specialized functions, including sech and tanh functions, can be solved using a variety of techniques, including this one. In the context of soliton equations or equations that explain wave-like processes, it is commonly used.

Suppose that an NLPDE is given by

$$F(\phi, \phi_\xi, \phi_\tau, \phi_{\xi\xi}, \phi_{\tau\tau}, \phi_{\xi\tau}, \ldots) = 0 \tag{2.205}$$

where $\phi = \phi(\xi, \tau)$ is an unknown function and F is a polynomial in and its partial derivatives, in which the highest-order derivatives and nonlinear terms are involved. In the following, we give the main steps of the sech–tanh expansion method:

Step 1: The traveling wave variable

$$\phi(\xi, \tau) = \phi(\chi) = l\xi - \rho\tau \tag{2.206}$$

where ρ and l are constants, permits us to reduce (2.205) into an ODE;

$$F(\phi, \phi', \phi'', \phi''', \ldots) = 0 \tag{2.207}$$

Step 2: Suppose that the solution of (2.206) can be expressed as follows:

$$\phi(\chi) = A_0 + \sum_{i=1}^{n} \mathrm{sech}^{i-1}(\chi)[A_i \tanh(\chi) + B_i \mathrm{sech}(\chi)] \tag{2.208}$$

where $A_i (i = 0, 1, 2, \ldots, n)$ and $B_i (i = 0, 1, 2, \ldots, n)$ are constants to be determined.

Step 3: The parameter "n" in (2.208) can be found by balancing the highest order derivatives term and the highest nonlinear term in (2.206);

(i) If n is a positive integer, then go to *Step 4*.
(ii) If n is not a positive integer, we put $\phi = v^m$ and then return to *Step 1*.

Step 4: Substituting (2.208) with the fixed parameter n into the obtained ODE. collecting all terms with the same power of $\mathrm{sech}^i \chi \tan^j \chi (i = 0, 1; s = 0, 1; j = 0, 1, 2, \ldots, n)$ to get a set of algebraic equations $c_{ijs}(A_s; B_s) = 0$, with respect to the unknowns $A_{0,\tau} A_i (i = 1, \ldots, n)$ and $B_i (i = 1, \ldots, n)$.

Step 5: Solve the set of algebraic equations to get A_0, A_i and B_i, and then we have the traveling wave solutions of the given nonlinear equation.

$$\begin{aligned} \phi_1 &= A_0 + A_1 \tanh(\chi) + B_1 \mathrm{sech}(\chi) \\ &\quad + A_2 \tanh(\chi)\mathrm{sech}(\chi) + B_2 \mathrm{sech}^2(\chi) \end{aligned} \tag{2.209}$$

Here consider a generalized KdV equation as $\frac{\partial \phi}{\partial \tau} + a\phi \frac{\partial \phi}{\partial x} + b \frac{\partial^3 \phi}{\partial x^3} = 0$.

Then after following Step 1 to Step 4 and and collecting all terms with the same powers of $\mathrm{sech}^i \chi \tanh^j \chi (i = 0, 1; j = 0, 1, \ldots, n)$ to zero together. Set to zero the coefficients of the unknowns A_0, A_1, A_2, B_1, B_2, and l. Solve the set of equations of $\operatorname{sech}^i\chi \tanh^j \chi$, By using Mathematica, we obtain the following solutions:

$$A_0 = -\frac{8Dl^2}{c}, \quad A_1 = 0, \quad A_2 = 0, \quad B_1 = 0,$$

$$B_2 = \frac{3bDl^2\rho}{c\sqrt{(a^2+ab+b^2)D^2l^6}}, \quad \gamma = -\frac{\sqrt{\rho}}{(a^2D^2l^6+abD^2l^6+b^2D^2l^6)^{1/4}}$$

So, the generalized solution can be written as

$$\begin{aligned}\phi_1 &= \frac{a^2\rho}{c(a^2+ab+b^2)l} + \frac{ab\rho}{c(a^2+ab+b^2)l} + \frac{b^2\rho}{c(a^2+ab+b^2)l} \\ &+ \frac{aDl^2\rho}{c\sqrt{(a^2+ab+b^2)D^2l^6}} + \frac{2bDl^2\rho}{c\sqrt{(a^2+ab+b^2)D^2l^6}} \\ &- \frac{\sqrt{\rho}}{(a^2D^2l^6+abD^2l^6+b^2D^2l^6)^{1/4}}\cos^2(\omega(\chi)) \end{aligned} \tag{2.210}$$

2.14 Jacobian Elliptic Function Expansion Method

The extended Jacobian elliptic function expansion method is a powerful mathematics technique used to find exact solutions for NLPDEs. It involves expanding solutions in terms of Jacobian elliptic functions such as Sine, Cosine, or other related functions.

Jacobian elliptic functions are solutions to the Jacobi elliptic differential equation, which are periodic functions that depend on two parameters: the modulus (m) and the argument (u). These functions are generalized trigonometric functions and can describe a wide range of wave-like phenomena.

The classical Jacobian elliptic function expansion includes: $\mathrm{Sn}(u,m)$: Sine amplitude, $\mathrm{Cn}(u,m)$: Cosine amplitude, $\mathrm{dn}(u,m)$: delta amplitude etc.

Algorithm:

Let us consider the NLPDE

$$P(u, u_t, u_x, u_{tt}, u_{xx}, \ldots) = 0 \tag{2.211}$$

where P is polynomial in $u(x,t)$ and its partial derivatives in which the highest order derivatives and nonlinear terms are involved. The key steps of this method are listed as follows.

Step 1:

Using the transformation $u = u(\xi),\ \xi = x - ct$, where c is the spped of the wave, one can reduce Eq. (2.211) to the following ODE

$$P(u, u', u''', \ldots) = 0 \tag{2.212}$$

where P is a polynomial in $u(\xi)$ and its total derivatives, while $' = \frac{d}{d\xi}$.

Step 2:

We consider its solution in the form

$$u(\xi) = a_0 + \sum_{j=1}^{n} f_i^{j-1}(\xi)[a_j f_j(\xi) + b_j g_j(\xi)], \quad i = 1, 2, 3, \ldots \tag{2.213}$$

where

$$\begin{aligned} &f_1(\xi) = sn\xi, \quad g_1(\xi) = cn\xi, \quad f_2(\xi) = sn\xi, \quad g_2(\xi) = dn\xi, \\ &f_3(\xi) = ns\xi, \quad g_3(\xi) = cs\xi, \quad f_4(\xi) = ns\xi, \quad g_4(\xi) = ds\xi, \\ &f_5(\xi) = sc\xi, \quad g_5(\xi) = nc\xi, \quad f_6(\xi) = sd\xi, \quad g_6(\xi) = nd\xi \end{aligned} \tag{2.214}$$

Here, $sn\xi$, $cn\xi$, $dn\xi$ are the Jacobian elliptic sine function, the Jacobian elliptic cosine function and the Jacobian elliptic function of the third kind and other Jacobian functions, denoted by Glaisher's symbols. Those are generated by these three kinds of functions, namely,

$$\begin{aligned} &ns\xi = \frac{1}{sn\xi}, \quad nc\xi = \frac{1}{cn\xi}, \quad nd\xi = \frac{1}{dn\xi}, \quad sc\xi = \frac{cn\xi}{sn\xi}, \\ &cs\xi = \frac{sn\xi}{cn\xi}, \quad ds\xi = \frac{dn\xi}{sn\xi}, \quad sd\xi = \frac{sn\xi}{dn\xi} \end{aligned}$$

They also have the relations

$$\begin{aligned} &sn^2\xi + cn^2\xi = 1, \quad dn^2\xi + m^2 sn^2\xi = 1, \quad ns^2\xi = 1 + cs^2\xi, \\ &ns^2\xi = m^2 + ds^2\xi, \quad sc^2\xi + 1 = nc^2\xi, \quad m^2 sd^2\xi + 1 = nd^2\xi \end{aligned} \tag{2.215}$$

where $0 < m < 1$. In addition, we have

$$\frac{d}{d\xi} sn\xi = cn\xi dn\xi, \quad \frac{d}{d\xi} cn\xi = -sn\xi dn\xi, \quad \frac{d}{d\xi} dn\xi = -m^2 sn\xi cn\xi \tag{2.216}$$

The derivatives of other Jacobian elliptic functions are obtained by using (2.216). If we define the degree of $u(\xi)$ as $D[u] = n$, it give rise to the degrees of other expression as

$$D\left[\frac{d^q u}{d\xi^q}\right] = n + q, \quad D\left[u^p \left(\frac{d^q u}{d\xi^q}\right)^8\right] = np + s(n + q) \tag{2.217}$$

We can find n in Eq. (2.213) by balancing the highest order linear and nonlinear terms in Eq. (2.212) in accordance with the criteria.

Furthermore, we observe that sn, cn, and dn degenerate as tanh, sech, and cosech, respectively, when $m \to 1$, but Eq. (2.217) degenerates as the following forms when

$$u(\xi) = a_0 + \sum_{j=1}^{N} \tanh^{j-1}(\xi)[a_j \tanh(\xi) + b_j \text{sech}(\xi)] \tag{2.218}$$

$$u(\xi) = a_0 + \sum_{j=1}^{N} \coth^{j-1}(\xi)[a_j coth(\xi) + b_j coth(\xi)] \tag{2.219}$$

$$u(\xi) = a_0 + \sum_{j=1}^{N} \tan^{j-1}(\xi)[a_j \tan(\xi) + b_j sec(\xi)] \tag{2.220}$$

$$u(\xi) = a_0 + \sum_{j=1}^{N} cot^{j-1}(\xi)[a_j cot(\xi) + b_j csc(\xi)] \tag{2.221}$$

Thus, the extended Jacobian elliptic function expansion approach is more general than the sine–cosine method, the Tanh-function method, and the Jacobian elliptic function expansion method.

mKdV equation:

Let us study the solution of the mKdV equation by using the extended Jacobian elliptic function method. The mKdV equation is

$$\frac{\partial u}{\partial t} + au^2 \frac{\partial u}{\partial x} + b \frac{\partial^2 u}{\partial x^2} = 0 \tag{2.222}$$

where a and b are constant.

Using the transformation $\xi = k(x - \lambda t)$, we get

$$bk^2 \frac{du}{d\xi} - \lambda \frac{du}{d\xi} + au^2 \frac{du}{d\xi} = 0 \tag{2.223}$$

Let us consider its solution of the form

$$u(\xi) = a_0 + \sum_{j=1}^{n} f_i^{j-1}(\xi)[a_j f_j(\xi) + b_j g_j(\xi)], \quad i = 1, 2, 3, 4, 5 \tag{2.224}$$

where $f_1(\xi)$, $f_2(\xi)$, $f_3(\xi)$, $f_4(\xi)$, $f_5(\xi)$, $g_1(\xi)$, $g_2(\xi)$, $g_3(\xi)$, $g_4(\xi)$, $g_5(\xi)$ are given in (2.214). If we obtain the solution of Eq. (2.223), then we can find the corresponding solution of Eq. (2.222).

Consider Eq. (2.223) has the formal solution

$$u = a_0 + \sum_{i=1}^{n} sd^{i-1}\xi(a_i sd\xi + b_i nd\xi) \tag{2.225}$$

where $sd\xi = sd(\xi, m)$ and $nd\xi = nd(\xi, m)$ are the Jacobian elliptic functions denoted by Glaisher's symbols. n is the integer. Balancing the highest order linearly derivative term $\frac{d^3u}{d\xi^3}$ and the nonlinear term $u^2\frac{du}{d\xi}$ in Eq. (2.223), we get $n = 1$. So, Eq. (2.223) has the solution

$$u = a_0 + a_1 sd(\xi) + b_1 nd(\xi) \tag{2.226}$$

where a_0, a_1, and b_1 are constants to be determined.

Substituting Eq. (2.226) into Eq. (2.223) and using symbolic computation we get a system of nonlinear algebraic equations as

$$\frac{a}{3}(a_1^3 + 3m^2a_1b_1^2) + 2bk^2a_1(m^4 - m^2) = 0 \tag{2.227}$$

$$\frac{a}{3}(m^2b_1^3 + 3a_1^2b_1) + 2bk^2b_1(m^4 - m^2) = 0 \tag{2.228}$$

$$\frac{a}{3}(3a_0a_1^2 + 3m^2a_0b_1^2) = 0 \tag{2.229}$$

$$2aa_0a_1b_1 = 0 \tag{2.230}$$

$$\frac{a}{3}(3a_0^2a_1 + 3a_1b_1^2) + a_1bk^2(2m^2 - 1) - \lambda a_1 = 0 \tag{2.231}$$

$$\frac{a}{3}b_1^3 + bk^2m^2b_1 - \lambda b_1 = 0 \tag{2.232}$$

$$\frac{a}{3}(a_0^3 + 3a_0b_1^2) - \lambda a_0 = 0 \tag{2.233}$$

from which we have the following solutions

$$a_0 = b_1 = 0, \quad a_1 = \pm m\sqrt{\frac{6(1-m^2)\lambda}{(2m^2-1)a}}, \quad k = \pm\sqrt{\frac{\lambda}{(2m^2-1)b}} \tag{2.234}$$

$$a_0 = a_1 = 0, \quad b_1 = \pm\sqrt{\frac{6(1-m^2)\lambda}{(2-m^2)a}}, \quad k = \pm\sqrt{\frac{\lambda}{(2-m^2)b}} \tag{2.235}$$

$$a_0 = 0, \quad a_1 = \pm m\sqrt{\frac{3(1-m^2)\lambda}{(m^2+1)a}}, \quad b_1 = \pm\frac{a_1}{m}, \quad k = \pm\sqrt{\frac{2\lambda}{(1+m^2)b}} \tag{2.236}$$

By using Eqs. (2.226) and (2.234)–(2.236), we can obtain the following Jacobian elliptic function solutions

$$u_1 = \pm m\sqrt{\frac{6(1-m^2)\lambda}{(2m^2-1)a}}\; sd\left[\sqrt{\frac{\lambda}{(2m^2-1)b}}(x-\lambda t)\right], \tag{2.237}$$

$$u_2 = \pm\sqrt{\frac{6(1-m^2)\lambda}{(2-m^2)a}}\; nd\left[\sqrt{\frac{\lambda}{(2-m^2)b}}(x-\lambda t)\right] \tag{2.238}$$

$$u_3 = \pm\sqrt{\frac{3(1-m^2)\lambda}{(m^2+1)a}}\left\{m\; sd\left[\sqrt{\frac{2\lambda}{(1+m^2)b}}(x-\lambda t)\right] \pm nd\left[\sqrt{\frac{2\lambda}{(1+m^2)b}}(x-\lambda t)\right]\right\} \tag{2.239}$$

Note: For particular cases

(i) When $m \to 1$ or 1, u_1 is a trivial solution of Eq. (2.222).
(ii) When $m \to 1$, $u_2 = 0$, while when $m \to 0$, $u_2 = \pm\sqrt{\frac{3\lambda}{a}}$.
(iii) When $m \to 1$, $u_3 = 0$, while when $m \to 0$, $u_3 = \pm\sqrt{\frac{3\lambda}{a}}$.

2.15 Modified Expansion Function Method

The Modified expansion function method is another powerful technique to solve NLPDE. It is an extension of the standard expansion function method, designed to offer a more flexible and efficient for solving NLPDEs. In essence, the modified expansion function method applies a series expansion or an approximate function for the unknown solution, which is then substituted into the governing differential equations. This method aims to modify the expansion functions in such a way that they better capture the solutions behavior.

Now, we consider an NLPDE

$$P(u, u_x, u_t, u_{xx}, u_{tt}, u_{tx}, \ldots) = 0 \tag{2.240}$$

where $u(x,t)$ is unknown function.

Step I: Consider the transformation

$$u(x,t) = U(\xi), \quad \xi = \lambda(x - ct) \tag{2.241}$$

where λ and c are constants. Substituting (2.241) into (2.240), we get the nonlinear ODE

$$N(U, U', U'', U''', \ldots) = 0 \tag{2.242}$$

Step II: Presume the traveling wave solution of Eq. (2.242) can be rewritten as

$$U(\xi) = \frac{\sum_{i=0}^{m} A_i [e^{-\vartheta(\xi)}]^i}{\sum_{j=0}^{n} B_j [e^{-\vartheta(\xi)}]^i} = \frac{A_0 + A_1 e^{-\vartheta} + \cdots + A_m e^{-m\vartheta}}{B_0 + B_1 e^{-\vartheta} + \cdots + B_n e^{-n\vartheta}} \tag{2.243}$$

where $A_i, B_j, (0 \leq i \leq m, 0 \leq j \leq n)$ are constants to be determined later, such that $A_m \neq 0$, $B_n \neq 0$. Also $\vartheta(\xi)$ satisfies the following ODE

$$\vartheta(\xi)'(\xi) = e^{-\vartheta(\xi)} + ae^{\vartheta(\xi)} + b \tag{2.244}$$

Equation (2.244) has the following families of solutions

Case 1: when $a \neq 0,\ b^2 - 4a > 0$,

$$\vartheta(\xi) = \ln\left(\frac{-\sqrt{b^2 - 4a}}{2a} \tanh\left(\frac{\sqrt{b^2 - 4a}}{2}(\xi + E)\right) - \frac{b}{2a}\right) \tag{2.245}$$

Case 2: when $a \neq 0,\ b^2 - 4a < 0$,

$$\vartheta(\xi) = \ln\left(\frac{\sqrt{-b^2 + 4a}}{2a} \tanh\left(\frac{\sqrt{-b^2 + 4a}}{2}(\xi + E)\right) - \frac{b}{2a}\right) \tag{2.246}$$

Case 3: when $a = 0,\ b \neq 0,\ b^2 - 4a > 0$,

$$\vartheta(\xi) = -\ln\left(\frac{b}{e^{b(\xi+E)} - 1}\right) \tag{2.247}$$

Case 4: when $a \neq 0,\ b \neq 0,\ b^2 - 4a = 0$

$$\vartheta(\xi) = -\ln\left(-\frac{2b(\xi + E) + 4}{b^2(\xi + E)}\right) \tag{2.248}$$

Case 5: when $a = 0,\ b = 0,\ b^2 - 4a = 0$

$$\vartheta(\xi) = -\ln(\xi + E) \tag{2.249}$$

such that $A_0, A_1, A_2, \ldots, A_m, B_0, B_1, B_2, \ldots, B_n, E, a, b$ are constants. By using the balance principle we can find the positive integer m and n between the highest order nonlinear term and the highest order linear term happening in Eq. (2.243).

Step III: Substituting Eq. (2.243) and (2.244) into Eq. (2.242), we get an algebraic equation system. This system has been solved by the computational program and then the solutions of the medium equal width equation have been obtained.

Modified BBM Equation

The medium equal width equation can be defined as follows:

$$u_t + 3u^2 u_x - du_{xxt} = 0 \tag{2.250}$$

Consider the following traveling wave transformation $u(x,t) = U(\xi)$, $\xi = \nu(x - ct)$. The following nonlinear ordinary differential equation is obtained,

$$cd\nu^2 U'' + U^3 - cU = 0 \tag{2.251}$$

If the balancing procedure is applied to Eq. (2.251), we have the following relationship $m = n + 1$. Choosing $n = 1$ we get $m = 2$, Eq. (2.243) is obtained for m and n values as follows;

$$U(\xi) = \frac{A_0 + A_1 e^{-\vartheta} + A_2 e^{-2\vartheta}}{B_0 + B_1 e^{-\vartheta}} \tag{2.252}$$

After all derivative terms from Eq. (2.252) have been found and the required arrangements have been done, we have found an algebraic equation system. With the aid of a computation program, we have solved the system to obtain the following coefficients.

$$A_0 = \frac{\sqrt{c}\lambda B_0}{\sqrt{\lambda^2 - 4\mu}}, \quad A_1 = \frac{\sqrt{c}(2B_0 + \lambda B_1)}{\sqrt{\lambda^2 - 4\mu}},$$

$$A_2 = \frac{\sqrt{c}B_1}{\sqrt{\lambda^2 - 4\mu}}, \quad k = \frac{2}{\sqrt{-d(\lambda^2 - 4\mu)}}$$

Case 1: when $\mu \neq 0,\ \lambda^2 - 4\mu > 0$,

$$u_1(x,t) = \frac{\sqrt{c}(\lambda^2 - 4\mu + \lambda\sqrt{\lambda^2 - 4\mu}\tanh[\psi])}{\sqrt{\lambda^2 - 4\mu}(\lambda + \sqrt{\lambda^2 - 4\mu}\tanh[\psi])} \tag{2.253}$$

where $\psi = \frac{1}{2}\sqrt{\lambda^2 - 4\mu}(\xi + E)$.

Case 2: when $\mu \neq 0,\ \lambda^2 - 4\mu < 0$,

$$u_2(x,t) = \frac{\sqrt{c}(\lambda^2 - 4\mu - \lambda\sqrt{-\lambda^2 + 4\mu}\tanh[\psi])}{\sqrt{\lambda^2 - 4\mu}(\lambda - \sqrt{-\lambda^2 + 4\mu}\tanh[\psi])} \tag{2.254}$$

where $\psi = \frac{1}{2}\sqrt{-\lambda^2 + 4\mu}(\xi + E)$.

Case 3: when $\mu = 0,\ \lambda \neq 0,\ \lambda^2 - 4\mu > 0,$

$$u_3(x,t) = \frac{\sqrt{c}\lambda \coth[\frac{1}{2}(\xi + E)]}{\sqrt{\lambda^2}} \tag{2.255}$$

The obtained coefficients have not given solutions for Family-4 and Family-5.

2.16 Generalized Riccati Equation Expansion Method

The generalized Riccati equation expansion Method is presented by a general ansatz, which is a direct and unified algebraic method for constructing multiple and more general traveling wave solutions of NLPDEs. It can be implemented in a computer system. Compared with most of the existing tanh methods, namely the extended tanh-function method, the modified extended tanh-function method and the generalized hyperbolic-function method, the proposed method is more powerful. By use of the method, one can not only successfully recover the previously known formal solutions but also construct new and more general formal solutions for some Nonlinear evolution equations.

We will describe the algorithm of the generalized Riccati equation expansion Method for find traveling wave solutions of nonlinear evolution equation.

Step I:

Let us consider NLPDEs for some physical field $u(x, y, t)$ as

$$P(u, u_t, u_x, u_{xt}, u_{xx}, \ldots) = 0 \tag{2.256}$$

we express the solutions of (2.256) by the new more general ansatz

$$u(x, y, t) = a_0 + \sum_{i=1}^{m} \left[a_i \phi^i(\xi) + b_i \phi^{i-1}(\xi)\sqrt{R + \phi^2(\xi)} + k_i \phi^{-i}(\xi) \right] \tag{2.257}$$

where m is an integer. It can be determined by balancing the highest-order derivative terms with the nonlinear terms in (2.256). R is a real constant. a_0, a_i, b_i, k_i $(i = 1, \ldots, m)$ and ξ are unknown functions of x, y, t. Also $\phi(\xi)$ satisfies the following nonlinear ODE

$$\frac{d\phi(\xi)}{d\xi} = R + \phi^2(\xi) \tag{2.258}$$

Step II:

Substituting (2.257) along with (2.258), multiplying the most simple common denominator in the obtained system, and setting the coefficients of $\phi^i(\xi)(\sqrt{R+\phi^2(\xi)})^n(j = 0,1,\ldots;n = 0,1)$, we obtain a set of over-determined system of PDEs with respect to unknown functions a_0, a_i, b_i, k_i $(i = 1,\ldots,m)$ and ξ.

Step III:

Solving this over determined PDEs, we would end up with the explicit expressions for a_0, a_i, b_i, k_i $(i = 1,\ldots,m)$ and ξ or the constraints among them.

Step IV:

It is well known that the general solution of (2.258) is:

(i) If $R < 0$,

$$\begin{aligned} \phi(\xi) &= -\sqrt{-R}\tanh(\sqrt{-R}\xi), \\ \phi(\xi) &= -\sqrt{-R}coth(\sqrt{-R}\xi) \end{aligned} \tag{2.259}$$

(ii) If $R = 0$,

$$\phi(\xi) = -\frac{1}{\xi} \tag{2.260}$$

(iii) If $R > 0$,

$$\begin{aligned} \phi(\xi) &= \sqrt{R}\tan(\sqrt{R}\xi), \\ \phi(\xi) &= -\sqrt{R}cot(\sqrt{R}\xi) \end{aligned} \tag{2.261}$$

Thus, according to (2.257), (2.259)–(2.261) and the conclusion in step III, we obtain many general solutions of Eq. (2.256).

The Cylindrical KdV Equation:

Let us consider the cylindrical KdV equation as

$$\frac{\partial u}{\partial t} + 6u\frac{\partial u}{\partial x} + \frac{\partial^3 u}{\partial x^3} + \frac{1}{2t}u = 0 \tag{2.262}$$

It can be written in following form

$$2t\left(\frac{\partial u}{\partial t} + 6u\frac{\partial u}{\partial x} + \frac{\partial^3 u}{\partial x^3}\right) + u = 0 \tag{2.263}$$

By balancing the highest-order contributions from both the linear and nonlinear terms in (2.263), we get $m = 2$.

According to the proposed method, we expand the solution in the form

$$u(x,t) = a_0(t)x + a_1(t)\phi(\xi) + a_2\phi^2(\xi) \tag{2.264}$$

where $\xi = a(t)x + b(t)$ and $\phi(\xi)$ satisfy (2.258).

Substituting (2.264) into (2.263) and collecting the coefficients of the polynomial of $\phi(\xi)$ and x, then setting each coefficients to zero, we obtain

$$24a_2a(2a^2 + a_2) = 0 \tag{2.265}$$

$$2ta_1a(a^2 + 3a_2) = 0 \tag{2.266}$$

$$2tRa_1(b_t + 2Ra^3) = 0 \tag{2.267}$$

$$12ta_0aa_1R + 12ta_0^2 + a_0 + 2ta_0 + 2ta_1Ra_t = 0 \tag{2.268}$$

$$2ta_{1t} + 4ta_2Rb_t + 12ta_1a_0 + 12ta_1^2aR + 32ta^3a_2R^2 + a_1 = 0 \tag{2.269}$$

$$4tRa_2(a_t + 6a_0a) = 0 \tag{2.270}$$

$$36ta_1aa_2R + 16ta^3a_1R + 2ta_1b_t + 12ta_2a_0 + a_2 + 2ta_{2t} = 0 \tag{2.271}$$

$$2ta_1(a_t + 6a_0a) = 0 \tag{2.272}$$

$$4t(6a_2^2aR + 3a_1^2a + 20a^3a_2R + a_2b_t) = 0 \tag{2.273}$$

$$4ta_2(a_t + 6a_0a) = 0 \tag{2.274}$$

where $a_t = \frac{\partial a}{\partial t}$ and $a_{it} = \frac{\partial a_i}{\partial t}$ $(i = 0, 1, 2)$.

Solving the powerful PDE tools package of Maple, solving the set of PDES (2.265)–(2.274), we can obtain the following nontrivial (i.e., $a \neq 0$) solutions:

$$a_1 = 0, \quad a_1 = \frac{1}{12t}, \quad a_2 = \frac{c_1}{t}, \quad b = -4Rc_1^2\sqrt{-\frac{2}{c_1t}} + c_2, \quad a = -\frac{\sqrt{-2c_1t}}{2t} \tag{2.275}$$

$$a_1 = 0, \quad a_1 = \frac{1}{12t}, \quad a_2 = \frac{c_1}{t}, \quad b = 4Rc_1^2\sqrt{-\frac{2}{c_1t}} + c_2, \quad a = \frac{\sqrt{-2c_1t}}{2t} \tag{2.276}$$

where c_1 and c_2 are arbitrary constants.

Thus, we can obtain the following solutions of the cylindrical KdV equation

$$u_1 = \frac{x}{12t} - \frac{c_1 R}{t} \tanh^2 \left[\pm \frac{x\sqrt{-2c_1 t}}{2t} \pm 4Rc_1^2 \sqrt{-\frac{2}{c_1 t}} + c_2 \right] \text{ when } R < 0 \tag{2.277}$$

$$u_2 = \frac{x}{12t} - \frac{c_1 R}{t} coth^2 \left[\pm \frac{x\sqrt{-2c_1 t}}{2t} \pm 4Rc_1^2 \sqrt{-\frac{2}{c_1 t}} + c_2 \right] \text{ when } R < 0 \tag{2.278}$$

$$u_3 = \frac{x}{12t} + \frac{c_1 R}{t} \tan^2 \left[\pm \frac{x\sqrt{-2c_1 t}}{2t} \pm 4Rc_1^2 \sqrt{-\frac{2}{c_1 t}} + c_2 \right] \text{ when } R > 0 \tag{2.279}$$

$$u_4 = \frac{x}{12t} - \frac{c_1 R}{t} cot^2 \left[\pm \frac{x\sqrt{-2c_1 t}}{2t} \pm 4Rc_1^2 \sqrt{-\frac{2}{c_1 t}} + c_2 \right] \text{ when } R > 0 \tag{2.280}$$

2.17 Generalized Kudryashov Method

The generalized Kudryashov method is a generalization of the standard Kudryashov method that finds exact solutions to NLPDEs and ODEs. This method is particularly effective for solving equations in fields such as fluid dynamics, plasma physics, and mathematical biology.

The basic steps of the generalized Kudryashov method are the following:

Consider the following the general NLPDE in the form

$$P(u, u_t, u_x, u_y, u_z, u_{xx}, u_{yy}, u_{zz}, u_{xt}, u_{xxx}, u_{xxt}, \ldots) = 0 \tag{2.281}$$

where P is a polynomial in $u(x, y, z, t, \ldots)$ and its partial derivatives, comprising its nonlinear terms and the highest-order derivative.

Step 1:

Combining the independent variables $x, y, z, t, \ldots,$ into one variable

$$u(x, y, z, t, \ldots) = u(\xi, X), \quad \xi = kx + ny + mz + \omega t + \cdots \tag{2.282}$$

where $k, n, m,$ and ω are arbitrary constants. Using the traveling wave transformation (2.282) into Eq. (2.281), we get

$$F(u, \omega u_\xi, n u_\xi, m u_\xi, k^2 u_{\xi\xi}, n^2 u_{\xi\xi}, m^2 u_{\xi\xi}, \ldots) = 0 \tag{2.283}$$

Step 2: We assume that Eq. (2.283) has the formal solution

$$u(\xi, X) = \frac{\sum_{i=0}^{M} a_i X^i(\xi)}{\sum_{j=0}^{N} b_j X^i(\xi)} \tag{2.284}$$

where $a_i (i \leq i \leq M)$ and $b_j (i \leq j \leq N)$ are constants to be determined, such that $a_m \neq 0$, $b_n \neq 0$. Here, highest-order derivative and the nonlinear terms in Eq. (2.283) are equated to obtain the value of M and N. $X(\xi)$ satisfy the ODE

$$X'(\xi) = X^2(\xi) - X(\xi) \tag{2.285}$$

The solution of Eq. (2.285) is

$$X(\xi) = \frac{1}{1 + Ce^{\xi}}$$

Step 3: Substituting Eq. (2.284) into Eq. (2.283) and then equating all the coefficients of the same power of $X(\xi)$ to 0. This procedure yields a system of algebraic equations. The solution of Eq. (2.281) can be acquired by solving the obtained system of equations.

Burgers' Equation:

Consider the Burger Equation

$$\frac{\partial u}{\partial t} + u\frac{\partial u}{\partial x} - \alpha\frac{\partial^2 u}{\partial x^2} = 0 \tag{2.286}$$

Using the traveling wave transformation $u(x,t) = u(\xi)$; $\xi = x - \lambda t$ into Eq. (2.286), we get

$$-\lambda u' + uu' - \alpha u'' = 0 \tag{2.287}$$

Integrating Eq. (2.287), we obtain

$$-\lambda u + \frac{u^2}{2} - \alpha u' + C_1 = 0 \tag{2.288}$$

where C_1 is the integrating constant. We suppose that Eq. (2.288) has the formal solution (2.284). Taking homogeneous balance between the highest order derivative term u' and highest nonlinear term u^2 in Eq. (2.288), we obtain $M = N + 1$. When $N = 1$, then $M = 2$. Thus, the solution for Eq. (2.288) takes the form

$$u = \frac{a_0 + a_1 X(\xi) + a_2 X^2(\xi)}{b_0 + b_1 X(\xi)} \tag{2.289}$$

where a_0, a_1, a_2, b_0 and b_1 are constant. Also, $a_2 \neq 0$ and $b_1 \neq 0$.

Substituting Eq. (2.289) into Eq. (2.288). Equating the coefficients of $X^j(\xi)$ to zero, $j \geq 0$, we obtain the following set of algebraic equations

$$a_2^2 + 2\alpha a_2 b_1 = 0 \tag{2.290}$$

$$-2a_1a_2 + 2\lambda a_2 b_1 - 2\alpha a_2 b_1 + 4\alpha a_2 b_0 = 0 \tag{2.291}$$

$$-a_1^2 - 4\alpha a_2 b_0 - 2C_1 b_1^2 + 2\lambda a_2 b_0 - 2a_0 a_2 - 2\alpha a_0 b_1 + \alpha a_1 b_0 + 2\lambda a_1 b_1 = 0 \tag{2.292}$$

$$2\lambda a_0 b_1 - 4C_1 b_0 b_1 + 2\lambda a_1 b_0 - 2\alpha a_1 b_0 - 2a_1 a_0 + 2\alpha a_0 b_1 \tag{2.293}$$

$$2\lambda a_0 b_0 - a_0^2 - 2C_1 b_0^2 = 0 \tag{2.294}$$

Solving Eqs. (2.290)–(2.294) by using maple, we find that the solution of Eq. (2.288) exists only in the following two cases:

Case-I: When

$$C_1 = \frac{a_1^2 - 4\alpha b_0 b_1 - 4\alpha a_1 b_0 + 4\alpha^2 b_0^2 + 2\alpha a_1 b_1}{2b_1^2},$$

$$\lambda = \frac{a_1 + \alpha b_1 - 2\alpha b_0}{b_1}, \quad a_0 = \frac{b_0(a_1 - 2\alpha b_0)}{b_1}$$

$$a_1 = \text{constant}, \quad a_2 = 2\alpha b_1, \quad b_0 = \text{constant}, \quad b_1 = \text{constant}$$

In this case the exact traveling wave solution of Eq. (2.288) is

$$u = \frac{b_0(a_1 - 2\alpha b_0)(1 + Ce^{\xi})^2 + a_1 b_1 (1 + Ce^{\xi}) + 2\alpha b_1^2}{b_1[b_1 + b_0(1 + Ce^{\xi})](1 + Ce^{\xi})} \tag{2.295}$$

Case II: When

$$C_1 = \frac{a_1^2 - 8\alpha a_1 b_0}{8b_0^2}, \quad \lambda = \frac{4\alpha b_0 - a_1}{2b_0}, \quad a_0 = \frac{-a_1}{2}$$

$$a_1 = \text{constant}, \quad a_2 = -4\alpha b_0, \quad b_0 = \text{constant}, \quad b_1 = -2b_0$$

In this case, the exact traveling wave solution of Eq. (2.288) is

$$u = \frac{2a_1(1 + Ce^{\xi}) - a_1(1 + Ce^{\xi})^2 - 8\alpha b_0}{2b_0(Ce^{\xi} - 1)(1 + Ce^{a\xi})} \tag{2.296}$$

2.17.1 *Burgers' equation*

Let us consider the Burgers equation as

$$\frac{\partial u}{\partial t} + u\frac{\partial u}{\partial x} - \alpha\frac{\partial^2 u}{\partial x^2} \tag{2.297}$$

The analytical solution of Eq. (2.297) is obtained by virtue of step 1 and step 2 of Section 2.3.

Step 1:

Consider the traveling wave transformation $u(x,t) = U(z),\ z = x - ct$ to covert Eq. (2.297) to the following ODE

$$-cU' + UU' - \alpha U'' = 0. \tag{2.298}$$

Step 2:

Integrating the above equation and using the boundary condition $U(\infty) = u_r$, $U(-\infty) = u_l$, with $U'(\infty) = 0$ and, $U'(-\infty) = 0$, we get the following equation

$$\frac{dU}{dz} = \frac{(U - u_r)(U - u_l)}{2\alpha} \tag{2.299}$$

If $0 < u_l < u_r$, then $dU/dz < 0$, since $u_l < U < u_r$. However, the solution U will not be contained within the defined boundary, and thus a solution cannot exist. On the other hand, if $0 < u_r < u_l$, we have $u_r < U < u_l$ and $dU/dz < 0$. Hence, a solution exists.

To get a particular type of traveling wave solution we consider $u_l = 1$ and $u_r = 0$, and integrate equation (2.299) over the finite interval, we obtained

$$2\alpha \int_{1/2}^{U} \frac{dm}{m^2 - m} = z \tag{2.300}$$

From the above equation, we get

$$U(z) = \frac{1}{1 + e^{\frac{z}{2\alpha}}} \tag{2.301}$$

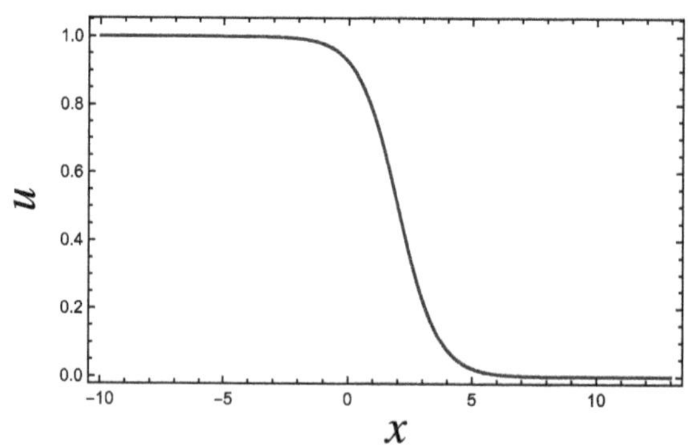

Figure 2.38. Solution of the Eq. (2.297) given in Eq. (2.302).

Therefore, the traveling wave solution in $u(x,t)$ is given by,

$$u(x,t) = \frac{1}{1+e^{\frac{x-ct}{2\alpha}}}. \tag{2.302}$$

2.17.2 *Sine-Gordon equation*

Let us consider the sine-Gordon equation as

$$\frac{\partial^2 u}{\partial X^2} - \frac{1}{c^2}\frac{\partial^2 u}{\partial T^2} = \sin u \tag{2.303}$$

In terms of the characteristic coordinate transformations

$$x = \pm\frac{1}{2}(X - cT), \quad t = \frac{1}{2}(X + cT) \tag{2.304}$$

Equation (2.303) can be written as

$$u_{xt} = \sin u \tag{2.305}$$

Thus, the two forms (2.303) and (2.305) of the sine-Gordon equation are interchangeable. The equation is invariant under $u \to u + 2n\pi$, where $n = \pm 1, \pm 2, \pm 3, \ldots$. The transformation $u \to u + (2n+1)\pi)$ leads to the replacement of $\sin u$ by $-\sin u$ in the equation.

Linearized solution:

If we linearized the nonlinear term $\sin u$ by u, then the linearized form of Eq. (2.303) is given by

$$u_{XX} - c^{-2}u_{TT} = u \tag{2.306}$$

Using a plane wave solution of the form $u(X,T) = Aexp[i(kX - \omega T)]$ in (2.306), we obtain the dispersion relation

$$\omega^2 = c^2(1 + k^2). \tag{2.307}$$

As a result, the equilibrium solution $u = 0$ of (2.303) is stable and ω is real for all real k. However, if $u = \bar{u} + \pi$ for a small perturbation $\bar{u}$, (2.303) may be linearized to get

$$\bar{u}_{XX} - c^{-2}\bar{u}_{TT} = -\bar{u} \tag{2.308}$$

Similarly, a plane wave solution of the form $\bar{u}(X,T) = Aexp[i(kX - \omega T)]$ also exists, provided the dispersion relation

$$\omega^2 = c^2(k^2 - 1) \tag{2.309}$$

is satisfied. Therefore, $\bar{u}(X,T)$ will expand exponentially in time T if $\omega^2 < 0$ (if $0 \leq k^2 < 1$), indicating that the equilibrium solution $u = \pi$ is unstable.

This makes sense because Eq. (2.303) reduces to the equation of a simple pendulum expressing limited oscillations when the term u_{XX} is omitted, or when $k = 0$. As a result, one crucial aspect of the equilibrium solutions of the sine-Gordon equation is the topic of stability or instability.

Nonlinear Waves:

Now, we discuss the soliton solutions of the sine-Gordon equation

$$\frac{\partial^2 u}{\partial x^2} - \frac{1}{c^2}\frac{\partial^2 u}{\partial t^2} = \sin u \tag{2.310}$$

Considering a solution of the form $u(x,t) = \phi(x - Ut) = \phi(\xi)$, by a wave-traveling transformation with a velocity U, we obtain an ODE for $\phi(\xi)$ as

$$\frac{d^2\phi}{d\xi^2} - V^2\frac{d^2 u}{d\xi^2} = \sin\phi \tag{2.311}$$

where $V^2 = \frac{U^2}{c^2}$. Dividing both sides by $(1 - V^2)$ and multiplying by $\phi'(\xi)$ gives

$$\phi'(\xi)\phi''(\xi) = \frac{\sin u}{1 - V^2}\phi'(\xi) \tag{2.312}$$

Or equivalently,

$$\frac{d}{d\xi}\left[\frac{1}{2}\left(\frac{d\phi}{d\xi}\right)^2 + \frac{\cos u}{1 - V^2}\right] = 0 \tag{2.313}$$

i.e.,

$$\frac{1}{2}\left(\frac{d\phi}{d\xi}\right)^2 + \frac{\cos u}{1 - V^2} = B = const \tag{2.314}$$

Solving for $\phi'(\xi)$ gives a first-order ODE for $\phi((\xi)$

$$\frac{d\phi}{d\xi} = \left(2A - \frac{2\cos\phi}{1 - V^2}\right)^{1/2} \tag{2.315}$$

We can separate the variables and, then, integrate to obtain

$$\int_{\phi_0}^{\phi} \frac{d\psi}{\sqrt{A - \cos\psi}} = \left(\frac{2}{1 - V^2}\right)^{1/2} \int_{\xi_0}^{xi} d\eta \tag{2.316}$$

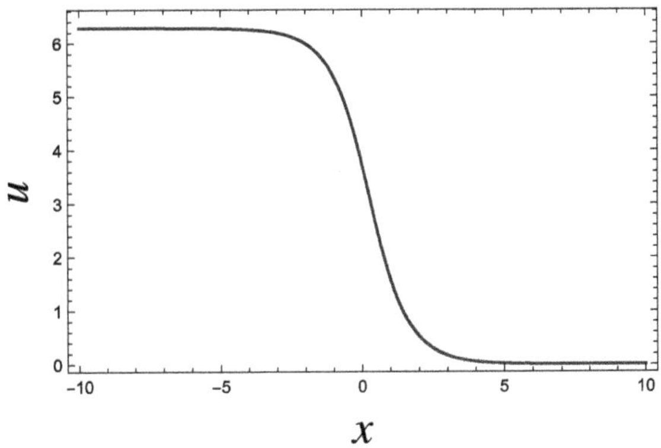

Figure 2.39. Solution of Eq. (11.7.8) given in Eq. (2.320).

where $A = B\sqrt{1-V^2}$. This result depends on the two parameters V (or $U = cV$), the velocity of the soliton, and B, an integrating constant. Thus, the stable solutions represent solitary waves, periodic waves, or a monotonically increasing function of ξ depending on the sign and magnitudes of U and A.

When $A = 1$, a solitary wave solution exists for any velocity, so that $0 < |V| < 1$ (or $0 < |U| < c$). Here, by using $1 - \cos\phi = 2\sin^2(\frac{1}{2}\phi)$ and $\frac{d}{d\phi}[\log\tan(\frac{1}{4}\phi)] = (2\sin\frac{1}{2}\phi)^{-1}$ and simplify both sides of (2.316), we get

$$\left(\frac{2}{1-V^2}\right)^{1/2}(\xi - \xi_0) = \sqrt{2}\int_{\phi_0}^{\phi}\frac{d\psi}{2\sin\frac{1}{2}\psi} = \sqrt{2}\log\left[\frac{\tan(\frac{1}{4}\phi)}{\tan(\frac{1}{4}\phi_0)}\right] \tag{2.317}$$

This gives the solution for $\phi(\xi)$ as

$$\phi(\xi) = 4\tan^{-1}\left(\alpha\exp\left(\frac{\xi-\xi_0}{\sqrt{1-V^2}}\right)\right) \tag{2.318}$$

where $\alpha = \tan(\frac{1}{4}\phi_0)$ and ξ_0 are constants of integration. For $\alpha = 1$ and $\xi_0 = 0$, a simple solitary wave solution is obtained as

$$u(x,t) = 4\tan^{-1}\left(\exp\left(\frac{x-Vt}{\sqrt{1-V^2}}\right)\right) \tag{2.319}$$

This is called the soliton (or kink) solution of the sine-Gordon equation. It represents a continuous profile with $u \to 0$ as $x \to -\infty$ and $u \to 2\pi$ as $x \to +\infty$. The soliton propagates in the positive x-direction with velocity V.

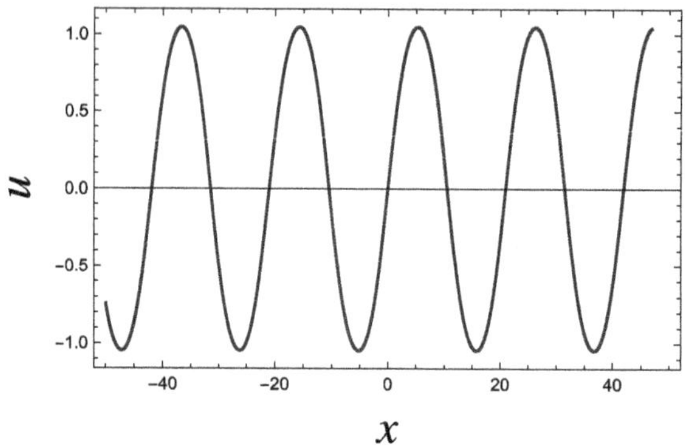

Figure 2.40. Graph of Eq. (2.324).

Similarly, one can obtain another solution obtained from (2.317) in the form

$$u(x,t) = 4\tan^{-1}\left(\exp\left(-\frac{x-Vt}{\sqrt{1-V^2}}\right)\right) = 4cot^{-1}\left(\exp\left(\frac{x-Vt}{\sqrt{1-V^2}}\right)\right) \tag{2.320}$$

This is called the antisoliton (or antikink). This profile travels in the negative x-direction with velocity V. It is also a continuous profile with $u \to 2\pi$ as $x \to -\infty$ and $u \to 0$ as $x \to +\infty$.

The partial derivatives of $u(x,t)$, that is, u_x and u_t, also represent solitary waves, which are given by

$$u_x(x,t) = \phi'(x-Ut), \quad u_t(x,t) = -U\phi'(x-Ut) \tag{2.321}$$

where $\phi'(\xi)$ is obtained from (2.318) as

$$\phi'(\xi) = \frac{4\tan\frac{1}{4}\phi}{\sqrt{1-V^2}(1+\tan^2(\frac{1}{4}\phi))} = \frac{2\text{sech}((x-Vt)/\sqrt{1-V^2})}{\sqrt{1-V^2}}$$

in which $\tan(\frac{1}{4}\phi)$ is replaced by $\exp(\frac{x-Vt}{\sqrt{1-V^2}})$.

If $A \neq 1$, other types of solutions can be obtained. In particular, when $V > 1$ and $|A| < 1$, the integral in (2.316) with $\phi_0 = 0$ and $\xi_0 = 0$ can be written as

$$\begin{aligned}\xi &= \left(\frac{1-V^2}{2}\right)^{1/2}\int_0^{\phi}\frac{d\psi}{\sqrt{(1-\cos\psi)-(1-A)}}\\ &= \left(\frac{1-V^2}{2}\right)^{1/2}\int_0^{\phi}\frac{d\psi}{\sqrt{2}\sqrt{\sin^2\frac{\psi}{2}-m^2}},\end{aligned}$$

where $m^2 = \frac{1}{2}(1 - A)$. By substituting $s = \frac{1}{m}\sin(\frac{\psi}{2})$, we get

$$= (V^2 - 1)^{1/2} \int_0^s \frac{ds}{\sqrt{1-s^2}\sqrt{1-s^2m^2}} \tag{2.322}$$

This can be expressed in terms of Jacobis elliptic function as

$$s = \frac{1}{m}\sin\left(\frac{\phi}{2}\right) = sn\left(\frac{\xi}{\sqrt{V^2-1}}, m\right) \tag{2.323}$$

where m is the modulus of Jacobis elliptic function $sn(z, m)$. Thus, the final solution is given by

$$u(x, r) = \phi(\xi) = 2\sin^{-1}\left[msn\left(\frac{\xi}{\sqrt{V^2-1}}, m\right)\right]. \tag{2.324}$$

2.18 Similarity Solution

Let us consider a PDE for $u(x, t)$ with $(x, t) \in R^2$. A new solution to the problem is obtained by variable transformation. It can be shown that $u(x - x_0, t - t_0)$ (where x_0 and t_0 are constants) and $u(-x, t)$ are also solutions if $u(x, t)$ is a solution to the heat equation $u_t = u_{xx}$. To obtain a new solution from an old one is simple and these solutions are called similarity solutions. To find a similarity solution one has to know symmetry transformation and related mathematics.

Symmetry transformation

Let $u(x, t)$ be a function and the independent variables can be transformed by $x' = X(x, t, u)$, $t' = T(x, t, u)$, and $u' = U(x, t, u)$ where X, T, U are smooth functions. A PDE is said to have a *symmetry* for a given solution $u(x, t)$ if $u'(x, t) = U(x, t, u(x, t, u), T(x, t, u))$ is also a solution. In other words, a symmetry maps solutions of an equation to another solution of the same equation.

Consider the equation $u_t = uu_{xx}$, and the transformation $x' = -x$, $t' = -t$ and $u' = -u$. Given a solution $u(x, t)$, it is easy to check that $u'(x, t) = -u(-x, -t)$ is also, simply by plugging it into the equation:

$$\frac{\partial}{\partial t}(-u(-x, -t)) = u_t(-x, -t)$$

and

$$(-u(-x,-t))\frac{\partial}{\partial x^2}(-u(-x,-t)) = u(-x,-t)u_{xx}(-x,-t)$$

Often a symmetry transformation takes one or more real parameters.

Similarity Solution:

It is well that if the PDEs are invariant under spatial translation, the obtained solutions are known as traveling waves. A similar idea works for a similarity solution. That the solution remains constant when space and time are rescaled. Similar to traveling waves, these kinds of solutions are possible only for specific equations. Provided the initial and boundary conditions satisfy the rescaling symmetry.

If a PDE has a symmetry transformation $(x, t) \to (x = L, t = t/L^\beta)$, then a solution of the PDE may be obtained in the form $u = f(\eta)$, is called a *similarity solution* where the similarity variable is $\eta = x/L^\beta$.

Let us consider a PDE in the form

$$f\left(\frac{\partial u}{\partial t}, \frac{\partial u}{\partial x}, \dots\right)$$

Obviously all solutions to PDEs are not similarity solutions. PDEs do not always have similar solutions. But when they exist, they can produce more general solutions. We consider the one-parameter stretching group

$$x' = \epsilon x, \quad t' = \epsilon^\beta t, \quad u' = \epsilon^\alpha u$$

with ϵ the group parameters, and α and β the family parameters of the group that label different groups of the family. The values of α and β are usually fixed by the initial and boundary conditions.

The solution to the PDE is a surface in the (x, t, u) space. Its general equation is

$$F(x, t, u) = 0$$

The equation is invariant to the stretching group

$$f(\epsilon x, \epsilon^\beta t, \epsilon^\alpha u) = 0$$

Differentiating it with respect to ϵ, then setting $\epsilon = 1$ leads to

$$xF_x + \beta t F_t + \alpha u F_u = 0$$

whose characteristic form is

$$\frac{dx}{x} = \frac{dt}{\beta t} = \frac{du}{\alpha u} = \frac{dF}{0}$$

Three integrals of this equation are

$$\xi = \frac{x}{t^{1/\beta}}, \quad f = \frac{u}{t^{\alpha/\beta}} \quad \text{and} \quad F$$

The most general solution is $F = F(\xi, f)$ where F is an arbitrary function. Since $F = 0$, we obtain an explicit solution in the form

$$u = t^{\alpha/\beta} f(\xi), \quad \text{with } \xi = \frac{x}{t^{1/\beta}}$$

When this form is substituted into the PDE, an ODE results. This ODE is called the principal equation. To solve the ODE is usually simpler to solve than the original PDE.

Asymptotic behaviour of similarity solutions: If the initial and boundary conditions impose some restriction that $a\alpha + b\beta = p$ (with a, b, and p constants), then the ODE is invariant to the associated stretching group $\xi' = \epsilon\xi$ and $f' = \epsilon^{p/a} f$.

If $p/a < 0$, then the principal equation admits an asymptotic solution $f = A\xi^{p/a}$. The value of A is found by substituting this form into the PDE.

The idea of similarity solutions is useful since it allows one to find a function f that normally satisfies an ODE and only has one independent variable, η.

Under the symmetry transformation, similarity solutions are invariant:

$$u(x/L, t/L^{\beta}) = f\left(\frac{x/L}{(t/L^{\beta})^{1/\beta}}\right) = f\left(\frac{x}{t^{\beta}}\right) = u(x,t)$$

The converse is also true. Every solution which is invariant under the symmetry transformation must be a function of the similarity variable η alone. This can be demonstrated with the one-to-one change of variables $\eta = \frac{x}{t^{1/\beta}}$, $\xi = xt$. If there is a solution which is invariant under the symmetry transformation, then it can be written in the new coordinates as $w(\eta, \xi)$. The transformation in terms of the new variables is $(\eta, \xi) \to (\eta, \xi/L^{1/\beta})$, so that

$$w(\eta, \xi) = w(\eta, \xi/L^{\beta+1})$$

Differentiating this expression with respect to L and setting $L = 1$ gives

$$w_\xi(\eta, \xi/L^{\beta+1})\xi = 0$$

which means that w must be a function of η alone.

2.18.1 *Heat equation*

Consider the heat equation

$$\partial_t u(x,t) - D\partial_{xx}u(x,t) = 0 \tag{2.325}$$

and introduce the dilation transformation

$$z = \epsilon^a x, \quad s = \epsilon^b t, \quad v(x,t) = \epsilon^c u(\epsilon^{-a}z, \epsilon^{-b}s) \tag{2.326}$$

Then $\partial_t u(x,t) = \epsilon^{-c}\partial_s v(z,s)s'(t) = \epsilon^{b-c}\partial_s v(z,s)$, etc., and so (2.325) becomes

$$\epsilon^{b-c}\partial_s v(z,s) - D\epsilon^{2a-c}\partial_{zz}v(z,s) = 0$$

Now, if $b - c = 2a - c$, (i.e., $b = 2a$) then Eq. (2.325) is invariant under this transformation; i.e., if $u(x,t)$, solves the heat equation in the variables x, t then for z, s, $v(z,s)$, given by (2.326), $v(z,s)$ solves the heat equation in the variables z, s. Note that

$$vs^{-c/b} = (\epsilon^c u)(\epsilon^b t)^{-c/b} = ut^{-c/b} \text{ and } \frac{z}{s^{a/b}} = \frac{\epsilon^a x}{(\epsilon^b t)^{a/b}} = \frac{x}{t^{a/b}}$$

i.e., both groupings of variables are invariant under the transformation (2.326) for all choices of a, b, c. This suggests that we look for a solution for (2.325) that is of the form

$$u = t^{c/b}y(\xi) \quad \text{for } \xi = \frac{x}{t^{a/b}} = \frac{x}{\sqrt{t}} \quad \text{since } a/b = 1/2 \tag{2.327}$$

i.e., $\partial_t u = t^{\frac{c}{2a}-1}(\frac{c}{2a}y(\xi) - \frac{\xi}{2}y'(\xi))$ and $\partial_{xx}u = t^{\frac{c}{2a}-1}y''(\xi)$. Then (2.325) becomes

$$t^{\frac{c}{2a}-1}\left[Dy''(\xi) - \frac{c}{2a}y(\xi) + \frac{\xi}{2}y'(\xi)\right] = 0 \tag{2.328}$$

i.e., the PDE has been reduced to an ODE. If (2.325) is satisfied for $x > 0$, $t > 0$ and if, in addition, $u(x,t)$ satisfies

$$u(x,0) = 0 \quad x > 0, \quad u(x,t) \to 0, \quad x > \infty, \quad \partial_x u(x,t) = Q, \quad t > 0,$$

then since $y(\xi) = t^{-c/2a}u(x,t)$ and $\xi = \frac{x}{\sqrt{t}}$, it follows that $\xi \to \infty$ if $x \to \infty$, or $t \to 0$ and $\xi = 0$ if $x = 0$ hence the first two side conditions reduce to

$y(\infty) = 0$. Now, $\partial_x u(t, 0) = t^{c/2a-1/2} y'(0)$ and this can equal the constant Q if and only if $c = a$. Then the initial boundary value problem for $u(x,t)$ reduces to the following problem for $y(\xi)$

$$Dy''(\xi) - \frac{1}{2}y(\xi) + \frac{\xi}{2}y'(\xi) = 0, \quad y'(0) = Q, \quad y(\xi) \to 0 \text{ as } \xi \to \infty \quad (2.329)$$

On the other hand, if the boundary condition at $x = 0$ is $u(0,t) = u_0$, then $u(0,t) = ut^{c/2a}y(0)$ can equal the constant u_0 if and only if $c = 0$. In this case, the initial boundary value problem for $u(x,t)$ reduces to

$$Dy''(\xi) + \frac{\xi}{2}y'(\xi) = 0, \quad y(0) = u_0, \quad y(\xi) \to 0 \text{ as } \xi \to \infty \qquad (2.330)$$

We can integrate this once to obtain

$$y'(\xi) = C_1 e^{-\xi^2/4D}$$

and

$$y(\xi) = C_1 \int_0^\xi e^{-\lambda^2/4D} d\lambda + C_2 = C_3 \quad erf\left(\frac{\xi}{\sqrt{4D}}\right) + C - 2$$

where $erf(x) = \frac{2}{\sqrt{\pi}} \int_0^x e^{-\eta^2} d\eta$.

Then the boundary conditions lead to

$$y(\xi) = u_0 - u_0 \; erf\left(\frac{\xi}{\sqrt{4D}}\right)$$

and

$$u(x,t) = u_0 - u_0 \; erf\left(\frac{x}{\sqrt{4Dt}}\right) = u_0 \; erfc\left(\frac{x}{\sqrt{4Dt}}\right)$$

where $erfc(s) = 1 - erf(s)$.

The solution of Eq. (2.329) is given by

$$y(\xi) = C_1\xi + C_2\left[2\sqrt{\pi D}e^{-\xi^2/4D} + \pi\xi \; erf\left(\frac{\xi}{\sqrt{4D}}\right)\right] \qquad (2.331)$$

and since

$$y'(\xi) = C_1 + C_2\left[\pi \; erf\left(\frac{\xi}{\sqrt{4D}}\right)\right]$$

The auxiliary conditions are satisfied for

$$\begin{aligned} y(\xi) &= Q\xi - \frac{Q}{\pi}\left[2\sqrt{\pi D}e^{-\xi^2/4D} + \pi\xi \; erf\left(\frac{\xi}{\sqrt{4D}}\right)\right] \\ &= Q\xi \; erfc\left(\frac{\xi}{\sqrt{4D}}\right) - 2Q\sqrt{\frac{D}{\pi}}e^{-\xi^2/4D} \end{aligned}$$

2.18.2 *Burgers' equation*

Let us consider the Burger's equation

$$\delta_t u(x,t) + u(x,t)\delta_x u(x,t) = 0 \tag{2.332}$$

Under the transformation $z = \epsilon^a x,\ s = \epsilon^b,\ v(z,s) = \epsilon^c u(\epsilon^{-a}z, \epsilon^{-b}s)$, this becomes

$$\epsilon^{b-c}\delta_s v(z,s) + \epsilon^{a-2c}v(z,s)\delta_z v(z,s) = 0 \tag{2.333}$$

and the equation is invariant if $c = a - b$. Then we look for a solution of the form $u = t^{a/b-1y(\xi)}$, $\xi = x/t^{a/b}$. Let $a/b - 1 = m$, m is the free parameter. So $u = t^m y(\xi)$, $\xi = x/t^{m+1}$. Therefore, $\delta_t = t^{m-1}[my(\xi) - (m+1)\xi y'(\xi)]$, $\delta_x = t^m y'(\xi)(t^{-m-1})$.

Equation (2.332) becomes

$$t^{m-1}[my(\xi) - (m+1)\xi y'(\xi) + y(\xi)y'(\xi)] = 0 \tag{2.334}$$

For $m = 0$, (2.334) reduces to

$$\begin{aligned} y(\xi)y'(\xi) - \xi y'(\xi) &= 0 \\ \implies y(\xi) - \xi &= 0 \\ \implies u(x,t) &= x/t. \end{aligned}$$

References

[1] I. N. Sneddon, *Elements of Partial differential equations*, Dover Publications, INC, Mineola, New York, 2006.

[2] B. B. Kadomtsev, and V. I. Petviashvili, *Sov. Phys. Dokl.* 15, 539–541 (1970).

[3] P. Chatterjee, K. Roy, and U. N. Ghosh, *Waves and Wave Interaction in Plasmas,* World Scientific Publishing Co. Pte. Ltd, Singapore, 2023.

[4] G. B. Whitham, *Linear and Nonlinear Waves*, John Wiley & sons, New York, 1973.

[5] Abdul-Majid Wazwaz, *Partial Differential Equation and Solitary Waves Theory*, Springer, Higher Education Press, Beijing, 2009.

[6] I. Shingareva and C. Lizarraga-celaya, *Solving Nonlinear Partial Differential Equations with Maple and Mathematica,* Springerwien New York, 2011.

[7] L. Debnath, *Nonlinear Partial Differential Equations for Scientists and Engineers*, Birkhauser, Boston, 2011.

[8] G. Baumann, Mathematica for theoretical Physics, Springer, New York, 2005.

[9] E. Tzirtzilakis, V. Mrinakis, C. Apokis, and T. Bountis, Soliton-Like Solutions of Higher Order Wave Equations of the Korteweg-de Vries Type. *Journal of Mathematical Physics*, 43, 6151 (2002).

[10] P. Rosenau and J. M. Hyman, Compactons: Solitons with Finite Wavelength. *Physical Review Letters*, 70, 564 (1993).

[11] P. Rosenau, Compact and Noncompact Dispersive Structures. *Physics Letters A*, 275, 193 (2000).

[12] A. M. Wazwwaz, Compactons Dispersive Structures for Variants of the K (m,n) and the KP Equations. Chaos, *Solitons and Fractals,* 15, 1053 (2002).

[13] A. M. Wazwwaz, *Compactons and Solitary Patterns Structures for Variants of the KdV and KP Equations. Applied Mathematics and Computation,* 139, 37 (2003).

[14] A. M. Wazwwaz, Dinstinct Variants of the KdV Equation with Compact and No compact Structures. *Applied Mathematics and Computation,* 150, 365 (2004).

[15] W. Malfliet, Math. *Meth. Appl. Sci.* 28, 2031 (2005).

[16] A. M. Wazwaz, The tanh-coth method for solitons and kink solutions for non-linear parabolic equations, *Appl. Math. and Comp.,* 188(2), 1467 (2007).

[17] M. Wang, X. Li, and J. Zhang, The (G'/G)-expansion method and travelling wave solutions of nonlinear evolution equations in mathematical physics, *Physics Letters A*, 372, 417(2008).

[18] E. M. E. Zayed, A modified (G'/G)- expansion method and its application for finding hyperbolic, trigonometric and rational function solutions of nonlinear evolution equations, *J. Info. Comput. Sci.,* 8, 003, 2013.

[19] A.-M. Wazwaz, A sine-cosine method for handling nonlinear wave equations, Mathematical and computer modelling 40, 499, 2004.

[20] Y. Zhen-Ya., New jacobian elliptic function solutions of modified KdV equation:II*, *Commun. Theor. Phys.,* 38, 400 (2002).

[21] Y. Zhen-Ya., New Jacobian Elliptic Function Solutions to Modified KdV Equation: I*, *Commun. Theor. Phys.,* 38, 143 (2002).

[22] C. Yong, L. Biao, and Z. Hong-Qing, Generalized Riccati equation expansion method and its application to the Bogoyavlenskii's generalized breaking soliton equation, *Chin. Phys. Soc.,* 9, 12, 2003.

[23] Y. Chen and Q. Wang, *A new Riccati equation rational expansion method and its application, Z. Naturforsch,* 60a, 1, (2005).

[24] H. Xuan and B. Li, *Symbolic computation and construction of soliton-like solutions of some nonlinear evolution equations, Z. Naturforsch,* 58a, 167, (2003).

[25] Md. R. Islam, H.-Or-Roshid, Application of generalized kudryashov method to the burger equation, *IJMTT,* 38 (2016).

Chapter 3

Approximate Analytical Solutions

3.1 Introduction

It's well known that the majority of nonlinear partial differential equation (NLPDE) lack explicit solutions, i.e., the solutions cannot be expressed in finite terms, in other words. Even if a solution can be found explicitly, it is frequently too complex to examine its main characteristics. Due to these challenges, the most time-consuming and challenging endeavor for academicians working with nonlinear models is investigating mathematical problems involving NLPDEs. When the homogeneous differential equations have forcing terms, damping terms, or both, they are not homogeneous. Non-homogeneous NLPDEs are generally not integrable. In science and engineering, however, these equations are commonly used to comprehend nonlinear models. In recent years, a number of analytical asymptotic and approximation techniques have been developed to discover approximate solutions of ODEs and PDEs due to growing interest in the creation of nonlinear models. These techniques include the following: the energy balance method, the variational approach method, the parameter expansion method, the amplitude–frequency formulation, the iteration perturbation method, the Adomian decomposition method (ADM), the variational iteration method (VIM), the homotopy perturbation method (HPM), the homotopy analysis method (HAM), the energy balance method, the variational approach method, and many more. Now, we are going to explain some simple but elegant techniques to explain the solutions of some important NLPDEs.

3.2 Conservation Laws and Integrals of the Motions

In physics, integrals of motion and conservation laws are closely connected ideas, especially in field theory and classical mechanics. Both describe quantities that, under particular circumstances and frequently as a result of system symmetries, stay constant over time. We want to find a technique to obtain the solution of a non-homogeneous NLPDE by using some well-known principles of the homogeneous NLPDE, by using conservation laws and integrals of motions of some homogeneous NLPDEs. By a conservation law we mean an equation of the form

$$T_t + X_x = 0 \tag{3.1}$$

where the conserved density T and the flux X are the functions of x, t, u and higher order derivatives of u with respect to x. For the most part, we only discuss conservation laws, where T and X are polynomials in u and derivatives of u w.r.t x alone. A conserved density is trivial if T is a x-derivative for all u. The triviality arises because if $T = F_x$, then we automatically have a conservation law

$$(F_x)_t + (-F_t)_x = 0 \tag{3.2}$$

where the t derivatives in the flux are eliminated by repeated use of the evolution equation.

Integrating (3.1) both sides with respect to 'x', we have

$$\begin{aligned}
&\int_{-\infty}^{\infty} T_t dx + \int_{-\infty}^{\infty} X_x dx = 0 \\
\Rightarrow &\int_{-\infty}^{\infty} T_t dx + \int_{-\infty}^{\infty} dX = 0 \\
\Rightarrow &\int_{-\infty}^{\infty} \frac{\partial T}{\partial t} dx + [X]_{-\infty}^{\infty} = 0 \\
\Rightarrow &\frac{d}{dt}\int_{-\infty}^{\infty} T dx = 0, \quad [As\ [X]_{-\infty}^{\infty} = 0] \\
\Rightarrow &\int_{-\infty}^{\infty} T dx = c \quad (c \text{ is a constant})
\end{aligned}$$

As an example, we consider the KdV equation

$$u_t - 6uu_x + u_{xxx} = 0 \tag{3.3}$$

u, u_x, u_{xx}, u_{xxx} all tend to zero as $x \to \pm\infty$. (i) Equation (3.3) can be rewritten as

$$(u)_t + (-3u^2 + u_{xx})_x = 0$$

Here, we have $T = u$, $X = -3u^2 + u_{xx}$ and $\int_{-\infty}^{\infty} T dx$=constant.

(ii) Multiplying Eq. (3.3) by $2u$ and rewriting the equation, we get

$$(u^2)_t + (-4u^3 + 2uu_{xx} - (u_x)^2)_x = 0$$

Here, we have $T = u^2$, $X = -4u^3 + 2uu_{xx} - (u_x)^2$ and $\int_{-\infty}^{\infty} u^2 dx$=constant.
(iii) Multiplying Eq. (3.3) by $(3u^2 - u_{xx})$ and rewriting the equation, we obtain

$$\left(u^3 + \frac{1}{2}(u_x)^2\right)_t + \left(-\frac{9}{2}u^4 + 3u^2u_{xx} - 6u(u_x)^2 + u_x u_{xxx} - \frac{1}{2}(u_{xx})^2\right)_x = 0$$

Here, $T = u^3 + \frac{1}{2}(u_x)^2$, $X = -\frac{9}{2}u^4 + 3u^2u_{xx} - 6u(u_x)^2 + u_x u_{xxx} - \frac{1}{2}(u_{xx})^2$, and $\int_{-\infty}^{\infty}(u^3)_t + \frac{1}{2}((u_x)^2)dx$ =constant.

Where the first simply restates the KdV equation in conservation form. The second is obtained after first multiplying the KdV equation by $2u$ and then rewriting it in the conservation form. The third conservation law is obtained by multiplying $(3u^2 - u_{xx})$ and performing suitable manipulations with differentiations. These conserved densities can be interpreted as mass, momentum, and energy of the KdV equation for some physical systems.

Furthermore, the existence of infinitely many conservation laws indicates that the KdV equation is a special equation of physical interest. Here, we consider a few conservation laws for the KdV equation.

Conserved quantities in KdV equation

Let us consider the general form of the KdV equation as

$$\frac{\partial u}{\partial \tau} + Au\frac{\partial u}{\partial \xi} + B\frac{\partial^3 u}{\partial \xi^3} = 0 \tag{3.4}$$

Now, we can show that the integral

$$I = \int_{-\infty}^{\infty} u^2 d\xi \tag{3.5}$$

is conserved.

Differentiating (3.5) under integration sign w.r.t. 'τ,' we get

$$\frac{dI}{d\tau} = \int_{-\infty}^{\infty} 2u\frac{\partial u}{\partial \tau} d\xi \tag{3.6}$$

Therefore,

$$\frac{\partial I}{\partial \tau} = 2\int_{-\infty}^{\infty} u\left(-Au\frac{\partial u}{\partial \xi} - B\frac{\partial^3 u}{\partial \xi^3}\right)d\xi \tag{3.7}$$

Now,

$$I_1 = -2A\int_{-\infty}^{\infty} u^2\frac{\partial u}{\partial \xi}d\xi = \left[-2A\frac{u^3}{3}\right]_{-\infty}^{\infty} = 0$$

Since, $u \to 0$ as $\xi \to \pm\infty$ and

$$\begin{aligned} I_2 = -2B\int_{-\infty}^{\infty} u\frac{\partial^3 u}{\partial \xi^3}d\xi &= -2B\left[\left(u\frac{\partial^2 u}{\partial \xi^2}\right)_{-\infty}^{\infty} - \int_{-\infty}^{\infty}\frac{\partial u}{\partial \xi}\frac{\partial^2 u}{\partial \xi^2}d\xi\right] \\ &= -2B\left[0 - \frac{1}{2}\int_{-\infty}^{\infty}\frac{d}{d\xi}\left(\frac{\partial u}{\partial \xi}\right)^2 d\xi\right] \\ &= -B\left[\left(\frac{\partial u}{\partial \xi}\right)^2\right]_{-\infty}^{\infty} = 0 \end{aligned}$$

From the property of solitons, we get $\frac{\partial \phi}{\partial \xi} \to 0$ as $\xi \to \pm\infty$. Therefore, we get $\frac{dI}{d\tau} = 0$, i.e., $I =$ constant. There exists an infinite number of conserved quantities like I.

3.3 Conservation Law of the Gardner Equation

Let us consider the generalized Gardner equation of the form

$$u_t + au^p u_x + bu^{2p}u_x + \mu u_{xxx} = 0 \tag{3.8}$$

First conservation law

The first conservation law, which represents the conservation of mass, can be obtained directly by rewriting (3.8) in the form

$$\begin{aligned} &(u)_t + \left(\frac{a}{p+1}u^{p+1} + \frac{b}{2p+1}u^{2p+1} + \mu u_{xx}\right)_x = 0 \\ &\Rightarrow T_t + X_x = 0 \end{aligned} \tag{3.9}$$

where $T = u$ and $X = \frac{a}{p+1}u^{p+1} + \frac{b}{2p+1} + \mu u_{xx}$.

Second conservation law

The second conservation law, which represents the conservation of momentum, is derived by multiplying Eq. (3.8) by $2u$. Then, (3.8) becomes

$$\begin{aligned}&2uu_t + 2au^{p+1}u_x + 2bu^{2p+1}u_x + 2\mu uu_{xxx} = 0\\&\Rightarrow (u^2)_t + \left(\frac{2a}{p+2}u^{p+2}\right)_x + \left(\frac{2b}{2p+2}u^{2p+2}\right)_x + (2\mu uu_{xx})_x\\&\quad - 2\mu u_x u_{xx} = 0\\&\Rightarrow (u^2)_t + \left(\frac{2a}{p+2}u^{p+2}\right)_x + \left(\frac{2b}{2p+2}u^{2p+2}\right)_x + (2\mu uu_{xx})_x\\&\quad - \left(\mu(u_x)^2\right)_x = 0\\&\Rightarrow [u^2]_t + \left[\frac{2a}{p+2}u^{p+2} + \frac{2b}{2p+2}u^{2p+2} + 2\mu uu_{xx} - \mu(u_x)^2\right]_x = 0\\&\Rightarrow T_t + X_x = 0\end{aligned}$$

where the conserved density is the moment $T = u^2$ and the associated conserved flux is $X = \frac{2a}{p+2}u^{p+2} + \frac{2b}{2p+2}u^{2p+2} + 2\mu uu_{xx} - \mu(u_x)^2$.

Third conservation law

First, we multiply Eq. (3.8) by $\left(\frac{2a}{p+1}u^{p+1} + \frac{2b}{2p+1}u^{2p+1}\right)$, then we obtain

$$\begin{aligned}&\frac{2a}{p+1}u^{p+1}u_t + \frac{2b}{2p+1}u^{2p+1}u_t + \frac{2a^2}{p+1}u^{2p+1}u_x + \frac{2ab}{2p+1}u^{3p+1}u_x\\&\quad + \frac{2ab}{p+1}u^{3p+1}u_x + \frac{2b^2}{2p+1}u^{4p+1}u_x + \frac{2a}{p+1}\mu u^{p+1}u_{xxx}\\&\quad + \frac{2b}{2p+1}\mu u^{2p+1}u_{xxx} = 0\end{aligned} \tag{3.10}$$

In the second step, we differentiate the generalized Gardner equation (3.8) with respect to the variable x

$$u_{xt} + apu^{p-1}u_x u_x + au^p u_{xx} + 2bpu^{2p-1}u_x u_x + bu^{2p}u_{xx} + \mu u_{xxxx} = 0 \tag{3.11}$$

Multiplying (3.11) by $-2\mu u_x$, we get

$$\begin{aligned}&-2\mu u_x u_{xt} - 2a\mu pu^{p-1}(u_x)^3 - 2\mu au^p u_x u_{xx} - 4\mu bpu^{2p-1}(u_x)^3\\&-2\mu bu^{2p}u_x u_{xx} - 2(\mu)^2 u_x u_{xxxx} = 0\end{aligned} \tag{3.12}$$

Adding (3.10) and (3.12), we have

$$\left[\frac{2au^{p+2}}{(p+1)(p+2)}+\frac{2bu^{2p+2}}{(2p+1)(2p+2)}-\mu(u_x)^2\right]_t+\left[\frac{a^2u^{2p+2}}{(p+1)^2}\right.$$
$$+\frac{b^2u^{4p+2}}{(2p+1)^2}+\frac{2abu^{3p+2}}{(2p+1)(3p+2)}+\frac{2abu^{3p+2}}{(p+1)(3p+2)}+\frac{2\mu au^{p+1}u_{xx}}{(p+1)}$$
$$+\frac{2\mu bu^{2p+1}u_{xx}}{(2p+1)}-2\mu au^p(u_x)^2-2\mu bu^{2p}(u_x)^2-2\mu^2u_xu_{xxx}$$
$$\left.+\mu^2(u_{xx})^2\right]_x=0$$
$$\text{i.e., } T_t+X_x=0 \tag{3.13}$$

where the conserved density is the Hamiltonian density given by $T = \frac{2au^{p+2}}{(p+1)(p+2)}+\frac{2bu^{2p+2}}{(2p+1)(2p+2)}-\mu(u_x)^2$ and associated conserved flux is given by $X = \frac{a^2u^{2p+2}}{(p+1)^2}+\frac{b^2u^{4p+2}}{(2p+1)^2}+\frac{2abu^{3p+2}}{(2p+1)(3p+2)}+\frac{2abu^{3p+2}}{(p+1)(3p+2)}+\frac{2\mu au^{p+1}u_{xx}}{(p+1)}+\frac{2\mu bu^{2p+1}u_{xx}}{(2p+1)}-2\mu au^p(u_x)^2-2\mu bu^{2p}(u_x)^2-2\mu^2u_xu_{xxx}+\mu^2(u_{xx})^2$.

3.4 Analytical Solution of KdV-Type Equations

It is well known that the KdV equation is completely integrable. We have already shown that it admits infinite conservation laws. However, if any additional term, like damping or forcing, is added to the KdV equation, the changed KdV equation will no longer be completely integrable. So, it is not possible to find an explicit expression of the damped or forced KdV equation. However, we seek an approximate solution for this type of equation. To do so, we assume the conservation principle of the damped or forced KdV equation is the same as that of the KdV equation for small values of damping and forcing. For small perturbations, we expect that the solution will remain close to the original soliton for some time. Therefore, the solution we seek will be roughly a soliton with a slowly changing shape and location plus a correction. Now, we find the solutions of the damped KdV, forced KdV, and damped-forced KdV equations.

Damped KdV equation

Consider the KdV equation with a damped coefficient c (very small) as

$$\frac{\partial u}{\partial t}+u\frac{\partial u}{\partial x}+\frac{\partial^3 u}{\partial x^3}+cu=0 \tag{3.14}$$

Let us assume that

$$I = \int_{-\infty}^{\infty} u^2\, dx$$

is conserved for (3.14). Now differentiating the equation with respect to t, we have

$$\begin{aligned}\frac{dI}{dt} &= \int_{-\infty}^{\infty} 2u\frac{\partial u}{\partial t}\, dx \\ &= -\int_{-\infty}^{\infty} 2u\left(u\frac{\partial u}{\partial x} + \frac{\partial^3 u}{\partial x^3} + cu\right) dx \\ &= I_1 + I_2 + I_3 \end{aligned} \tag{3.15}$$

where

$$I_1 = -\int_{-\infty}^{\infty} 2u^2\frac{\partial u}{\partial x}\, dx = -2\left[\frac{u^3}{3}\right]_{-\infty}^{\infty} = 0$$

$$\begin{aligned} I_2 &= -\int_{-\infty}^{\infty} 2u\frac{\partial^3 u}{\partial x^3}\, dx = -2\left[u\frac{\partial^2 u}{\partial x^2}\right]_{-\infty}^{\infty} + 2\int_{-\infty}^{\infty} \frac{\partial u}{\partial x}\frac{\partial^2 u}{\partial x^2}\, dx \\ &= \int_{-\infty}^{\infty} \frac{d}{dx}\left(\frac{\partial u}{\partial x}\right)^2 dx \\ &= \left[\left(\frac{\partial u}{\partial x}\right)^2\right]_{-\infty}^{\infty} \\ &= 0 \end{aligned}$$

$$I_3 = -2\int_{-\infty}^{\infty} cu^2\, dx$$

From Eq. (3.15), we get

$$\begin{aligned} \frac{dI}{dt} &= -2\int_{-\infty}^{\infty} cu^2\, dx \\ \Rightarrow \frac{dI}{dt} &= -2cI \\ \Rightarrow \frac{dI}{I} &= -2cdt \\ \Rightarrow \frac{1}{2c}\log I &= -t + c^* \end{aligned} \tag{3.16}$$

Initially, when $t = t_0$, $I = I_0$, we can find $c^* = \frac{\log I_0}{2c} + t_0$. Putting the value of c^* in the above equation, we get

$$I = I_0 \exp[-2c(t - t_0)] \tag{3.17}$$

which is the solution of Eq. (3.17).

Now, let us assume that the solitary wave solution of Eq. (3.14) is of the form

$$u = u_m(t)\ \mathrm{sech}^2\left(\frac{x - M(t)t}{\omega(t)}\right) \tag{3.18}$$

where u_m, M, ω are functions of t.

So,

$$I = \int_{-\infty}^{\infty} u_m^2(t)\ \mathrm{sech}^4\left(\frac{x - M(t)t}{\omega(t)}\right) dx = \frac{4}{3}\omega(t)u_m^2(t) \tag{3.19}$$

At $t = 0$, we get $I_0 = \frac{4}{3}\omega(t_0)u_m^2(t_0)$. From (3.17) and (3.19), we get

$$I = \frac{4}{3}\omega(t_o)u_m^2(t_0)e^{-2c(t-t_0)} \tag{3.20}$$

where $u_m(t_0) = 3M_0$ and $\omega(t_0) = \sqrt{\frac{4}{M_0}}$. So,

$$\begin{aligned}
&\frac{4}{3}\omega(t)u_m^2(t) = \frac{4}{3}\omega(t_o)u_m^2(t_0)e^{-2c(t-t_0)} \\
&\Rightarrow \frac{\frac{4}{3}\omega(t)u_m^2(t)}{\frac{4}{3}\omega(t_o)u_m^2(t_0)} = e^{-2c(t-t_0)} \\
&\Rightarrow \frac{\frac{4}{3}(3M)^2\sqrt{\frac{4}{M}}}{\frac{4}{3}(3M_0)^2\sqrt{\frac{4}{M_0}}} = e^{-2c(t-t_0)} \\
&\Rightarrow \left(\frac{M}{M_0}\right)^{\frac{3}{2}} = e^{-2c(t-t_0)} \\
&\Rightarrow M = M_0 e^{-\frac{4}{3}c(t-t_0)}
\end{aligned}$$

Therefore, the solution of (3.14) is considered of the form

$$u = u_m(t)\ \mathrm{sech}^2\left(\frac{x - M(t)t}{\omega(t)}\right)$$

where $u_m(t) = 3M(t)$, $\omega(t) = \sqrt{\frac{4}{M(t)}}$, and $M(t) = M_0 e^{-\frac{4}{3}c(t-t_0)}$. Here, the velocity of the solitary wave is proportional to the exponential value of negative time (t). Also, the amplitude and width of the solitary wave structure are proportional to the velocity of the wave and inversely proportional to the square root of the velocity of the wave, respectively.

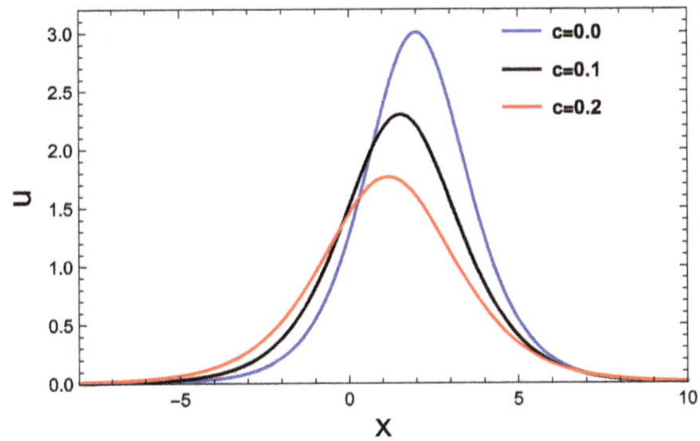

Forced KdV equation

Consider the KdV equation with the forced term $f_0 \cos(\omega t)$ as

$$\frac{\partial u}{\partial t} + u\frac{\partial u}{\partial x} + \frac{\partial^3 u}{\partial x^3} = f_0 \cos(\omega t) \tag{3.21}$$

For the small value of f_0, we consider the solution of (3.21) to be

$$u = u_m(t)\ \mathrm{sech}^2\left(\frac{x - M(t)t}{\omega(t)}\right) \tag{3.22}$$

where $u_m(t) = 3M$ and $\omega(t) = \sqrt{\frac{4}{M}}$.

Let

$$\begin{aligned} I &= \int_{-\infty}^{\infty} u^2\, dx \\ \therefore \frac{dI}{dt} &= \int_{-\infty}^{\infty} 2u\frac{\partial u}{\partial t}\, dx \\ &= \int_{-\infty}^{\infty} 2u\left(-u\frac{\partial u}{\partial x} - \frac{\partial^3 u}{\partial x^3} + f_0\cos(\omega t)\right) dx \\ &= I_1 + I_2 + I_3 \end{aligned} \tag{3.23}$$

where

$$\begin{aligned} I_1 &= 0,\ \ I_2 = 0 \\ I_3 &= \int_{-\infty}^{\infty} 2uf_0\cos(\omega t)\, dx = 2f_0\cos(\omega t)\int_{-\infty}^{\infty} u\, dx \end{aligned}$$

$$\Rightarrow I_3 = 2f_0\cos(\omega t)\int_{-\infty}^{\infty} u_m(t)\ \mathrm{sech}^2\left(\frac{x-M(t)t}{\omega(t)}\right)dx$$

$$\Rightarrow I_3 = 4f_0\cos(\omega t)u_m(t)\omega(t)$$

$$\therefore \frac{dI}{dt} = 4f_0\cos(\omega t)u_m(t)\omega(t) \tag{3.24}$$

We also have

$$\begin{aligned}
I &= \int_{-\infty}^{\infty} u^2\,dx \\
&= \int_{-\infty}^{\infty} u_m^2(t)\ \mathrm{sech}^4\left(\frac{x-M(t)t}{\omega(t)}\right)dx \\
&= \frac{4}{3}u_m^2(t)\omega(t) \\
&= \frac{4}{3}(3M(t))^2\sqrt{\frac{4}{M(t)}}\left[\text{since } u_m(t) = 3M(t), \omega(t) = \sqrt{\frac{4}{M(t)}}\right] \\
&= 24M^{\frac{3}{2}}(t)
\end{aligned} \tag{3.25}$$

Now, from (3.24) and (3.25),

$$\frac{d}{dt}\left(24M^{\frac{3}{2}}(t)\right) = 4f_0\cos(\omega t)u_m(t)\omega(t)$$

$$\Rightarrow 32M^{\frac{3}{2}}(t)\frac{dM(t)}{dt} = 4f_0\cos(\omega t)3M(t)\sqrt{\frac{4}{M(t)}}$$

$$\Rightarrow \frac{dM(t)}{dt} = f_0\cos(\omega t)\frac{2}{3}$$

Integrating we have

$$M(t) = \frac{2}{3\omega}f_0\sin(\omega t) + c$$

Using the initial condition $t = 0, M(t) = M_0$ in the above equation we get $c = M_0$. Therefore,

$$M(t) = \frac{2}{3\omega}f_0\sin(\omega t) + M_0$$

Solution of the forced KdV Eq. (3.21) is of the form

$$u = u_m(t)\,\mathrm{sech}^2\left(\frac{x - M(t)t}{\omega(t)}\right)$$

where $u_m(t) = 3M(t)$, $\omega(t) = \sqrt{\frac{4}{M(t)}}$, and $M(t) = \frac{2}{3\omega} f_0 \sin(\omega t) + M_0$.

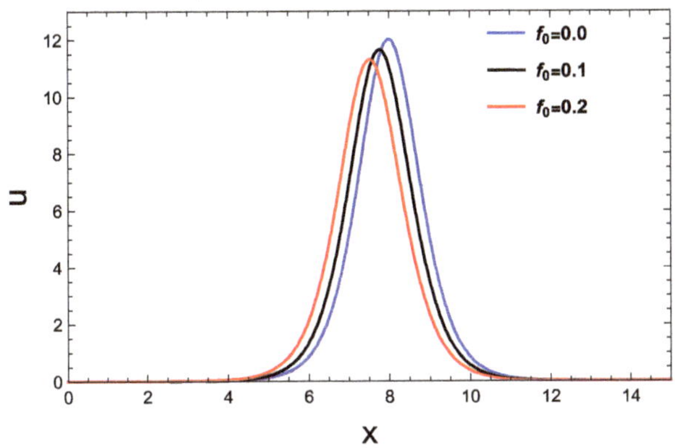

Damped forced KdV equation

Consider the KdV equation with a damping and forcing term as

$$\frac{\partial u}{\partial t} + u\frac{\partial u}{\partial x} + \frac{\partial^3 u}{\partial x^3} + cu = f_0 \cos(\omega t) \tag{3.26}$$

For the small values of forcing and damping term, we assume the solution of Eq. (3.26) is of the form

$$u = u_m(t)\,\mathrm{sech}^2\left(\frac{x - M(t)t}{\omega(t)}\right) \tag{3.27}$$

where $u_m(t) = 3M$ and $\omega(t) = \sqrt{\frac{4}{M}}$ are the amplitude and width of the solitary wave solution.

Let

$$I = \int_{-\infty}^{\infty} u^2\, dx$$

So,

$$\begin{aligned}\frac{dI}{dt} &= \int_{-\infty}^{\infty} 2u\frac{\partial u}{\partial t}\,dx \\ &= \int_{-\infty}^{\infty} 2u\left(-u\frac{\partial u}{\partial x} - \frac{\partial^3 u}{\partial x^3} - cu + f_0\cos(\omega t)\right)dx \\ &= I_1 + I_2 + I_3 + I_4 \end{aligned} \tag{3.28}$$

where

$$I_1 = 0, \quad I_2 = 0, \quad I_3 = -2\int_{-\infty}^{\infty} cu^2\,dx = -2cI$$

$$I_4 = \int_{-\infty}^{\infty} 2uf_0\cos(\omega t)\,dx = 4f_0\cos(\omega t)u_m(t)\omega(t)$$

From Eq. (3.28), we have

$$\frac{dI}{dt} + 2cI = 4f_0\cos(\omega t)u_m(t)\omega(t) \tag{3.29}$$

We also have

$$I = \int_{-\infty}^{\infty} u^2\,dx = \int_{-\infty}^{\infty} u_m^2(t)\,\mathrm{sech}^4\left(\frac{x - M(t)t}{\omega(t)}\right)dx = 24M^{\frac{3}{2}}(t) \tag{3.30}$$

Now, from (3.29) and (3.30), we get

$$M(t)e^{\frac{4}{3}ct} = \frac{2}{3}f_0\int e^{\frac{4}{3}ct}\cos(\omega t)\,dt = \frac{2}{3}f_0I_* + c_0 \tag{3.31}$$

Now,

$$\begin{aligned} I_* &= \int e^{\frac{4}{3}ct}\cos(\omega t)\,dt \\ &= \frac{e^{\frac{4}{3}ct}}{(\frac{4c}{3})^2 + \omega^2}\left(\frac{4c}{3}\cos(\omega t) + \omega\sin(\omega t)\right) \\ &= \frac{9e^{\frac{4}{3}ct}}{16c^2 + 9\omega^2}\left(\frac{4c}{3}\cos(\omega t) + \omega\sin(\omega t)\right) \end{aligned}$$

Substituting the value of I_* in Eq. (3.31), we obtain

$$M(t)e^{\frac{4}{3}ct} = \frac{2}{3}f_0\frac{9e^{\frac{4}{3}ct}}{16c^2+9\omega^2}\left(\frac{4c}{3}\cos(\omega t)+\omega\sin(\omega t)\right)+c_0 \quad (3.32)$$

When $t=0$, $M(t)=M_0$, then

$$c_0 = -\frac{8f_0c}{16c^2+9\omega^2}+M_0$$

From (3.32),

$$M(t) = \left[M_0 - \frac{8f_0c}{16c^2+9\omega^2}\right]e^{-\frac{4}{3}ct}+\frac{6f_0}{16c^2+9\omega^2}\left(\frac{4c}{3}\cos(\omega t)+\omega\sin(\omega t)\right) \quad (3.33)$$

So, the solution of the damped forced KdV Eq. (3.26) is of the form

$$u = u_m(t)\ \mathrm{sech}^2\left(\frac{x-M(t)t}{\omega(t)}\right)$$

where $u_m(t)=3M(t)$, $\omega(t)=\sqrt{\frac{4}{M(t)}}$ and $M(t)$ is of the form (3.33).

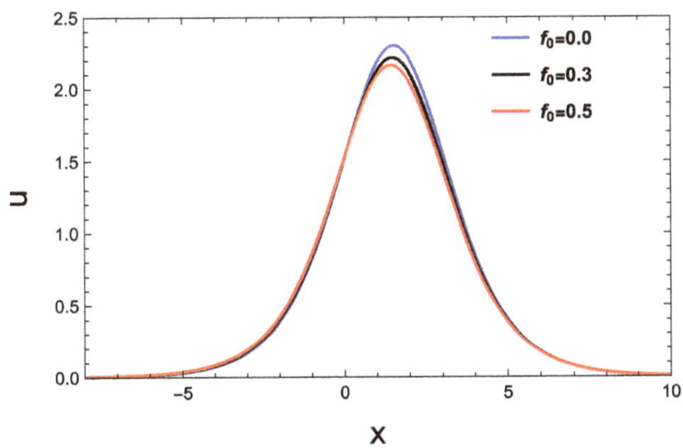

Damped MKdV equation

Consider the MKdV equation with a damped term c (very small) as

$$\frac{\partial u}{\partial t} + u^2\frac{\partial u}{\partial x} + \frac{\partial^3 u}{\partial x^3} + cu = 0 \tag{3.34}$$

Let

$$I = \int_{-\infty}^{\infty} u^2\,dx$$

Therefore,

$$\begin{aligned}\frac{dI}{dt} &= \int_{-\infty}^{\infty} 2u\frac{\partial u}{\partial t}\,dx = -\int_{-\infty}^{\infty} 2u\left(u^2\frac{\partial u}{\partial x} + \frac{\partial^3 u}{\partial x^3} + cu\right)dx \\ &= I_1 + I_2 + I_3 \end{aligned}\tag{3.35}$$

where

$$I_1 = 0, \quad I_2 = 0, \quad I_3 = -2\int_{-\infty}^{\infty} cu^2\,dx$$

From Eq. (3.35), we get

$$\frac{1}{2c}\log I = -t + c^* \tag{3.36}$$

Initially, when $t = t_0, I = I_0$, we can find $c^* = \frac{\log I_0}{2c} + t_0$. Putting the value of c^* in the above equation, we get

$$I = I_0 \exp[-2c(t - t_0)] \tag{3.37}$$

which is the solution of Eq. (3.37).

Now, let us assume that the solitary wave solution of Eq. (3.34) is of the form

$$u = u_m(t)\,\operatorname{sech}\left(\frac{x - M(t)t}{\omega(t)}\right) \tag{3.38}$$

where u_m, M, ω are functions of t.

Therefore,

$$I = \int_{-\infty}^{\infty} u^2\,dx = \int_{-\infty}^{\infty} u_m^2(t)\,\operatorname{sech}^2\left(\frac{x - M(t)t}{\omega(t)}\right)dx = 2\omega(t)u_m^2(t) \tag{3.39}$$

At $t = 0$, we get $I_0 = 2\omega(t_0)u_m^2(t_0)$. From (3.37) and (3.39), we get

$$I = 2\omega(t_o)u_m^2(t_0)e^{-2c(t-t_0)} \tag{3.40}$$

where $u_m(t_0) = 3M_0$ and $\omega(t_0) = \sqrt{\frac{4}{M_0}}$. So,

$$\begin{aligned} 2\omega(t)u_m^2(t) &= 2\omega(t_o)u_m^2(t_0)e^{-2c(t-t_0)} \\ \Rightarrow \frac{2\omega(t)u_m^2(t)}{2\omega(t_o)u_m^2(t_0)} &= e^{-2c(t-t_0)} \\ \Rightarrow M &= M_0 e^{-\frac{4}{3}c(t-t_0)} \end{aligned}$$

Therefore, the solution of (3.34) is of the form

$$u = u_m(t)\ \text{sech}^2\left(\frac{x - M(t)t}{\omega(t)}\right)$$

where $u_m(t) = 3M(t)$, $\omega(t) = \sqrt{\frac{4}{M(t)}}$ and $M(t) = M_0 e^{-\frac{4}{3}c(t-t_0)}$.

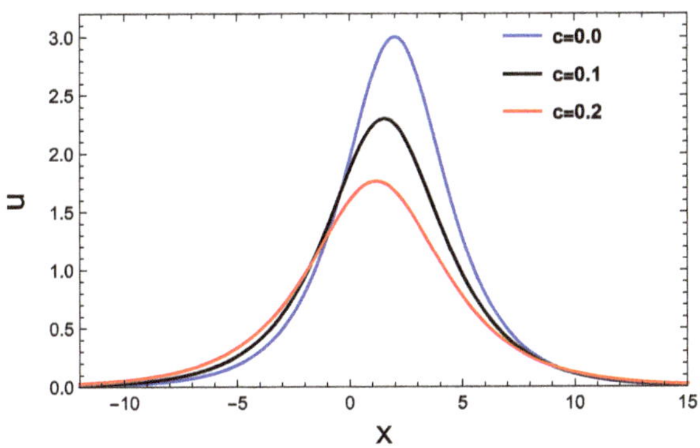

Forced mKdV equation

Consider the MKdV equation with the forced term $f_0 \cos(\omega t)$ as

$$\frac{\partial u}{\partial t} + u^2\frac{\partial u}{\partial x} + \frac{\partial^3 u}{\partial x^3} = f_0 \cos(\omega t) \tag{3.41}$$

Now, consider the solution of (3.41)

$$u = u_m(t)\ \mathrm{sech}\left(\frac{x - M(t)t}{\omega(t)}\right) \tag{3.42}$$

where $u_m(t) = 3M$ and $\omega(t) = \sqrt{\frac{4}{M}}$.

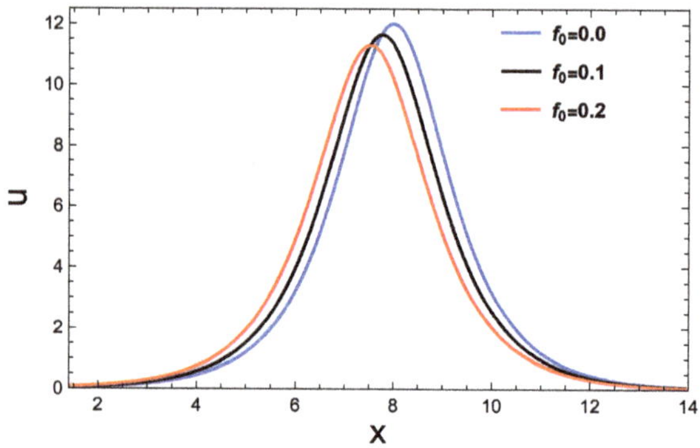

Let

$$I = \int_{-\infty}^{\infty} u^2 dx$$

Therefore,

$$\frac{dI}{dt} = \int_{-\infty}^{\infty} 2u\left(-u^2\frac{\partial u}{\partial x} - \frac{\partial^3 u}{\partial x^3} + f_0\cos(\omega t)\right) dx = I_1 + I_2 + I_3, \tag{3.43}$$

where $I_1 = 0,\ I_2 = 0,\ I_3 = -2f_0\pi\cos(\omega t)u_m(t)\omega(t)$.

Therefore,

$$\frac{dI}{dt} = -2f_0\pi\cos(\omega t)u_m(t)\omega(t). \tag{3.44}$$

We also have

$$I = \int_{-\infty}^{\infty} u_m^2(t)\ \mathrm{sech}^2\left(\frac{x - M(t)t}{\omega(t)}\right) dx = 36M^{\frac{3}{2}}(t). \tag{3.45}$$

Now, from (3.44) and (3.45), we get

$$\frac{dM(t)}{dt} = -\frac{2}{9}f_0\pi\cos(\omega t)$$

Integrating we have

$$M(t) = -\frac{2}{9\omega} f_0 \pi \sin(\omega t) + c$$

Using the initial condition $t = 0, M(t) = M_0$ in the above equation, we get $c = M_0$

$$\therefore M(t) = -\frac{2}{9\omega} f_0 \pi \sin(\omega t) + M_0$$

Therefore, the solution of the forced MKdV Eq. (3.41) is of the form

$$u = u_m(t) \text{ sech} \left(\frac{x - M(t)t}{\omega(t)} \right)$$

where $u_m(t) = 3M(t),\ \ \omega(t) = \sqrt{\frac{4}{M(t)}}$, and $M(t) = -\frac{2}{9\omega} f_0 \pi \sin(\omega t) + M_0$.

Damped forced mKdV equation

Consider the mKdV equation with a damping and forcing term as

$$\frac{\partial u}{\partial t} + u^2 \frac{\partial u}{\partial x} + \frac{\partial^3 u}{\partial x^3} + cu = f_0 \cos(\omega t) \tag{3.46}$$

Now, we assume the solution of Eq. (3.46) is of the form

$$u = u_m(t) \text{ sech} \left(\frac{x - M(t)t}{\omega(t)} \right) \tag{3.47}$$

where $u_m(t) = 3M$ and $\omega(t) = \sqrt{\frac{4}{M}}$.

Let

$$I = \int_{-\infty}^{\infty} u^2 dx$$

Therefore,

$$\begin{aligned} \frac{dI}{dt} &= \int_{-\infty}^{\infty} 2u \frac{\partial u}{\partial t}\, dx \\ &= \int_{-\infty}^{\infty} 2u \left(-u^2 \frac{\partial u}{\partial x} - \frac{\partial^3 u}{\partial x^3} - cu + f_0 \cos(\omega t) \right) dx \\ &= I_1 + I_2 + I_3 + I_4 \end{aligned} \tag{3.48}$$

where

$$I_1 = 0, \quad I_2 = 0, \quad I_3 = -2cI, \quad I_4 = -2 f_0 \pi \cos(\omega t) u_m(t) \omega(t)$$

From Eq. (3.48), we have

$$\frac{dI}{dt} + 2cI = -2f_0\pi \cos(\omega t) u_m(t)\omega(t) \tag{3.49}$$

We also have

$$I = 36M^{\frac{3}{2}}(t) \tag{3.50}$$

Now, from (3.49) and (3.50), we get

$$M(t)e^{\frac{4}{3}ct} = -\frac{2}{9}f_0\pi \int e^{\frac{4}{3}ct}\cos(\omega t)\,dt = -\frac{2}{9}f_0\pi I_* + c_0 \tag{3.51}$$

Now,

$$\begin{aligned} I_* &= \int e^{\frac{4}{3}ct}\cos(\omega t)\,dt \\ &= \frac{9e^{\frac{4}{3}ct}}{16c^2 + 9\omega^2}\left(\frac{4c}{3}\cos(\omega t) + \omega \sin(\omega t)\right) \end{aligned}$$

From (3.51),

$$M(t)e^{\frac{4}{3}ct} = -\frac{2}{9}f_0\pi \frac{9e^{\frac{4}{3}ct}}{16c^2 + 9\omega^2}\left(\frac{4c}{3}\cos(\omega t) + \omega \sin(\omega t)\right) + c_0 \tag{3.52}$$

when $t = 0$, $M(t) = M_0$ then, $c_0 = \frac{\frac{8}{3}f_0\pi c}{16c^2+9\omega^2} + M_0$. From (3.52),

$$M(t) = \left[M_0 + \frac{\frac{8}{3}f_0\pi c}{16c^2 + 9\omega^2}\right]e^{-\frac{4}{3}ct} - \frac{2f_0\pi}{16c^2 + 9\omega^2}\left(\frac{4c}{3}\cos(\omega t) + \omega \sin(\omega t)\right) \tag{3.53}$$

Solution of the damped forced Eq. (3.46) is of the form

$$u = u_m(t)\ \mathrm{sech}\left(\frac{x - M(t)t}{\omega(t)}\right)$$

where $u_m(t) = 3M(t)$, $\omega(t) = \sqrt{\frac{4}{M(t)}}$, and $M(t)$ is of the form (3.53).

Damped SKdV equation

Consider the SKdV equation with the damped term c (very small) as

$$\frac{\partial u}{\partial t} + \sqrt{u}\frac{\partial u}{\partial x} + \frac{\partial^3 u}{\partial x^3} + cu = 0 \tag{3.54}$$

Let

$$I = \int_{-\infty}^{\infty} u^2 dx$$

Therefore,

$$\begin{aligned}\frac{dI}{dt} &= \int_{-\infty}^{\infty} 2u\frac{\partial u}{\partial t}\, dx \\ &= -\int_{-\infty}^{\infty} 2u\left(\sqrt{u}\frac{\partial u}{\partial x} + \frac{\partial^3 u}{\partial x^3} + cu\right) dx \\ &= I_1 + I_2 + I_3 \end{aligned}\tag{3.55}$$

where

$$I_1 = 0,\ I_2 = 0,\ I_3 = -2cI$$

So,

$$\frac{1}{2c}\log I = -t + c^* \tag{3.56}$$

Initially, when $t = t_0, I = I_0$, we can find $c^* = \frac{\log I_0}{2c} + t_0$. Putting the value of c^* in the above equation, we get

$$I = I_0 \exp[-2c(t - t_0)] \tag{3.57}$$

which is the solution of Eq. (3.56).

Now, let us assume that the solitary wave solution of Eq. (3.54) is of the form

$$u = u_m(t)\ \text{sech}^4\left(\frac{x - M(t)t}{\omega(t)}\right) \tag{3.58}$$

where u_m, M, ω are functions of t.

Therefore,

$$I = \int_{-\infty}^{\infty} u_m^2(t)\, \operatorname{sech}^8\left(\frac{x - M(t)t}{\omega(t)}\right) dx = \frac{32}{35}\omega(t)u_m^2(t) \tag{3.59}$$

At $t = 0$, we get $I_0 = \frac{32}{35}\omega(t_0)u_m^2(t_0)$. From (3.57) and (3.59), we obtain

$$I = \frac{32}{35}\omega(t_o)u_m^2(t_0)e^{-2c(t-t_0)} \tag{3.60}$$

where $u_m(t_0) = 3M_0$ and $\omega(t_0) = \sqrt{\frac{4}{M_0}}$. So,

$$\frac{32}{35}\omega(t)u_m^2(t) = \frac{32}{35}\omega(t_o)u_m^2(t_0)e^{-2c(t-t_0)}$$
$$\Rightarrow M = M_0 e^{-\frac{4}{3}c(t-t_0)}$$

Therefore, the solution of (3.54) is of the form

$$u = u_m(t)\, \operatorname{sech}^4\left(\frac{x - M(t)t}{\omega(t)}\right)$$

where $u_m(t) = 3M(t)$, $\omega(t) = \sqrt{\frac{4}{M(t)}}$, and $M(t) = M_0 e^{-\frac{4}{3}c(t-t_0)}$.

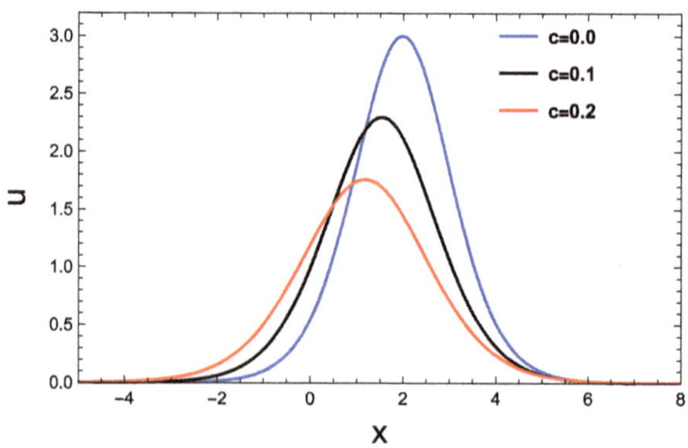

Forced SKdV equation

Consider the SKdV equation with the forced term $f_0 \cos(\omega t)$ as

$$\frac{\partial u}{\partial t} + \sqrt{u}\frac{\partial u}{\partial x} + \frac{\partial^3 u}{\partial x^3} = f_0 \cos(\omega t) \tag{3.61}$$

Now, consider the solution of Eq. (3.61)

$$u = u_m(t)\ \mathrm{sech}^4\left(\frac{x - M(t)t}{\omega(t)}\right) \tag{3.62}$$

where $u_m(t) = 3M$ and $\omega(t) = \sqrt{\frac{4}{M}}$.

Let

$$I = \int_{-\infty}^{\infty} u^2 dx$$

Therefore,

$$\frac{dI}{dt} = \int_{-\infty}^{\infty} 2u\left(- \sqrt{u}\frac{\partial u}{\partial x} - \frac{\partial^3 u}{\partial x^3} + f_0 \cos(\omega t)\right) dx = I_1 + I_2 + I_3 \tag{3.63}$$

where

$$I_1 = 0, \quad I_2 = 0, \quad I_3 = \frac{8}{3} f_0 \cos(\omega t) u_m(t)\omega(t) \tag{3.64}$$

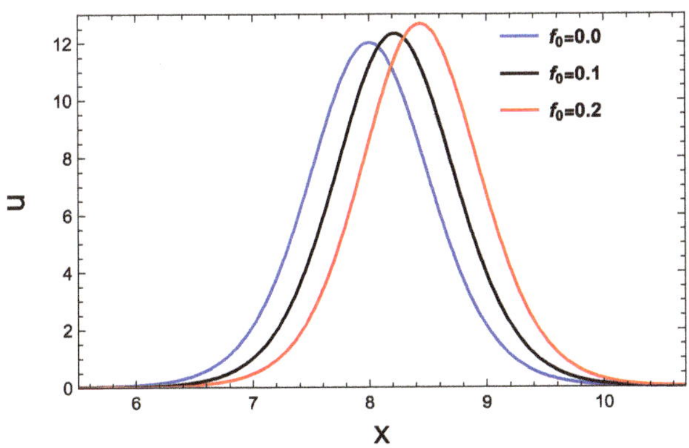

We also have

$$I = \frac{576}{35} M^{\frac{3}{2}}(t) \tag{3.65}$$

Now, from (3.63) and (3.64), we get

$$M(t) = \frac{35}{54\omega} f_0 \sin(\omega t) + c$$

Using the initial condition $t = 0, M(t) = M_0$ in the above equation we get $c = M_0$.

Therefore,

$$M(t) = \frac{35}{54\omega} f_0 \sin(\omega t) + M_0$$

So, the solution of the forced Eq. (3.61) is of the form

$$u = u_m(t)\ \text{sech}^4\left(\frac{x - M(t)t}{\omega(t)}\right)$$

where $u_m(t) = 3M(t),\ \ \omega(t) = \sqrt{\frac{4}{M(t)}}$, and $M(t) = \frac{35}{54\omega} f_0 \sin(\omega t) + M_0$.

Damped forced SKdV equation

Consider the SKdV equation with a damping and forcing term as

$$\frac{\partial u}{\partial t} + \sqrt{u}\frac{\partial u}{\partial x} + \frac{\partial^3 u}{\partial x^3} + cu = f_0 \cos(\omega t) \tag{3.66}$$

Now, we assume the solution of Eq. (3.66) is of the form

$$u = u_m(t)\ \text{sech}^4\left(\frac{x - M(t)t}{\omega(t)}\right) \tag{3.67}$$

where $u_m(t) = 3M$ and $\omega(t) = \sqrt{\frac{4}{M}}$.

Let

$$I = \int_{-\infty}^{\infty} u^2 dx$$

Therefore,

$$\begin{aligned}\frac{dI}{dt} &= \int_{-\infty}^{\infty} 2u\left(-\sqrt{u}\frac{\partial u}{\partial x} - \frac{\partial^3 u}{\partial x^3} - cu + f_0 \cos(\omega t)\right) dx \\ &= I_1 + I_2 + I_3 + I_4 \end{aligned} \tag{3.68}$$

where

$$I_1 = 0, \quad I_2 = 0, \quad I_3 = -2cI, \quad I_4 = \frac{8}{3} f_0 \cos(\omega t) u_m(t) \omega(t)$$

From Eq. (3.68), we have

$$\frac{dI}{dt} + 2cI = \frac{8}{3} f_0 \cos(\omega t) u_m(t) \omega(t) \tag{3.69}$$

We also have

$$I = \frac{576}{35} M^{\frac{3}{2}}(t) \tag{3.70}$$

Now, from Eqs. (3.69) and (3.70), we get

$$M(t) e^{\frac{4}{3}ct} = \frac{35}{54} f_0 \int e^{\frac{4}{3}ct} \cos(\omega t)\, dt = \frac{35}{54} f_0 I_* + c_0 \tag{3.71}$$

Now,

$$\begin{aligned} I_* &= \int e^{\frac{4}{3}ct} \cos(\omega t)\, dt \\ &= \frac{e^{\frac{4}{3}ct}}{(\frac{4c}{3})^2 + \omega^2} \left(\frac{4c}{3} \cos(\omega t) + \omega \sin(\omega t) \right) \\ &= \frac{9 e^{\frac{4}{3}ct}}{16c^2 + 9\omega^2} \left(\frac{4c}{3} \cos(\omega t) + \omega \sin(\omega t) \right) \end{aligned}$$

From Eq. (3.71), we get

$$M(t) e^{\frac{4}{3}ct} = \frac{35}{54} f_0 \frac{9 e^{\frac{4}{3}ct}}{16c^2 + 9\omega^2} \left(\frac{4c}{3} \cos(\omega t) + \omega \sin(\omega t) \right) + c_0 \tag{3.72}$$

when $t = 0$, $M(t) = M_0$, then $c_0 = M_0 - \frac{70 f_0 c}{9(16c^2 + 9\omega^2)}$. From Eq. (3.72), we get

$$\begin{aligned} M(t) &= \left[M_0 - \frac{70 f_0 c}{9(16c^2 + 9\omega^2)} \right] e^{-\frac{4}{3}ct} \\ &\quad + \frac{35 f_0}{6(16c^2 + 9\omega^2)} \left(\frac{4c}{3} \cos(\omega t) + \omega \sin(\omega t) \right) \end{aligned} \tag{3.73}$$

So, the solution of the damped forced Eq. (3.66) is of the form

$$u = u_m(t)\ \mathrm{sech}^4\left(\frac{x - M(t)t}{\omega(t)}\right)$$

where $u_m(t) = 3M(t)$, $\omega(t) = \sqrt{\frac{4}{M(t)}}$, and $M(t)$ is of the form (3.73).

Damped modified Gardner equation

Consider the modified Gardner equation with a damped term c (very small) as

$$\frac{\partial u}{\partial t} + u^3\frac{\partial u}{\partial x} + \frac{\partial^3 u}{\partial x^3} + cu = 0 \tag{3.74}$$

Let

$$I = \int_{-\infty}^{\infty} u^2 dx$$

Therefore,

$$\frac{dI}{dt} = -\int_{-\infty}^{\infty} 2u\left(u^3\frac{\partial u}{\partial x} + \frac{\partial^3 u}{\partial x^3} + cu\right)dx = I_1 + I_2 + I_3 \tag{3.75}$$

where

$$I_1 = 0, \quad I_2 = 0, \quad I_3 = -2\int_{-\infty}^{\infty} cu^2 dx$$

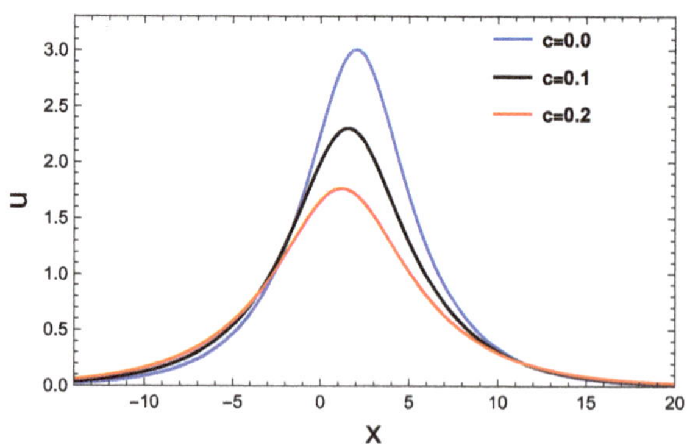

From Eq. (3.75), we get

$$\frac{1}{2c}\log I = -t + c^* \tag{3.76}$$

Initially, when $t = t_0, I = I_0$, we can find $c^* = \frac{\log I_0}{2c} + t_0$. Putting the value of c^* in the above equation, we get

$$I = I_0 \exp[-2c(t - t_0)] \tag{3.77}$$

which is the solution of Eq. (3.76).

Now, let us assume that the solitary wave solution of Eq. (3.74) is of the form

$$u = u_m(t) \operatorname{sech}^{\frac{2}{3}} \left(\frac{x - M(t)t}{\omega(t)} \right) \tag{3.78}$$

where u_m, M, ω are functions of t. Therefore,

$$I = u_m^2(t)\omega(t)\beta \left(\frac{2}{3}, \frac{1}{2} \right) \tag{3.79}$$

At $t = 0$, $I_0 = u_m^2(t_0)\omega(t_0)\beta(\frac{2}{3}, \frac{1}{2})$. Therefore,

$$I = u_m^2(t_0)\omega(t_0)\beta \left(\frac{2}{3}, \frac{1}{2} \right) e^{-2c(t-t_0)} \tag{3.80}$$

So, we get from Eqs. (3.79) and (3.80),

$$M = M_0 e^{-\frac{4}{3}c(t-t_0)}$$

Therefore, the solution of (3.74) is of the form

$$u = u_m(t) \operatorname{sech}^{\frac{2}{3}} \left(\frac{x - M(t)t}{\omega(t)} \right)$$

where $u_m(t) = 3M(t)$, $\omega(t) = \sqrt{\frac{4}{M(t)}}$, and $M(t) = M_0 e^{-\frac{4}{3}c(t-t_0)}$.

Forced modified Gardner equation

Consider the equation with the forced term $f_0 \cos(\omega t)$ as

$$\frac{\partial u}{\partial t} + u^3 \frac{\partial u}{\partial x} + \frac{\partial^3 u}{\partial x^3} = f_0 \cos(\omega t) \tag{3.81}$$

Now, consider the solution of Eq. (3.81)

$$u = u_m(t) \operatorname{sech}^{\frac{2}{3}} \left(\frac{x - M(t)t}{\omega(t)} \right) \tag{3.82}$$

where $u_m(t) = 3M$ and $\omega(t) = \sqrt{\frac{4}{M}}$.

Let

$$I = \int_{-\infty}^{\infty} u^2 dx$$

Therefore,

$$\frac{dI}{dt} = \int_{-\infty}^{\infty} 2u(-u^3\frac{\partial u}{\partial x} - \frac{\partial^3 u}{\partial x^3} + f_0\cos(\omega t))dx = I_1 + I_2 + I_3 \quad (3.83)$$

where

$$I_1 = 0, \quad I_2 = 0, \quad I_3 = 2f_0\cos(\omega t)u_m(t)\omega(t)\beta\left(\frac{1}{3}, \frac{1}{2}\right)$$

From Eq. (3.83),

$$\frac{dI}{dt} = 2f_0\cos(\omega t)u_m(t)\omega(t)\beta\left(\frac{1}{3}, \frac{1}{2}\right) \quad (3.84)$$

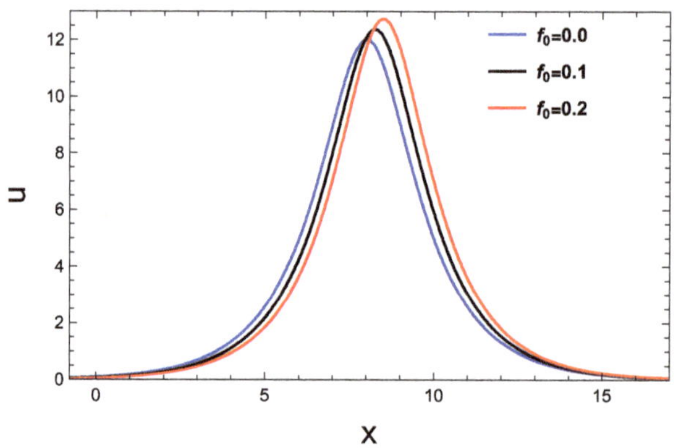

Therefore,

$$I = u_m^2(t)\omega(t)\beta\left(\frac{2}{3}, \frac{1}{2}\right) \quad (3.85)$$

Now, from Eqs. (3.84) and (3.85),

$$M = \frac{4}{9}f_0\frac{\sin(\omega t)}{\omega}\frac{\beta(\frac{1}{3}, \frac{1}{2})}{\beta\left(\frac{2}{3}, \frac{1}{2}\right)} + c^*$$

When t=0, $M(t) = M_0, c^* = M_0$,

$$M(t) = \frac{4}{9}f_0\frac{\sin(\omega t)}{\omega}\frac{\beta(\frac{1}{3}, \frac{1}{2})}{\beta\left(\frac{2}{3}, \frac{1}{2}\right)} + M_0 \quad (3.86)$$

Therefore, the solution of the forced Eq. (3.81) is of the form

$$u = u_m(t)\ \mathrm{sech}^{2/3}\left(\frac{x - M(t)t}{\omega(t)}\right)$$

where $u_m(t) = 3M(t)$, $\omega(t) = \sqrt{\frac{4}{M(t)}}$, and $M(t)$ is of the form (3.86).

Damped forced modified Gardner equation

Consider the equation with the damped and forced term as

$$\frac{\partial u}{\partial t} + u^3\frac{\partial u}{\partial x} + \frac{\partial^3 u}{\partial x^3} + cu = f_0\cos(\omega t) \tag{3.87}$$

Now, we assume the solution of Eq. (3.87) is of the form

$$u = u_m(t)\ \mathrm{sech}^{\frac{2}{3}}\left(\frac{x - M(t)t}{\omega(t)}\right) \tag{3.88}$$

where $u_m(t) = 3M$ and $\omega(t) = \sqrt{\frac{4}{M}}$.

Let

$$I = \int_{-\infty}^{\infty} u^2 dx$$

Therefore,

$$\frac{dI}{dt} = \int_{-\infty}^{\infty} 2u\left(-u^3\frac{\partial u}{\partial x} - \frac{\partial^3 u}{\partial x^3} - cu + f_0\cos(\omega t)\right)dx = I_1 + I_2 + I_3 + I_4 \tag{3.89}$$

where

$$I_1 = 0, \quad I_2 = 0, \quad I_3 = -2cI, \quad I_4 = 2f_0\cos(\omega t)u_m(t)\omega(t)\beta\left(\frac{1}{3}, \frac{1}{2}\right)$$

From Eq. (3.89), we have

$$\frac{dI}{dt} + 2cI = 2f_0\cos(\omega t)u_m(t)\omega(t)\beta\left(\frac{1}{3}, \frac{1}{2}\right) \tag{3.90}$$

Therefore,

$$I = u_m^2(t)\omega(t)\beta\left(\frac{2}{3}, \frac{1}{2}\right) = 18M^{\frac{3}{2}}(t)\beta\left(\frac{2}{3}, \frac{1}{2}\right) \tag{3.91}$$

Now, from Eqs. (3.90) and (3.91),

$$M(t)e^{\frac{4}{3}ct} = \frac{4}{9}\frac{\beta(\frac{1}{3},\frac{1}{2})}{\beta\left(\frac{2}{3},\frac{1}{2}\right)}f_0\int e^{\frac{4}{3}ct}\cos(\omega t)dt = \frac{4}{9}f_0\frac{\beta\left(\frac{1}{3},\frac{1}{2}\right)}{\beta(\frac{2}{3},\frac{1}{2})}I_* + c_0 \tag{3.92}$$

Now,

$$\begin{aligned}I_* &= \int e^{\frac{4}{3}ct}\cos(\omega t)\,dt\\ &= \frac{9e^{\frac{4}{3}ct}}{16c^2+9\omega^2}\left(\frac{4c}{3}\cos(\omega t)+\omega\sin(\omega t)\right)\end{aligned}$$

From Eq. (3.92),

$$M(t)e^{\frac{4}{3}ct} = \frac{4}{9}f_0\frac{\beta\left(\frac{1}{3},\frac{1}{2}\right)}{\beta\left(\frac{2}{3},\frac{1}{2}\right)}\frac{9e^{\frac{4}{3}ct}}{16c^2+9\omega^2}\left(\frac{4c}{3}\cos(\omega t)+\omega\sin(\omega t)\right)+c_0 \tag{3.93}$$

when $t=0$, $M(t)=M_0$, then $c_0 = M_0 - \frac{16f_0c}{3(16c^2+9\omega^2)}\frac{\beta(\frac{1}{3},\frac{1}{2})}{\beta(\frac{2}{3},\frac{1}{2})}$. Therefore,

$$\begin{aligned}M(t) &= \left[M_0 - \frac{16f_0c}{3(16c^2+9\omega^2)}\frac{\beta(\frac{1}{3},\frac{1}{2})}{\beta\left(\frac{2}{3},\frac{1}{2}\right)}\right]e^{-\frac{4}{3}ct}\\ &\quad + \frac{4f_0}{16c^2+9\omega^2}\left(\frac{4c}{3}\cos(\omega t)+\omega\sin(\omega t)\right)\frac{\beta(\frac{1}{3},\frac{1}{2})}{\beta\left(\frac{2}{3},\frac{1}{2}\right)}\end{aligned} \tag{3.94}$$

Solution of the damped forced Eq. (3.87) is of the form

$$u = u_m(t)\,\operatorname{sech}^{\frac{2}{3}}\left(\frac{x-M(t)t}{\omega(t)}\right)$$

where $u_m(t) = 3M(t)$, $\omega(t) = \sqrt{\frac{4}{M(t)}}$, and $M(t)$ is of the form (3.94).

Damped MSKdV equation

Consider the MSKdV equation with a damped term c (very small) as

$$\frac{\partial u}{\partial t} + u^{\frac{3}{2}}\frac{\partial u}{\partial x} + \frac{\partial^3 u}{\partial x^3} + cu = 0 \tag{3.95}$$

Let

$$I = \int_{-\infty}^{\infty} u^2 dx$$

Therefore,

$$\frac{dI}{dt} = -\int_{-\infty}^{\infty} 2u \left(u^{\frac{3}{2}} \frac{\partial u}{\partial x} + \frac{\partial^3 u}{\partial x^3} + cu \right) dx = I_1 + I_2 + I_3 \qquad (3.96)$$

where

$$I_1 = 0, \quad I_2 = 0, \quad I_3 = -2 \int_{-\infty}^{\infty} cu^2 dx$$

From Eq. (3.96), we get

$$\frac{1}{2c} \log I = -t + c^* \qquad (3.97)$$

Initially, when $t = t_0, I = I_0$, we can find $c^* = \frac{\log I_0}{2c} + t_0$. Putting the value of c^* in the above equation, we get

$$I = I_0 \exp[-2c(t - t_0)] \qquad (3.98)$$

which is the solution of Eq. (3.97).

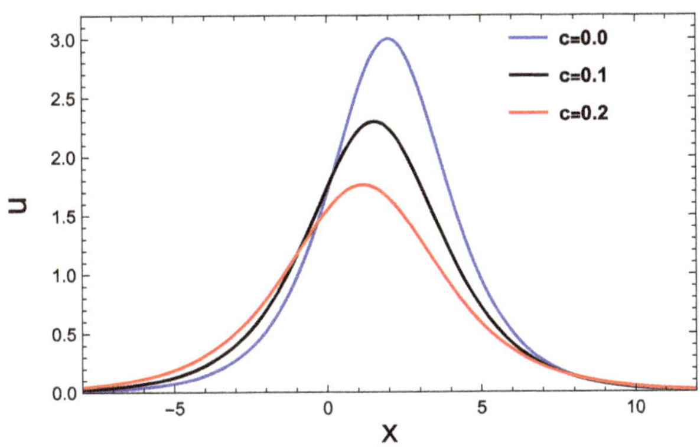

Now, let us assume that the solitary wave solution of Eq. (3.95) is of the form

$$u = u_m(t) \operatorname{sech}^{\frac{4}{3}} \left(\frac{x - M(t)t}{\omega(t)} \right) \qquad (3.99)$$

where u_m, M, ω are functions of t.

Therefore,

$$I = u_m^2(t)\omega(t)\beta\left(\frac{4}{3}, \frac{1}{2}\right) \tag{3.100}$$

At $t = 0$, $I_0 = u_m^2(t_0)\omega(t_0)\beta(\frac{4}{3}, \frac{1}{2})$. So,

$$I = u_m^2(t_0)\omega(t_0)\beta\left(\frac{4}{3}, \frac{1}{2}\right)e^{-2c(t-t_0)} \tag{3.101}$$

From Eqs. (3.100) and (3.101), we get

$$M = M_0 e^{-\frac{4}{3}c(t-t_0)}$$

Solution of (3.95) is of the form

$$u = u_m(t)\ \text{sech}^{\frac{4}{3}}\left(\frac{x - M(t)t}{\omega(t)}\right)$$

where $u_m(t) = 3M(t)$, $\omega(t) = \sqrt{\frac{4}{M(t)}}$, and $M(t) = M_0 e^{-\frac{4}{3}c(t-t_0)}$.

Forced MSKdV equation

Consider the MSKdV equation with the forced term $f_0\cos(\omega t)$ as

$$\frac{\partial u}{\partial t} + u^{\frac{3}{2}}\frac{\partial u}{\partial x} + \frac{\partial^3 u}{\partial x^3} = f_0\cos(\omega t) \tag{3.102}$$

Now, consider the solution of Eq. (3.102)

$$u = u_m(t)\ \text{sech}^{\frac{4}{3}}\left(\frac{x - M(t)t}{\omega(t)}\right) \tag{3.103}$$

where $u_m(t) = 3M$ and $\omega(t) = \sqrt{\frac{4}{M}}$.

Let

$$I = \int_{-\infty}^{\infty} u^2 dx$$

Therefore,

$$\frac{dI}{dt} = \int_{-\infty}^{\infty} 2u\left(-u^{\frac{3}{2}}\frac{\partial u}{\partial x} - \frac{\partial^3 u}{\partial x^3} + f_0\cos(\omega t)\right)dx = I_1 + I_2 + I_3 \tag{3.104}$$

where

$$I_1 = 0, \quad I_2 = 0, \quad I_3 = 2f_0\cos(\omega t)u_m(t)\omega(t)\beta\left(\frac{2}{3}, \frac{1}{2}\right)$$

From Eq. (3.104),

$$\frac{dI}{dt} = 2f_0 \cos(\omega t) u_m(t)\omega(t)\beta\left(\frac{2}{3}, \frac{1}{2}\right) \tag{3.105}$$

Therefore,

$$I = u_m^2(t)\omega(t)\beta\left(\frac{4}{3}, \frac{1}{2}\right) \tag{3.106}$$

Now, from Eqs. (3.105) and (3.106), we get

$$M = \frac{4}{9} f_0 \frac{\sin(\omega t)}{\omega} \frac{\beta\left(\frac{2}{3}, \frac{1}{2}\right)}{\beta(\frac{4}{3}, \frac{1}{2})} + c^*$$

When $t = 0$, $M(t) = M_0, c^* = M_0$ and therefore,

$$M(t) = \frac{4}{9} f_0 \frac{\sin(\omega t)}{\omega} \frac{\beta\left(\frac{2}{3}, \frac{1}{2}\right)}{\beta(\frac{4}{3}, \frac{1}{2})} + M_0 \tag{3.107}$$

Solution of the forced Eq. (3.102) is of the form

$$u = u_m(t)\ \text{sech}^{4/3}\left(\frac{x - M(t)t}{\omega(t)}\right)$$

where $u_m(t) = 3M(t),\ \omega(t) = \sqrt{\frac{4}{M(t)}}$, and $M(t)$ is of the form (3.107).

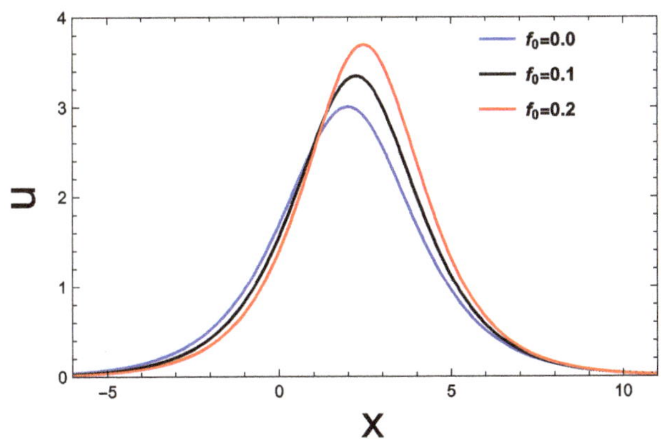

Damped forced MSKdV equation

Consider the MSKdV equation with a damped and forced term as

$$\frac{\partial u}{\partial t} + u^{\frac{3}{2}}\frac{\partial u}{\partial x} + \frac{\partial^3 u}{\partial x^3} + cu = f_0\cos(\omega t) \tag{3.108}$$

Now, we assume the solution of Eq. (3.108) is of the form

$$u = u_m(t)\,\operatorname{sech}^{\frac{4}{3}}\left(\frac{x - M(t)t}{\omega(t)}\right) \tag{3.109}$$

where $u_m(t) = 3M$ and $\omega(t) = \sqrt{\frac{4}{M}}$.

Let

$$I = \int_{-\infty}^{\infty} u^2 dx$$

Therefore,

$$\frac{dI}{dt} = \int_{-\infty}^{\infty} 2u\left(-u^{\frac{3}{2}}\frac{\partial u}{\partial x} - \frac{\partial^3 u}{\partial x^3} - cu + f_0\cos(\omega t)\right)dx = I_1 + I_2 + I_3 + I_4 \tag{3.110}$$

where

$$I_1 = 0, \quad I_2 = 0, \quad I_3 = -2cI, \quad I_4 = 2f_0\cos(\omega t)u_m(t)\omega(t)\beta\left(\frac{2}{3}, \frac{1}{2}\right)$$

From Eq. (3.110), we have

$$\frac{dI}{dt} + 2cI = 2f_0\cos(\omega t)u_m(t)\omega(t)\beta\left(\frac{2}{3}, \frac{1}{2}\right) \tag{3.111}$$

We also have

$$I = u_m^2(t)\omega(t)\beta\left(\frac{4}{3}, \frac{1}{2}\right) = 18M^{\frac{3}{2}}(t)\beta\left(\frac{4}{3}, \frac{1}{2}\right) \tag{3.112}$$

Now, from Eqs. (3.111) and (3.112), we get

$$M(t)e^{\frac{4}{3}ct} = \frac{4}{9}f_0\frac{\beta\left(\frac{2}{3}, \frac{1}{2}\right)}{\beta\left(\frac{4}{3}, \frac{1}{2}\right)}\frac{9e^{\frac{4}{3}ct}}{16c^2 + 9\omega^2}\left(\frac{4c}{3}\cos(\omega t) + \omega\sin(\omega t)\right) + c_0 \tag{3.113}$$

when $t = 0$, $M(t) = M_0$, then $c_0 = M_0 - \frac{16f_0c}{3(16c^2+9\omega^2)}\frac{\beta(\frac{2}{3},\frac{1}{2})}{\beta(\frac{4}{3},\frac{1}{2})}$. From Eq. (3.113), we get

$$M(t) = \left[M_0 - \frac{16f_0c}{3(16c^2 + 9\omega^2)}\frac{\beta\left(\frac{2}{3}, \frac{1}{2}\right)}{\beta(\frac{4}{3}, \frac{1}{2})}\right]e^{-\frac{4}{3}ct} + \frac{4f_0}{16c^2 + 9\omega^2}\left(\frac{4c}{3}\cos(\omega t) + \omega\sin(\omega t)\right)\frac{\beta\left(\frac{2}{3}, \frac{1}{2}\right)}{\beta(\frac{4}{3}, \frac{1}{2})} \tag{3.114}$$

Solution of the damped forced Eq. (3.108) is of the form

$$u = u_m(t)\ \mathrm{sech}^{\frac{4}{3}}\left(\frac{x - M(t)t}{\omega(t)}\right)$$

where $u_m(t) = 3M(t)$, $\omega(t) = \sqrt{\frac{4}{M(t)}}$, and $M(t)$ is of the form (3.114).

Generalized damped ZK equation

Consider the generalized damped Zakharov–Kuznetsov (GDZK) equation as

$$\frac{\partial u}{\partial t} + au^n\frac{\partial u}{\partial x} + b\frac{\partial}{\partial x}\left(\frac{\partial^2 u}{\partial x^2} + \frac{\partial^2 u}{\partial y^2}\right) + Au = 0 \tag{3.115}$$

where A is the damping coefficient.

Now, we assume the solution of Eq. (3.115) is of the form

$$u = \left[u_m \mathrm{sech}^2\left(\frac{n}{2}\sqrt{\frac{c(t)}{2b}}\xi\right)\right]^{1/n} \tag{3.116}$$

where $u_m = \frac{c(t)(n+1)(n+2)}{2a}$.

Let

$$I = \frac{2^{\frac{4}{n}-1}c(t)^{\frac{2}{n}-\frac{1}{2}}\beta(\frac{2}{n},\frac{2}{n})\left(\frac{(n+1)(n+2)}{2a}\right)^{\frac{2}{n}}}{n\sqrt{\frac{1}{2b}}} \tag{3.117}$$

Therefore,

$$\frac{dI}{dt} = \frac{2^{\frac{4}{n}-1}c(t)^{\frac{2}{n}-\frac{1}{2}-1}\beta\left(\frac{2}{n},\frac{2}{n}\right)\left(\frac{(n+1)(n+2)}{2a}\right)^{\frac{2}{n}}\left(\frac{2}{n}-\frac{1}{2}\right)}{n\sqrt{\frac{1}{2b}}}\frac{d}{dt}(c(t)) \tag{3.118}$$

So,

$$\begin{aligned}\frac{dI}{dt} &= -2\int_{-\infty}^{\infty} u(au^n u_x + b(u_{xx}+u_{yy})_x + Au)d\xi \\ &= -2A\frac{2^{\frac{4}{n}-1}c(t)^{\frac{2}{n}-\frac{1}{2}}\beta\left(\frac{2}{n},\frac{2}{n}\right)\left(\frac{(n+1)(n+2)}{2a}\right)^{\frac{2}{n}}}{n\sqrt{\frac{1}{2b}}}\end{aligned} \tag{3.119}$$

Since $\int_{\infty}^{\infty} u^n u u_x d\xi = 0$ and $\int_{\infty}^{\infty} u^n u(u_{xx} + u_{yy})_x d\xi = 0$. Considering Eqs. (3.118) and (3.119), we get a differential equation for $c(t)$ as

$$\frac{1}{c(t)}\frac{d}{dt}(c(t)) = -\frac{2A}{\left(\frac{2}{n}-\frac{1}{2}\right)} \tag{3.120}$$

So, the solution for Eq. (3.115) is of the form

$$u = \left[u_m \text{sech}^2\left(\frac{n}{2}\sqrt{\frac{c(t)}{2b}}\xi\right)\right]^{1/n}$$

where $c(t) = c_0 e^{-\frac{4An}{4-n}t}$ with $c = c_0$ at $t = 0$.

Generalized forced ZK equation

Consider the generalized forced Zakharov–Kuznetsov (GFZK) equation as

$$\frac{\partial u}{\partial t} + au^n\frac{\partial u}{\partial x} + b\frac{\partial}{\partial x}\left(\frac{\partial^2 u}{\partial x^2} + \frac{\partial^2 u}{\partial y^2}\right) = f\cos(\omega t) \tag{3.121}$$

Following the same formulation as stated, the solution of (3.121) may be written as

$$u = \left[u_m \text{sech}^2\left(\frac{n}{2}\sqrt{\frac{c(t)}{2b}}\xi\right)\right]^{1/n} \tag{3.122}$$

where $u_m = \frac{c(t)(n+1)(n+2)}{2a}$ and $c(t)^{\frac{1}{n}} = \frac{\beta(\frac{1}{n},\frac{1}{n})f\cos(\omega t)}{\omega(4-n)2^{\frac{2}{n}}\beta(\frac{2}{n},\frac{2}{n})\left(\frac{(n+1)(n+2)}{2a}\right)^{\frac{2}{n}}} + c_0^{\frac{1}{n}}$.

3.5 Adomian Decomposition Method (ADM)

Now, we are going to explain the Adomian decomposition method (ADM) that demonstrates how it may be applied to obtain near solutions in a series of an NLPDE that converges quickly. The method relies on selecting a differential operator, which might be linear or nonlinear, ordinary or partial. It enables the calculation of a convergent power series, which is the series solution of the equation.

George Adomian first used this method to solve ODEs and PDEs in the early 1980s. "Adomian Decomposition Method (ADM)" is the name given to this technique. It is an advanced method for solving NLPDEs and other nonlinear equations that make use of specific polynomials. This method provides effective algorithms for analytical approximations of real-world problems comprising of applied science and engineering. Additionally, it enables us to solve both boundary value problems (BVPs) and initial value problems (IVPs) that are either linear or nonlinear. There are no restrictions on this technique, such as linearization, perturbation, and *ad hoc* assumptions. Furthermore, the use of Green's functions — which are frequently overly complex and unpredictable — is not necessary for the ADM. A set of Adomian polynomials is used to handle NLPDEs in place of nonlinearity, and as these polynomials contribute to the solution's series components, their evaluation is required. Lastly, direct substitution makes it simple to confirm the accuracy of the approximate solutions that were found.

3.5.1 *Basic algorithm*

We consider a general nonlinear differential equation of the form

$$F[u(t)] = g(t) \tag{3.123}$$

where F, which consists of both linear and nonlinear terms, is the generic nonlinear differential operator. First, we decompose the linear terms of (3.123) into $L + R$, where R is a remainder of the linear operator and L is invertible (often the highest order derivative). Equation (3.123) can thus be expressed as

$$Lu + Ru + Nu = g \tag{3.124}$$

where Nu represents the nonlinear terms. By solving Eq. (3.124) for Lu gives $Lu = g - Ru - Nu$. Since L is invertible, an equivalent expression can be written as

$$L^{-1}Lu = L^{-1}g - L^{-1}(Ru) - L^{-1}(Nu) \tag{3.125}$$

If Eq. (3.123) corresponds to an IVP, then the integral operator L^{-1} of (3.125) may be considered as the definite integral from t_0 to t. Solving Eq. (3.125) for u yields

$$u = u(t_0) + L^{-1}g - L^{-1}(Ru) - L^{-1}(Nu) \tag{3.126}$$

If L is a second-order operator, then L^{-1} is a two-fold integration operator. By solving Eq. (3.125), we have

$$u = u(t_0) + u(t_0)(t - t_0) + L^{-1}g - L^{-1}(Ru) - L^{-1}(Nu) \tag{3.127}$$

Further, if Eq. (3.123) corresponds to the BVP, then the indefinite integrations are used and the constants are evaluated from the given boundary conditions. Therefore, solving Eq. (3.125) for u yields

$$u = A + Bt + L^{-1}g - L^{-1}(Ru) - L^{-1}(Nu) \tag{3.128}$$

ADM assumes the solution u that can be expanded into infinite series as

$$u = \sum_{n=0}^{\infty} u_n \tag{3.129}$$

Additionally, if the nonlinear term Nu is analytic, it will be expressed as

$$Nu = \sum_{n=0}^{\infty} A_n(u_0, u_1, \ldots, u_n) \tag{3.130}$$

where A_n are Adomian polynomials. Therefore, the solution of Eq. (3.128) is given by

$$\sum_{n=0}^{\infty} u_n = u_0 - L^{-1}\left(R\sum_{n=0}^{\infty} u_n\right) - L^{-1}\left(\sum_{n=0}^{\infty} A_n\right) \tag{3.131}$$

with u_0 identified as $u_0 = A + Bt + L^{-1}g$.

Expanding the series (3.131), we get

$$u_0 + u_1 + u_2 + \cdots = u_0 - L^{-1}Ru_0 - L^{-1}Ru_1 - \cdots - L^{-1}A_0 - L^{-1}A_1 - \cdots$$

Now, comparing the various coefficients, we get the following recurrence relations:

$$\left.\begin{array}{rl} u_1 = & -L^{-1}Ru_0 - L^{-1}A_0 \\ u_2 = & -L^{-1}Ru_1 - L^{-1}A_1 \\ \ldots & \qquad\qquad \ldots\ldots \\ u_{n+1} = & -L^{-1}Ru_n - L^{-1}A_n \\ \ldots & \qquad\qquad \ldots\ldots \\ \ldots & \qquad\qquad \ldots\ldots \end{array}\right\} \tag{3.132}$$

The Adomian polynomials can generally be produced in a number of ways. However, the recurrence formula in this case is provided by

$$A_n = \frac{1}{n!}\left[\frac{d^n}{d\lambda^n}\left[N\left(\sum_{i=0}^{n} \lambda^i u_i\right)\right]\right]_{\lambda=0}, \quad n = 0, 1, 2, \ldots \tag{3.133}$$

Using (3.128) and (3.133), we can compute all the components u_n and hence $u = \sum_{n=0}^{\infty} u_n$. If the series converges, the n-term partial sum $u = \sum_{n=0}^{n-1} u_i$ will be the approximate solution of Eq. (3.123) since $\lim_{n\to\infty} u_n = \sum_{i=0}^{\infty} u_i = u$ by definition.

3.5.2 *Solution of the KdV equation*

Considering the standard KdV equation

$$\phi_\tau + 6\phi\phi_\xi + \phi_{\xi\xi\xi} = 0 \tag{3.134}$$

with initial condition $\phi(\xi, 0) = f(\xi)$. An operator form of this equation can be written as

$$L_\tau(\phi) + 6N\phi + L_{\xi\xi\xi}(\phi) = 0 \tag{3.135}$$

where $L_\tau \equiv \partial/(\partial\tau)$ is the linear operator and $N\phi = \phi\phi_\xi$ is the nonlinear operator with $L_{\xi\xi\xi} \equiv \partial^3/(\partial\xi^3)$. It is assumed that L_τ^{-1} is an integral operator given by $L_\tau^{-1} \equiv \int_0^\tau (\cdot) d\tau$. Operating with the integral operator L_τ^{-1} on both sides of Eq. (3.135) we obtained

$$L_\tau^{-1} L_\tau(\phi) = -6L_\tau^{-1}(N\phi) - L_\tau^{-1} L_{\xi\xi\xi}(\phi) \tag{3.136}$$

Therefore, it follows that

$$\phi(\xi, \tau) = \phi(\xi, 0) - 6L_t^{-1}(N\phi) - L_t^{-1} L_{\xi\xi\xi}(\phi) \tag{3.137}$$

According to the ADM, the solution can be expressed as

$$\phi(\xi, \tau) = \sum_{n=0}^{\infty} \phi_n(\xi, \tau) \tag{3.138}$$

and the nonlinear part by

$$N(\phi) = -\sum_{n=0}^{\infty} A_n \tag{3.139}$$

where

$$A_n = \frac{1}{n!} \frac{d^n}{d\lambda^n} \left[N \left(\sum_{i=0}^{\infty} \lambda^i \phi_i \right) \right]_{\lambda=0}, \quad \text{for} \quad n \in N \tag{3.140}$$

After substituting the above values in Eq. (3.138), we obtained

$$\phi(\xi, \tau) = f(\xi) - \int_0^\tau \left[L \left(\sum_{n=0}^{\infty} \phi_n(\xi, t) \right) + 6 \sum_{n=0}^{\infty} A_n \right] dt \tag{3.141}$$

For the KdV equation, the Adomian polynomials can be expressed as follows:

$$A_n = \sum_{i=0}^{n} \phi_i \frac{\partial \phi_{n-i}}{\partial \xi} \tag{3.142}$$

Thus, we can infer $\phi_n(\xi, \tau)$, specifically

$$\begin{cases} \phi_0(\xi, \tau) = f(\xi) \\ \phi_{n+1}(\xi, \tau) = -\int_0^\tau \left[L_{\xi\xi\xi}\phi_n + A_n \right] dt, \quad n \geq 0 \end{cases} \tag{3.143}$$

If $f(\xi) = 2\,\mathrm{sech}^2(\xi)$, then by using the recurrence formula (3.143), one can obtain

$$\begin{aligned}
\phi_0 &= 2\,\mathrm{sech}^2(\xi),\ A_0 = -8\,\mathrm{sech}^4(\xi)\tanh(\xi)\\
\phi_1 &= 16\tau\,\mathrm{sech}^2(\xi)\tanh(\xi),\ A_1 = -32\tau(-3+2\cosh(2\xi))\,\mathrm{sech}^6(\xi)\\
\phi_2 &= 32\tau^2(-2+\cosh(2\xi))\,\mathrm{sech}^4(\xi)\\
A_2 &= 128\tau^2(13\sinh(\xi)-2\sinh(3\xi))\,\mathrm{sech}^7(\xi)\\
\phi_3 &= \frac{128}{3}\tau^3(-11\sinh(\xi)+\sinh(3\xi))\,\mathrm{sech}^5(\xi)
\end{aligned} \tag{3.144}$$

3.5.3 *Solution of the damped KdV equation*

Considering the damped KdV equation

$$\phi_\tau^{(1)} + A\phi^{(1)}\phi_\xi^{(1)} + B\phi_{\xi\xi\xi}^{(1)} + D\phi^{(1)} = 0 \tag{3.145}$$

This equation can be rewritten in operator form as

$$L\phi^{(1)} + AN(\phi^{(1)}) + BS\phi_1 + D\phi^{(1)} = 0 \tag{3.146}$$

with initial condition $\phi^{(1)}(\xi,0) = \psi(\xi),\ \xi \in R$, where $L = \frac{\partial}{\partial\tau}$ and $S = \frac{\partial^3}{\partial\xi^3}$ are differential operators and the nonlinear term $N(\phi^{(1)}) = \phi_1\frac{\partial\phi^{(1)}}{\partial\xi}$. $\phi^{(1)}(\xi,\tau)$ can be expressed as

$$\phi^{(1)}(\xi,\tau) = \sum_{n=0}^{\infty}\phi_n^{(1)}(\xi,\tau) \tag{3.147}$$

Also, $N(\phi^{(1)})$ can be decomposed by an infinite series of polynomials given by

$$N(\phi^{(1)}(\xi,\tau)) = \sum_{n=0}^{\infty}A_n(\phi_0^{(1)},\phi_1^{(1)},\phi_2^{(1)},\ldots,\phi_n^{(1)}) \tag{3.148}$$

where Adomian polynomials A_n's are defined as

$$A_n = \frac{1}{n!}\frac{d^n}{d\lambda^n}\left[N\left(\sum_{i=0}^{n}\lambda^i\phi_i^{(1)}\right)\right]_{\lambda=0},\quad n = 0,1,2,3,\ldots \tag{3.149}$$

The inverse operator L^{-1} is an integral operator given by

$$L^{-1}(\cdot) = \int_0^\tau(\cdot)d\tau \tag{3.150}$$

Applying L^{-1} on Eq. (3.159) and using the initial condition, we get

$$\phi^{(1)}(\xi,\tau) = \psi(\xi) - L^{-1}(AN(\phi^{(1)}) + BS\phi^{(1)} + D\phi^{(1)}) \qquad (3.151)$$

Substituting Eqs. (3.159) and (3.160) into Eq. (3.164), we obtain

$$\sum_{n=0}^{\infty}\phi_n^{(1)}(\xi,\tau) = \psi(\xi) - L^{-1}\left(A\sum_{n=0}^{\infty}A_n + BS\left(\sum_{n=0}^{\infty}\phi_n^{(1)}(\xi,\tau)\right)\right.$$
$$\left. + D\sum_{n=0}^{\infty}\phi_n^{(1)}(\xi,\tau)\right) \qquad (3.152)$$

Taking $\phi_0^{(1)} = \psi(\xi)$, we get the following recurrence relation:

$$\phi_{(n+1)}^{(1)} = -L^{-1}(AA_n + BS\phi_n^{(1)} + D\phi_n^{(1)}), \quad n \geq 0 \qquad (3.153)$$

where A_n are Adomian polynomials representing the nonlinear term. In this case, $N(\phi^{(1)}) = \phi^{(1)}\frac{\partial\phi^{(1)}}{\partial\xi}$, which gives

$$A_n = \sum_{k=0}^{n}\phi_n^{(1)}\frac{\partial\phi_{(n-k)}^{(1)}}{\partial\xi}, \quad n \geq 0 \qquad (3.154)$$

Using the solution of the standard KdV equation as the initial approximation of the damped KdV Eq. (3.145) as $\psi(\xi) = A_m\,\mathrm{sech}^2(\xi/W)$, where $A_m = \frac{3M}{A}$, $W = 2\sqrt{\frac{B}{M}}$ (M is the traveling wave's speed), and applying Eqs. (3.153) and (3.154), one can obtain

$$\phi_1^{(1)} = \frac{1}{\left(W^3\cosh^5\left(\frac{\xi}{W}\right)\right)}\left(A_m\tau\left(24\sinh\left(\frac{\xi}{W}\right) + D\cosh^3\left(\frac{\xi}{W}\right)\right.\right.$$
$$\left.\left. - 8B\sinh\left(\frac{\xi}{W}\right)\cosh^2\left(\frac{\xi}{W}\right) - 2AA_mW^2\sinh\left(\frac{\xi}{W}\right)\right)\right) \qquad (3.155)$$

$$\phi_2^{(1)} = -\left(2520A_mB^2t^2 + 7A^2A_m^3W^4t^2 - 2A_mt^2\cosh^2\left(\frac{\xi}{W}\right)(3A^2A_m^2W^4\right.$$
$$\left. - AA_mBW^2 + 1680B^2\right) - \frac{A_mt^2\cosh\left(\frac{\xi}{W}\right)^6(64B^2 + C^2W^6)}{2}$$

$$- 294AA_m^2BW^2t^2 + 16A_mBt^2\cosh^4\left(\frac{\xi}{W}\right)(63B - 5AA_mW^2)$$
$$+ 8A_mBCW^3t^2\sinh\left(\frac{\xi}{W}\right)\cosh^5\left(\frac{\xi}{W}\right) - 3A_mDW^3t^2\sinh\left(\frac{\xi}{W}\right)$$
$$\times \cosh^3\left(\frac{\xi}{W}\right)(8B - AA_mW^2)\Big)\Big/\left(W^6\cosh^8\left(\frac{\xi}{W}\right)\right) \quad (3.156)$$

and so on. The solution of Eq. (3.145) can be obtained by inserting ϕ_{11}, ϕ_{12}, etc., into Eq. (3.147).

3.5.4 *Solution of the damped mKdV equation*

Considering the damped modified KdV equation

$$\phi_1\tau + A_1\phi_1^2\phi_1\xi + B_1\phi_1\xi\xi\xi + D_1\phi_1 = 0 \quad (3.157)$$

This equation can can be expressed in operator form as

$$\mathcal{L}(\phi_1) + A_1\mathcal{N}(\phi_1) + B_1\mathcal{S}(\phi_1) + D_1\phi_1 = 0 \quad (3.158)$$
$$\phi_1(\xi, 0) = \psi(\xi), \quad \xi \in R$$

where $\mathcal{L} = \frac{\partial}{\partial\tau}$ and $\mathcal{S} = \frac{\partial^3}{\partial\xi^3}$ are differential operators and $\mathcal{N}(\phi_1) = \phi_1^2\frac{\partial\phi_1}{\partial\xi}$ is the nonlinear term. $\phi_1(\xi, \tau)$ can be expressed as

$$\phi_1(\xi, \tau) = \sum_{n=0}^{\infty}\phi_1{}^{(n)}(\xi, \tau) \quad (3.159)$$

and an infinite number of the polynomials can be used to decompose the nonlinear term $\mathcal{N}(\phi_1)$ as

$$\mathcal{N}(\phi_1(\xi, \tau)) = \sum_{n=0}^{\infty}\mathcal{A}_n(\phi_1^{(0)}, \phi_1^{(1)}, \phi_1^{(2)}, \ldots, \phi_1^{(n)}) \quad (3.160)$$

where Adomian polynomials $\mathcal{A}_n$'s are described as

$$\mathcal{A}_n = \frac{1}{n!}\frac{d^n}{d\lambda^n}\left[\mathcal{N}\left(\sum_{i=0}^{\infty}\lambda^i\phi_1^{(i)}\right)\right]_{\lambda=0}, \quad n = 0, 1, 2, 3, \ldots \quad (3.161)$$

The inverse integral operator $\mathcal{L}^{-1}$ is define as

$$\mathcal{L}^{-1}(\cdot) = \int_0^{\tau}(\cdot)d\tau \quad (3.162)$$

After applying $\mathcal{L}^{-1}$ on Eq. (3.159) with the initial condition, we obtain

$$\phi_1(\xi, \tau) = \psi(\xi) - \mathcal{L}^{-1}(A_1\mathcal{N}(\phi_1) + B_1\mathcal{S}(\phi_1) + D_1\phi_1) \quad (3.163)$$

Substituting Eqs. (3.159) and (3.160) into Eq. (3.163), we obtain

$$\sum_{n=0}^{\infty} \phi_1^{(n)}(\xi,\tau) = \psi(\xi) - \mathcal{L}^{-1}\left(A_1 \sum_{n=0}^{\infty} \mathcal{A}_n + B_1 \mathcal{S}\left(\sum_{n=0}^{\infty} \phi_1^{(n)}(\xi,\tau)\right)\right.$$
$$\left. + D_1 \sum_{n=0}^{\infty} \phi_1^{(n)}(\xi,\tau)\right) \tag{3.164}$$

Taking $\phi_1^{(0)} = \psi(\xi)$, we get the recurrence relation

$$\phi_1^{(n+1)} = -\mathcal{L}^{-1}(A\mathcal{A}_n + B\mathcal{S}\phi_1^{(n)} + D\phi_1^{(n)}), \quad n \geq 0 \tag{3.165}$$

where the Adomian polynomials $\mathcal{A}_n$ represent the nonlinear term. In the case $\mathcal{N}(\phi_1) = \phi_1^2 \frac{\partial \phi_1}{\partial \xi}$ which gives

$$\mathcal{A}_n = \sum_{k=0}^{n} \left(\sum_{j=0}^{k} \phi_1^{(j)} \phi_1^{(k-j)}\right) \frac{\partial}{\partial \xi} \phi_1^{(n-k)}, \quad n \geq 0 \tag{3.166}$$

If $\psi(\xi)$ is known, then by using the recurrence formula (3.165)–(3.166), one can obtain the solution ϕ_1 easily.

3.6 Homotopy Analysis Method (HAM)

The Homotopy Analysis Method (HAM) is another analytical technique for solving nonlinear problems in differential equations, both ODEs and PDEs. Unlike traditional perturbation methods, HAM does not require small parameters in the equation and provides a way to control and ensure the convergence of the solution series. Liao developed HAM as early as 1992.

Principles of HAM

(i) *Construction of the Homotopy*: The essence of HAM lies in constructing a homotopy, which is a continuous transformation from an initial guess to the exact solution.
(ii) *Homotopy Parameter*: HAM introduces an auxiliary parameter, often denoted as h, that controls the convergence region and the rate of convergence of the series solution.

Steps for Applying HAM to PDEs

Step I: Identify the PDE:

Consider the general NLPDE

$$\mathcal{N}[y(x,t)] = 0 \tag{3.167}$$

where $\mathcal{N}$ is the nonlinear operator and $y(x,t)$ is an unknown function with initial condition

$$y(x,t) = y_0(x,t) \quad \text{at} \quad t = 0$$

Step II: Construction of the zeroth-order deformation equation:

Let us create a homotopy that continuously deforms from an initial guess $y_0(x,t)$ to the actual solution $y(x,t)$ as

$$(1-p)\mathcal{L}[w(x,t;p) - y_0(x,t)] = phH(x,t)\mathcal{N}[w(x,t;p)] \tag{3.168}$$

where $p \in [0,1]$ is the embedding parameter, $h \neq 0$ is an auxiliary parameter, $H(x,t) \neq 0$ is an auxiliary function, $\mathcal{L}$ is a linear auxiliary operator, and $w(x,t;p)$ is the approximate solution of the given problem. For $p = 0$, Eq. (3.168) changes to

$$\mathcal{L}[w(x,t;0) - y_0(x,t)] = 0 \tag{3.169}$$

and in view of the linearity of the operator $\mathcal{L}$, the zeroth-order deformation equation is given as

$$w(x,t;0) = y_0(x,t) \tag{3.170}$$

The homotopy parameter p is considered in such a way that as it grows continuously in $[0,1]$, $w(x,t;p)$ changes from the initial guess $y_0(x,t)$ to the solution $y(x,t)$. Accordingly, for $p = 1$, Eq. (3.168) gives

$$w(x,t;1) = y(x,t) \tag{3.171}$$

Step III: Construction of the kth-order deformation equation:

Taylor's series expansion of $w(x,t;p)$ about the point $p = 0$ gives

$$w(x,t;p) = y_0(x,t) + \sum_{k=1}^{\infty} y_k(x,t)p^m \tag{3.172}$$

where

$$y_k(x,t) = \frac{1}{k!}\frac{\partial^k w(x,t;p)}{\partial p^k}\bigg|_{p=0} \tag{3.173}$$

The series (3.172) converges for $p = 1$ when $H(x,t)$, h, $\mathcal{L}$, and $y_0(x,t)$ are chosen correctly. Thus,

$$w(x,t;1) = y_0(x,t) + \sum_{k=1}^{\infty} y_k(x,t) \tag{3.174}$$

By virtue of Eq. (3.171),

$$y(x,t) = y_0(x,t) + \sum_{k=1}^{\infty} y_k(x,t) \tag{3.175}$$

which is a solution of the original nonlinear differential Eq. (3.167).

Let us define a vector

$$\vec{y}_n(x,t) = \{y_0(x,t), y_1(x,t), y_2(x,t), \ldots, y_n(x,t)\} \tag{3.176}$$

Differentiating Eq. (3.168) k times with regard to p and dividing the result by $k!$ and then putting $p = 0$ gives the kth-order deformation equation

$$\mathcal{L}[y_k(x,t) - \chi_k y_{k-1}(x,t)] = hH(x,t)\mathscr{R}_k(\vec{y}_{k-1}) \tag{3.177}$$

where

$$\mathscr{R}_k(\vec{y}_{k-1}) = \frac{1}{(k-1)!}\frac{\partial^{k-1}\mathcal{N}[w(x,t;p)]}{\partial p^{k-1}}\bigg|_{p=0} \tag{3.178}$$

and

$$\chi_k = \begin{cases} 0 & k \le 1 \\ 1 & k > 1 \end{cases} \tag{3.179}$$

Step IV: To solve the deformation equations:

Using the initial guess and the boundary conditions, one can easily obtain $y_k(x,t)$ for $k = 1, 2, \ldots$ from Eq. (3.177) as

$$y_k(x,t) = \chi_k y_{k-1}(x,t)] + h\mathcal{L}^{-1}[H(x,t)\mathscr{R}_k(\vec{y}_{k-1})] \tag{3.180}$$

by the help of symbolic computation software such as Maple or Mathematica. After determining $y_k(x,t)$, for $k = 1, 2, \ldots$, an approximate solution to the problem given in Eq. (3.167) is

$$y(x,t) \approx y_0(x,t) + \sum_{k=1}^{\infty} y_k(x,t). \tag{3.181}$$

3.6.1 *Approximate solution of KdV equation*

In this section, we apply the HAM to the KdV equation to obtain approximate solutions. Consider the standard KdV equation of the form

$$\phi_\tau + 6\phi\phi_\xi + \phi_{\xi\xi\xi} = 0 \tag{3.182}$$

with initial condition $\phi_0(\xi,\tau) = \phi(\xi,0) = -2\frac{e^\xi}{(1+e^\xi)^2}$. Here, we choose

$$\mathcal{L}\phi(\xi,\tau;p) = \frac{\partial\phi(\xi,\tau;p)}{\partial\tau}$$

as the auxiliary linear operator satisfying

$$\mathcal{L}[c] = 0$$

where c is a constant. We consider the auxiliary function

$$H(\xi,\tau) = 1$$

zeroth-order deformation problem

$$(1-p)\mathcal{L}[w(\xi,\tau;p) - \phi_0(\xi,\tau)] = phH(\xi,\tau)\mathcal{N}[w(\xi,\tau;p)]$$

$$\phi_0(\xi,\tau) = -2\frac{e^\xi}{(1+e^\xi)^2} \tag{3.183}$$

$$\mathcal{N}[\phi(\xi,\tau;p)] = \frac{\partial\phi(\xi,\tau;p)}{\partial\tau} - 6\phi(\xi,\tau;p)\frac{\partial\phi(\xi,\tau;p)}{\partial\xi} + \frac{\partial^3\phi(\xi,\tau;p)}{\partial\xi^3}$$

and the mth-order deformation problem as

$$\begin{aligned}&\mathcal{L}[\phi_m(\xi,\tau) - \chi_m\phi_m(\xi,\tau)]\\ &\quad = ph\left[\frac{\partial\phi_{m-1}}{\partial\tau} - 6\sum_{i=0}^{m-1}\phi_i\frac{\partial\phi_{m-1-i}}{\partial\xi} + \frac{\partial^3\phi_{m-1}}{\partial\xi^3}\right]\end{aligned} \tag{3.184}$$

$$\phi_m(\xi,0) = 0 \quad (m \geq 1) \tag{3.185}$$

We can use software Mathematica for solving the set of linear Eq. (3.184) with condition (3.185). It is found that the solution in a series form is given by

$$\begin{aligned}\phi(\xi,\tau) &= -2\frac{e^\xi}{(1+e^\xi)^2} + \left(2e^\xi(-1+e^\xi)h\tau\log[e]\left(12e^\xi\right.\right.\\ &\quad \left.\left.+\log[e]^2 - 10e^\xi\log[e]^2 + e^{2\xi}\log[e]^2\right)/\left(1+e^\xi\right)^5\right) + \cdots\end{aligned} \tag{3.186}$$

3.6.2 Approximate solution of DMKdV equation

We consider the damped modified KdV(DMKdV) equation as

$$\frac{\partial\phi^{(1)}}{\partial\tau}+D(\phi^{(1)})^2\frac{\partial\phi^{(1)}}{\partial\xi}+B\frac{\partial^3\phi^{(1)}}{\partial\xi^3}+C\phi^{(1)}=0 \tag{3.187}$$

The DMKdV equation is known to be non-integrable and to have no analytic solution. As a consequence, it will be numerically solved using HAM. The initial guess is taken as

$$\phi_0^{(1)}(\xi,\tau)=A_m\ \text{sech}\left(\frac{\xi}{W}\right) \tag{3.188}$$

which is the one soliton solution of the MKdV equation at $\tau=0$

$$\frac{\partial\phi^{(1)}}{\partial\tau}+D(\phi^{(1)})^2\frac{\partial\phi^{(1)}}{\partial\xi}+B\frac{\partial^3\phi^{(1)}}{\partial\xi^3}=0 \tag{3.189}$$

where $A_m=\sqrt{6U/D}$ gives the amplitude and $W=\sqrt{B/U}$ gives the width of the soliton (U being the Mach number).

To obtain the solution of the DMKdV Eq. (3.187), consider the linear operator as

$$\mathcal{L}[\psi(\xi,\tau;p)]=\frac{\partial w(\xi,\tau;p)}{\partial\tau} \tag{3.190}$$

which has the property $\mathcal{L}[\phi_0^{(1)}(\xi,\tau)]=0$, where $\phi_0^{(1)}(\xi,\tau)$ is the initial guess. Also, assume that $\mathcal{L}^{-1}$ exists and is defined as

$$\mathcal{L}^{-1}=\int_0^\tau(.)d\tau$$

Define the nonlinear operator as

$$\mathcal{N}[w(\xi,\tau;p)]=\frac{\partial w(\xi,\tau;p)}{\partial\tau}+Dw^2(\xi,\tau;p)\frac{\partial w(\xi,\tau;p)}{\partial\xi}+B\frac{\partial^3 w(\xi,\tau;p)}{\partial\xi^3}+Cw(\xi,\tau;p) \tag{3.191}$$

Thus, the zeroth-order deformation equation takes the form

$$(1-p)\mathcal{L}[w(\xi,\tau;p)-\phi_0^{(1)}(\xi,\tau)]=phH(\xi,\tau)\mathcal{N}[w(\xi,\tau;p)] \tag{3.192}$$

Equation (3.192) gives $w(\xi,\tau;0)=\phi_0^{(1)}(\xi,\tau)$ and $w(\xi,\tau;1)=\phi^{(1)}(\xi,\tau)$ for $p=0$ and $p=1$, respectively. The kth-order deformation equation can be written as

$$\mathcal{L}[\phi_k^{(1)}(\xi,\tau)-\chi_k\phi_{k-1}^{(1)}(\xi,\tau)]=hH(\xi,\tau)\mathscr{R}_k(\vec{\phi}_{k-1}^{(1)}) \tag{3.193}$$

where

$$\vec{\phi}_n^{(1)}=\{\phi_0^{(1)},\phi_1^{(1)},\phi_2^{(1)},\ldots,\phi_n^{(1)}\} \tag{3.194}$$

and

$$\begin{aligned}\mathscr{R}_k(\vec{\phi}_{k-1}^{(1)})={}&\frac{\partial\phi_{k-1}^{(1)}(\xi,\tau)}{\partial\tau}+D\sum_{n=0}^{k-1}\left(\sum_{j=0}^{n}\phi_j^{(1)}(\xi,\tau)\phi_{n-j}^{(1)}(\xi,\tau)\right)\\&\times\frac{\partial\phi_{k-1-n}^{(1)}(\xi,\tau)}{\partial\xi}+B\frac{\partial^3\phi_{k-1}^{(1)}(\xi,\tau)}{\partial\xi^3}+C\phi_{k-1}^{(1)}(\xi,\tau)\end{aligned} \tag{3.195}$$

Taking $H(\xi,\tau)=1$, Eq. (3.193) becomes

$$\phi_k^{(1)}(\xi,\tau)=\chi_k\phi_{k-1}^{(1)}(\xi,\tau)]+h\mathcal{L}^{-1}[\mathscr{R}_k(\vec{\phi}_{k-1}^{(1)})] \tag{3.196}$$

The approximate solution of the DMKdV Eq. (3.187) by HAM is

$$\begin{aligned}\phi^{(1)}={}&\frac{1}{2w^6}\Bigg(A_mh\tau\,\mathrm{sech}^2\left(\frac{\xi}{w}\right)\Bigg(-2h\tau\,\mathrm{sech}^6\left(\frac{\xi}{w}\right)\\&\times\left(A_m^2D^2w^4-22A_mBDw^2+136B^2\right)+12h\tau\tanh^2\left(\frac{\xi}{w}\right)\mathrm{sech}^4\left(\frac{\xi}{w}\right)\\&\times\left(A_m^2D^2w^4-32A_mBDw^2+240B^2\right)\\&-2\tanh\left(\frac{\xi}{w}\right)\mathrm{sech}^2\left(\frac{\xi}{w}\right)\Bigg(16Bh\tau\tanh^3\left(\frac{\xi}{w}\right)\left(57B-5A_mDw^2\right)\\&+A_mDw^5\left(3Ch\tau+2h+2\right)-16Bw^3\left(Ch\tau+h+1\right)\Bigg)\\&+\Bigg(\Bigg(\left(Cw^3-8B\tanh^3\left(\frac{\xi}{w}\right)\right)\Bigg(w^3(h(C\tau+2)+2)\\&-8Bh\tau\tanh^3\left(\frac{\xi}{w}\right)\Bigg)\Bigg)\Bigg)\Bigg)+\cdots\end{aligned}$$

3.6.3 *Approximate solution of Burgers equation*

We consider the standard Burgers equation as

$$\phi_\tau+\phi\phi_\xi-\phi_{\xi\xi}=0 \tag{3.197}$$

with initial solution $\phi_0(\xi,\tau)=\frac{1}{2}-\frac{1}{2}\tanh\frac{1}{4}(\xi)$ at $\tau=0$.

Here, we choose

$$\mathcal{L}\phi(\xi,\tau;p) = \frac{\partial\phi(\xi,\tau;p)}{\partial\tau}$$

as the auxiliary linear operator satisfying

$$\mathcal{L}[c] = 0$$

where c is a constant.

We consider auxiliary function

$$H(\xi,\tau) = 1,$$

zeroth-order deformation problem

$$\begin{aligned}
&(1-p)\mathcal{L}[\phi(\xi,\tau;p) - \phi_0(\xi,\tau)] = phH(\xi,\tau)\mathcal{N}[\phi(\xi,\tau;p)] \\
&\phi_0(\xi,\tau) =) = \frac{1}{2} - \frac{1}{2}\tanh\frac{1}{4}(\xi), \qquad (3.198)\\
&\mathcal{N}[\phi(\xi,\tau;p)] = \frac{\partial\phi(\xi,\tau;p)}{\partial\tau} + \phi(\xi,\tau;p)\frac{\partial\phi(\xi,\tau;p)}{\partial\xi} - \frac{\partial^2\phi(\xi,\tau;p)}{\partial\xi^2}
\end{aligned}$$

and the mth-order deformation problem as

$$\mathcal{L}[\phi_m(\xi,\tau) - \chi_m\phi_m(\xi,\tau)] = ph\left[\frac{\partial\phi_{m-1}}{\partial\tau} + \sum_{i=0}^{m-1}\phi_i\frac{\partial\phi_{m-1-i}}{\partial\xi} - \frac{\partial^2\phi_{m-1}}{\partial\xi^2}\right] \tag{3.199}$$

$$\phi_m(\xi,0) = 0 \quad (m \geq 1) \tag{3.200}$$

We can use software Mathematica for solving the set of linear Eq. (3.199) with condition (3.200). It is found that the solution in a series form is given by

$$\phi(\xi,\tau) = \frac{1}{2} - \frac{-1+e^{x/2}}{2(1+e^{x/2})} - \frac{e^{\xi/2}ht\log[e](2-\log[e]+e^{x/2}\log[e])}{4(1+e^{\xi/2})^3} + \cdots \tag{3.201}$$

3.7 Homotopy Perturbation Method (HPM)

In 1998, Liao's early idea to construct the one-parameter family of equations was introduced by He's so-called homotopy perturbation method (HPM), which is only a special case of the homotopy analysis method. But this homotopy perturbation method is almost the same as Liao's early one-parameter family equation which can be found in the work of

Liao. The author has proved that these two methods lead to the same solutions of the considered equation for high-order approximations. This is also mainly because of the Taylor series properties. In He's idea, the modification of HAM is done and a "homotopy perturbation method" is used.

Consider the following nonlinear differential equation:

$$N[y(\vec{x}, t)] = 0$$

In this equation, N is a nonlinear operator, $\vec{x}$ is a vector of spatial variables, t stands for time, and y is an unknown function, respectively. The zeroth-order deformation equation was proposed by He

$$(1 - p)\mathcal{L}[w(\vec{x}, t; p) - y_0(\vec{x}, t)] + \mathcal{N}[w(\vec{x}, t; p)] = 0 \tag{3.202}$$

where $p \in [0, 1]$ is the embedding parameter, $\mathcal{L}$ is a linear auxiliary operator, and $y_0(\vec{x}, t)$ is an initial estimate of $y(\vec{x}, t)$ such that $\mathcal{L}[y_0(\vec{x}, t)] = 0$. For $p = 0$, Eq. (3.202) changes to

$$w(\vec{x}, t; 0) = y_0(\vec{x}, t) \tag{3.203}$$

and for $p = 1$, Eq. (3.202) gives

$$w(\vec{x}, t; 1) = y(\vec{x}, t) \tag{3.204}$$

As the homotopy parameter p grows continuously in $[0, 1]$, $w(\vec{x}, t; p)$ changes from the initial guess $y_0(\vec{x}, t)$ to the solution $y(\vec{x}, t)$.

Taylor's series expansion of $w(\vec{x}, t; p)$ about the point $p = 0$ gives

$$w(x, t; p) = y_0(x, t) + \sum_{k=1}^{\infty} y_k(\vec{x}, t)p^k \tag{3.205}$$

where

$$y_k(\vec{x}, t) = \frac{1}{k!} \frac{\partial^k w(\vec{x}, t; p)}{\partial p^k}\bigg|_{p=0} \tag{3.206}$$

The series (3.172) converges for $p = 1$ when $H(x, t)$, h, $\mathcal{L}$, and $y_0(x, t)$ are chosen correctly. The approximations provided by the "homotopy perturbation method" are indeed a special case of those provided by the homotopy analysis method when $H(\vec{x}, t) = 1$ and $h = -1$ and the same starting guess and auxiliary linear operator are selected.

3.7.1 *Solution of KdV equation*

Considering the KdV equation

$$\phi_\tau + \alpha\phi\phi_\xi + \phi_{\xi\xi\xi} = 0 \tag{3.207}$$

with initial condition $\phi(\xi, 0) = f(\xi)$, where α is constant. First, we construct a homotopy as follows:

$$(1-p)\left[\dot{Y} - \dot{\phi_0}\right] + p\left[\dot{Y} + \alpha YY' + Y'''\right] = 0 \tag{3.208}$$

where $\dot{Y} = \frac{\partial Y}{\partial \tau}$, $Y' = \frac{\partial Y}{\partial \xi}$, $Y''' = \frac{\partial^3 Y}{\partial \xi^3}$, and $p \in [0,1]$. With the initial approximation

$$Y_0 = \phi_0 = \frac{3\lambda}{\alpha}\left(1 - \tanh^2\left(\frac{\sqrt{\lambda}}{2}\xi\right)\right)$$

suppose the solution of Eq. (3.208) has the form

$$\begin{aligned} Y &= Y_0 + pY_1 + p^2Y_2 + p^3Y_3 + \cdots \\ &= \sum_{n=0}^{\infty} p^n Y_n(\xi, \tau) \end{aligned} \tag{3.209}$$

Substituting Eq. (3.209) into Eq. (3.208) and arranging the coefficients of "p" powers, we have

$$p^0 : \dot{Y_0} - \dot{\phi_0} = 0 \tag{3.210}$$

$$p^1 : \dot{Y_1} + \dot{\phi_0} + \alpha Y_0 Y_0' + Y_0''' = 0 \tag{3.211}$$

$$p^2 : \dot{Y_2} + Y_1 + \alpha Y_0 Y_1' + \alpha Y_1 Y_0' + Y_1''' = 0 \tag{3.212}$$

$$p^3 : \dot{Y_3} + Y_1 + \alpha Y_0 Y_2' + \alpha Y_1 Y_1' + \alpha Y_2 Y_0' + Y_2''' = 0 \tag{3.213}$$

$$\vdots$$

and finally using Mathematica, the solutions of the equation can be obtained as follows:

$$Y_0 = \frac{3\lambda}{\alpha} - \frac{3\lambda}{\alpha}\tanh^2\left(\frac{\sqrt{\lambda}}{2}x\right) \tag{3.214}$$

$$Y_1 = \frac{3\tau\lambda^{\frac{5}{2}}}{\alpha}\,\mathrm{sech}^2\left(\frac{\sqrt{\lambda}}{2}x\right)\tanh\left(\frac{\sqrt{\lambda}}{2}x\right) \tag{3.215}$$

$$Y_2 = \frac{3\tau^2\lambda^4}{4\alpha}\,\text{sech}^2\left(\frac{\sqrt{\lambda}}{2}x\right)\left(-2+\cosh\left(x\sqrt{\lambda}\right)\right) \tag{3.216}$$

$$\vdots$$

The above terms of the series (Eq. (3.209)) could be calculated. When the series (Eq. (3.209)) is considered with the terms (Eqs. (3.210)–(3.213)) and supposing $p = 1$, an approximate series solution of the considered KdV equation is obtained as follows:

$$\begin{aligned}\phi(\xi,\tau) &= \frac{3\lambda}{\alpha} - \frac{3\lambda}{\alpha}\tanh^2\left(\frac{\sqrt{\lambda}}{2}x\right) + \frac{3\tau\lambda^{\frac{5}{2}}}{\alpha}\,\text{sech}^2\left(\frac{\sqrt{\lambda}}{2}x\right)\tanh\left(\frac{\sqrt{\lambda}}{2}x\right)\\ &\quad + \frac{3\tau^2\lambda^4}{4\alpha}\,\text{sech}^2\left(\frac{\sqrt{\lambda}}{2}x\right)\left(-2+\cosh\left(x\sqrt{\lambda}\right)\right) + \cdots\end{aligned} \tag{3.217}$$

3.7.2 *Solution of mKdV equation*

Considering the modified KdV equation

$$\phi_\tau + 6\phi^2\phi_\xi + \phi_{\xi\xi\xi} = 0 \tag{3.218}$$

with initial condition $\phi(\xi,0) = f(\xi)$, where α is constant. For the Homotopy perturbation method's implementation, first, we construct a homotopy as follows:

$$(1-p)\left[\dot{Y} - \dot{\phi_0}\right] + p\left[\dot{Y} + 6Y^2Y' + Y'''\right] = 0 \tag{3.219}$$

where $\dot{Y} = \frac{\partial Y}{\partial \tau}, Y' = \frac{\partial Y}{\partial \xi}, Y''' = \frac{\partial^3 Y}{\partial \xi^3}$, and $p \in [0,1]$. With the initial approximation

$$Y_0 = \phi_0 = \sqrt{c}\,\text{sech}\left[k + \sqrt{c}\xi\right]$$

suppose the solution of Eq. (3.219) has the form

$$\begin{aligned}Y &= Y_0 + pY_1 + p^2Y_2 + p^3Y_3 + \cdots\\ &= \sum_{n=0}^{\infty} p^n Y_n(\xi,\tau)\end{aligned} \tag{3.220}$$

Substituting Eq. (3.220) into Eq. (3.219) and arranging the coefficients of "p" powers, we have

$$p^0 : \dot{Y}_0 - \dot{\phi}_0 = 0 \tag{3.221}$$

$$p^1 : \dot{Y}_1 + \dot{\phi}_0 + 6Y_0^2 Y_0' + Y_0''' = 0 \tag{3.222}$$

$$p^2 : \dot{Y}_2 + 12Y_0 Y_1 Y_0' + 6Y_0^2 Y_1' + Y_1''' = 0 \tag{3.223}$$

$$p^3 : \dot{Y}_3 + 12Y_0 Y_2 Y_0' + 6Y_1^2 Y_0' + 12Y_0 Y_1 Y_1' + Y_2''' = 0 \tag{3.224}$$

$$\vdots$$

and finally using Mathematica, the solutions of the equation can be obtained as follows:

$$\begin{aligned}\phi(\xi,\tau) &= \sqrt{c}\,\text{sech}(k+\sqrt{c}\xi) + \frac{1}{4}c^{\frac{7}{2}}\tau^2(-3+\cosh(2(k+\sqrt{c}\xi)))\\ &\times \text{sech}^3(k+\sqrt{c}\xi) + \frac{1}{24}c^5\tau^3\,\text{sech}^4(k+\sqrt{c}\xi)\\ &\times(-23\sinh(k+\sqrt{c}\xi)+\sinh(3(k+\sqrt{c}\xi))) + c^2\tau\,\text{sech}(k+\sqrt{c}\xi)\\ &\times \tanh(k+\sqrt{c}\xi) + \cdots\end{aligned}$$

3.8 Variational Iteration Method (VIM)

He was the first to suggest the variational iteration method (VIM). With this approach, differential equations are handled without the need to make constrictive assumptions that could alter the solutions' physical structure. A broad class of nonlinear problems with approximations that quickly converge to precise solutions is suggested by the VIM.

We consider an ODE

$$Ly + N(y) = f(x) \tag{3.225}$$

where L and N are linear and nonlinear differential operators, respectively, and $f(x)$ is an given inhomogeneous term. In the VIM, a correction functional of Eq. (3.225) is defined in the following form:

$$y_{n+1}(x) = y_n(x) + \int_0^x \lambda(\tau)(Ly_n(\tau) + N(\tilde{y_n}(\tau)) - f(\tau))d\tau \tag{3.226}$$

where $\lambda(\tau)$ is a general Lagrange multiplier, which can be identified using the variational theory. Furthermore, $y_n(x)$ is the nth approximation of $y(x)$ and $\tilde{y_n}(x)$ is considered as a restricted variation, i.e., $\delta\tilde{y_n}(x) = 0$.

By imposing the variation and by considering the restricted variation, Eq. (3.226) is reduced to

$$\begin{aligned}\delta y_{n+1}(x) &= \delta y_n(x) + \delta\left(\int_0^x \lambda(\tau) L y_n(\tau) d\tau\right) \\ &= \delta y_n(x) + \left[\lambda(\tau)\int_0^\tau L\delta y_n(\xi)d\xi\right]_{\tau=0}^{\tau=x} \\ &\quad - \int_0^x \lambda'(\tau)\left(\int_0^\tau L\delta y_n(\xi)d\xi\right)d\tau\end{aligned} \tag{3.227}$$

Obviously, in (3.227), we used the integration by parts, which is based on the following formula:

$$\int \lambda(\tau) y_n'(\tau) d\tau = \lambda(\tau) y_n(\tau) - \int \lambda'(\tau) y_n(\tau) d\tau \tag{3.228}$$

Now, by applying the stationary condition for (3.227), the optimum value of the Lagrange multiplier $\lambda(\tau)$ can be identified. Once $\lambda(\tau)$ is obtained, the solution of Eq. (4.57) can be determined by calculating the successive approximation $y_n(x),\ n = 0, 1, 2, \ldots$, using the formula

$$y_{n+1}(x) = y_n(x) + \int_0^x \lambda(\tau)(L y_n(\tau) + N(y_n(\tau)) - f(\tau))d\tau \tag{3.229}$$

where $y_0(x)$ is the starting function, which has to be prescribed by the user.

It is shown that the approximate solution $y_n(x)$ of the the exact solution $y(x)$ can be achieved using any selected function $y_0(x)$. Consequently, the approximate solution is given as the limit $y(x) = \lim_{n\to\infty} y_n(x)$. In other words, the correction functional (3.226) will give a sequence of approximations and the exact solution is obtained at the limit of the successive approximations. In general, it is difficult to calculate this limit. Consequently, an accurate solution can be obtained by considering a large value for n. This value depends on the interval I where a good approximation of the solution is desired.

Example

Let us consider the following IVP:

$$y'(x) + y(x)^2 = 0, \quad y(0) = 1 \tag{3.230}$$

The corresponding exact solution is

$$y(x) = (1+x)^{-1} = 1 - x + x^2 - x^3 + x^4 - x^5 + x^6 + \cdots \tag{3.231}$$

Here, we have

$$Ly = y', \quad N(y) = y^2, \quad f(x) = 0$$

To determine the Lagrange multiplier, we insert these expression into (3.226) and obtain

$$y_{n+1}(x) = y_n(x) + \int_0^x \lambda(\tau)\left(\frac{dy_n(\tau)}{d\tau} + \tilde{y_n(\tau)^2}\right) d\tau \tag{3.232}$$

Making the above correction functional stationary w.r.t. y_n, noting that $\delta\tilde{y_n}(x) = 0$ and $\delta y_n(0) = 0$, it follows with (3.227)

$$\begin{aligned}
\delta y_{n+1}(x) &= \delta y_n(x) + \delta\left(\int_0^x \lambda(\tau)\frac{dy_n(\tau)}{d\tau} d\tau\right) \\
&= \delta y_n(x) + \left[\lambda(\tau)\int_0^\tau \frac{d\delta y_n(\xi)}{d\xi} d\xi\right]_{\tau=0}^{\tau=x} \\
&\quad - \int_0^x \lambda'(\tau)\left(\int_0^\tau \frac{d\delta y_n(\xi)}{d\xi} d\xi\right) d\tau. \\
&= \delta y_n(x) + \lambda(\tau)\delta y_n(\tau)\mid_{\tau=x} - \int_0^x \lambda'(\tau)\delta y_n(\tau) d\tau \\
&= 0
\end{aligned}$$

Thus, we obtained the equations

$$1 + \lambda(x) = 0 \quad \text{and} \quad \lambda'(x) = 0 \tag{3.233}$$

Now, substitute the solution $\lambda = -1$ of (3.233) into (3.232). It results in the following successive iteration formula:

$$y_{n+1}(x) = y_n(x) - \int_0^x \left(\frac{dy_n(\tau)}{d\tau} + y_n(\tau)^2\right) d\tau \tag{3.234}$$

We have to choose a starting function $y_0(x)$, which satisfies the given initial condition $y(0) = 1$. Starting with $y_0(x) \equiv 1$, we compute the successive approximations

$$\begin{aligned}
y_0(x) &= 1 \\
y_1(x) &= 1 - x \\
y_2(x) &= 1 - x + x^2 - \frac{1}{3}x^3 \\
y_3(x) &= 1 - x + x^2 - \frac{1}{3}x^3 + \frac{2}{3}x^4 - \frac{1}{5}x^5 + \frac{1}{9}x^6 - \frac{1}{63}x^7
\end{aligned}$$

Comparing the iterates with Taylor series of the solution (see (3.231)), we see that in $y_3(x)$ the first four terms are correct.

In the above example, the linear operation $L = \frac{d}{dx}$. More generally, let us assume that $L = \frac{d^m}{dx^m}$, $m \geq 1$.

The corresponding optimal values of the Lagrange multipliers are given. It holds

$$\begin{aligned} &\lambda = -1, && \text{for } m = 1 \\ &\lambda = \tau - 1, && \text{for } m = 2 \\ &\lambda = \frac{(-1)^m}{(m-1)!}(\tau - 1), && \text{for } m \geq 1 \end{aligned} \tag{3.235}$$

Substituting (3.235) into the corresponding functional (3.226), we get the following iteration formula:

$$y_{n+1}(x) = y_n(x) + \int_0^x \frac{(-1)^m}{(m-1)!}(\tau - x)^{m-1}(Ly_n(\tau) + N(y_n(\tau)) - f(\tau))d\tau \tag{3.236}$$

where $y_0(x)$ must be given by the user.

3.8.1 *Solution of the KdV equation*

Let us consider the standard KdV equation

$$\phi_\tau + 6\phi\phi_\xi + \phi_{\xi\xi\xi} = 0 \tag{3.237}$$

with initial condition $\phi(\xi, 0) = 6\xi$. To obtain the solution of Eq. (3.237) by VIM, we construct the correction functional as

$$\begin{aligned} \phi_{n+1}(\xi, \tau) = \phi_n(\xi, \tau) + \int_0^\tau \lambda(t)\Bigg(&\frac{\partial \phi_n(\xi, t)}{\partial t} - 6\tilde{\phi_n}(\xi, t)\frac{\partial \tilde{\phi_n}(\xi, t)}{\partial \xi} \\ &+ \frac{\partial^3 \tilde{\phi_n}(\xi, t)}{\partial \xi^3}\Bigg) dt \end{aligned} \tag{3.238}$$

where λ is the general Lagrange multiplier whose optimal value is found using variational theory and ϕ_0 is an initial solution with or without unknown parameters. In the case of no unknown parameters, ϕ_0 should satisfy initial boundary conditions, and $\tilde{\phi_n}$ is the restricted variation (He 1999b), therefore, $\delta\tilde{\phi_n} = 0$. Taking variation to the both sides of Eq. (3.238) gives

$$\delta\phi_{n+1}(\xi,\tau) = \delta\phi_n(\xi,\tau) + \int_0^\tau \delta\lambda(t)\left(\frac{\partial\phi_n(\xi,t)}{\partial t} - 6\tilde{\phi}_n(\xi,t)\frac{\partial\tilde{\phi}_n(\xi,t)}{\partial\xi} + \frac{\partial^3\tilde{\phi}_n(\xi,t)}{\partial\xi^3}\right)dt$$

$$\Rightarrow \delta\phi_{n+1}(\xi,\tau) = \delta\phi_n(\xi,\tau) + \int_0^\tau \delta\lambda(t)\left(\frac{\partial\phi_n(\xi,t)}{\partial t}\right)dt$$

Simplifying by integration by parts, we get

$$\delta\phi_{n+1}(\xi,\tau) = \delta\phi_n(\xi,\tau) + \delta\lambda(t) - \int_0^\tau \lambda'(t)\delta\phi_n(\xi,t)dt$$

Making the correction functional stationary $\delta\phi_{n+1}(\xi,\tau) = 0$, we have

$$0 = \delta\phi_n(\xi,\tau) + \delta\lambda(t) - \int_0^\tau \lambda'(t)\delta\phi_n(\xi,t)dt$$

Therefore, the stationary points are $1+\lambda = 0$ and $\lambda' = 0$, and solving them we obtained $\lambda = -1$. Hence, the iterative formula becomes

$$\phi_{n+1}(\xi,\tau) = \phi_n(\xi,\tau) - \int_0^\tau\left(\frac{\partial\phi_n(\xi,t)}{\partial t} - 6\tilde{\phi}_n(\xi,t)\frac{\partial\tilde{\phi}_n(\xi,t)}{\partial\xi} + \frac{\partial^3\tilde{\phi}_n(\xi,t)}{\partial\xi^3}\right)dt \tag{3.239}$$

From the initial condition, we have $\phi_0 = 6\xi$ and at $n = 0$, Eq. (3.239) becomes

$$\phi_1(\xi,\tau) = \phi_0(\xi,\tau) - \int_0^\tau\left(\frac{\partial\phi_0(\xi,t)}{\partial t} - 6\tilde{\phi}_n(\xi,t)\frac{\partial\tilde{\phi}_0(\xi,t)}{\partial\xi} + \frac{\partial^3\tilde{\phi}_0(\xi,t)}{\partial\xi^3}\right)dt$$
$$= 6\xi(1+36\tau)$$

When $n = 1$,

$$\phi_2(\xi,\tau) = \phi_1(\xi,\tau) - \int_0^\tau\left(\frac{\partial\phi_1(\xi,t)}{\partial t} - 6\tilde{\phi}_n(\xi,t)\frac{\partial\tilde{\phi}_1(\xi,t)}{\partial\xi} + \frac{\partial^3\tilde{\phi}_1(\xi,t)}{\partial\xi^3}\right)dt$$
$$= 6\xi(1+36\tau+(36\tau)^2$$

Consequently, the successive approximations are obtained

$$\begin{aligned}\phi_1(\xi,\tau) &= 6\xi(1+36\tau)\\ \phi_2(\xi,\tau) &= 6\xi(1+36\tau+(36\tau)^2)\\ \phi_3(\xi,\tau) &= 6\xi(1+36\tau+(36\tau)^2+(36\tau)^3)\end{aligned}$$

Hence, the closed form and exact solution is given as

$$\begin{aligned}\phi(\xi,\tau) &= 6\xi\left(1+36\tau+(36\tau)^2+(36\tau)^3+(36\tau)^4+\cdots\right)\\ &= \frac{6\xi}{1-36\tau}\end{aligned} \tag{3.240}$$

3.8.2 *Solution of the Burgers equation*

Let us consider the standard Burgers equation

$$\phi_\tau+\phi\phi_\xi-\phi_{\xi\xi}=0 \tag{3.241}$$

with initial condition $\phi(\xi,0)=\xi$. To obtain the solution of Eq. (3.241) by VIM, we construct the correction functional as

$$\begin{aligned}\phi_{n+1}(\xi,\tau) &= \phi_n(\xi,\tau)\\ &\quad+\int_0^\tau \lambda(t)\left(\frac{\partial\phi_n(\xi,t)}{\partial t}+\tilde{\phi}_n(\xi,t)\frac{\partial\tilde{\phi}_n(\xi,t)}{\partial\xi}-\frac{\partial^2\tilde{\phi}_n(\xi,t)}{\partial\xi^2}\right)dt\end{aligned} \tag{3.242}$$

Proceeding as before, we find $\lambda=-1$. Substituting this value of the Lagrangian multiplier into functional (3.242) gives the iteration formula

$$\begin{aligned}\phi_{n+1}(\xi,\tau) &= \phi_n(\xi,\tau)\\ &\quad-\int_0^\tau \left(\frac{\partial\phi_n(\xi,t)}{\partial t}+\tilde{\phi}_n(\xi,t)\frac{\partial\tilde{\phi}_n(\xi,t)}{\partial\xi}-\frac{\partial^2\tilde{\phi}_n(\xi,t)}{\partial\xi^2}\right)dt\end{aligned} \tag{3.243}$$

Selecting the initial value $\phi_0(\xi,0)=\xi$ for $\phi_0(\xi,\tau)$ and using (3.243), we obtain the following successive approximations:

$$\begin{aligned}\phi_0(\xi,\tau) &= \xi\\ \phi_1(\xi,\tau) &= \xi(1-\tau)\\ \phi_2(\xi,\tau) &= \xi\left(1-\tau+\tau^2-\frac{1}{3}\tau^3\right)\end{aligned}$$

$$\phi_3(\xi,\tau) = \xi(1-\tau+\tau^2-\tau^3+\tau^4) + \text{other terms}$$

$$\vdots$$

$$\phi_n(\xi,\tau) = \xi(1-\tau+\tau^2-\tau^3+\tau^4-\tau^5+\cdots)$$

Recall that

$$\phi = \lim_{n\to\infty} \phi_n \tag{3.244}$$

Hence, the exact solution is

$$\phi(\xi,\tau) = \frac{\xi}{1+\tau}. \tag{3.245}$$

3.8.3 *Solution of the mKdV equation*

We now consider the mKdV equation

$$\phi_\tau + 6\phi^2\phi_\xi + \phi_{\xi\xi\xi} = 0 \tag{3.246}$$

with the initial condition $\phi(\xi,0) = mkcn[k\xi,m]$, where k is an arbitrary constant, cn is a doubly periodic function, and m is the modulus of Jacobi elliptic functions $(0 < m < 1)$.

To solve Eq. (3.246) by means of the variational iteration method, we construct a correction functional which reads as

$$\phi_{n+1}(\xi,\tau) = \phi_n(\xi,\tau) + \int_0^\tau \lambda(t)\left(\frac{\partial\phi_n(\xi,t)}{\partial t} + 6\tilde{\phi}_n^2(\xi,t)\frac{\partial\tilde{\phi}_n(\xi,t)}{\partial\xi} + \frac{\partial^3\tilde{\phi}_n(\xi,t)}{\partial\xi^3}\right)dt \tag{3.247}$$

where $\delta\tilde{\phi}_n$ is considered as a restricted variation. Its stationary conditions can be obtained as follows:

$$\lambda'(t) = 0 \tag{3.248}$$

$$1 + \lambda(t)|_{t=\tau} = 0 \tag{3.249}$$

where Eq. (3.248) is called Lagrange–Euler equation and Eq. (3.249) natural boundary condition.

The Lagrange multiplier, therefore, can be obtained as $\lambda = -1$, and the following variational iteration formula can be obtained as

$$\phi_{n+1}(\xi,\tau) = \phi_n(\xi,\tau) - \int_0^\tau \left(\frac{\partial\phi_n(\xi,t)}{\partial t} + 6\tilde{\phi}_n^2(\xi,t)\frac{\partial\tilde{\phi}_n(\xi,t)}{\partial\xi} + \frac{\partial^3\tilde{\phi}_n(\xi,t)}{\partial\xi^3}\right)dt \tag{3.250}$$

We start with an initial approximation $\phi(\xi,0) = mkcn[k\xi,m]$, by the above iteration (3.250), we can obtain directly the other components as

$$\begin{aligned}
\phi_1(\xi,\tau) &= mk^4 sn[k\xi,m]dn[k\xi,m](2m^2-1)\tau \\
\phi_2(\xi,\tau) &= \frac{1}{2}mk^7 cn[k\xi,m]\left(8m^6 sn^2[k\xi,m] - 8m^4 sn^2[k\xi,m]\right. \\
&\quad \left. + 2m^2 sn^2[k\xi,m] - 4m^4 + 4m^2 - 1\right)\tau^2 \\
\phi_3(\xi,\tau) &= \frac{1}{6}k^{10} msn[k\xi,m]dn[k\xi,m]\left(48m^8 sn^2[k\xi,m] - 72m^6\right. \\
&\quad \times sn^2[k\xi,m] - 32m^8 + 40m^6 + 36m^4 sn^2[k\xi,m] - 12m^4 \\
&\quad \left. - 6m^2 sn^2[k\xi,m] - 2m^2 + 1\right)\tau^3
\end{aligned}$$

and so on, in the same manner using the rest of components of the iteration formula (3.250) which were obtained by the Mathematica Package. Consequently, we have the solution of (3.246) in a series form

$$\begin{aligned}
\phi(\xi,\tau) &= mkcn[k\xi,m] + mk^4 sn[k\xi,m]dn[k\xi,m](2m^2-1)\tau\frac{1}{2}k^7 \\
&\quad \times mcn[k\xi,m]\left(8m^6 sn^2[k\xi,m] - 8m^4 sn^2[k\xi,m]\right. \\
&\quad \left. + 2m^2 sn^2[k\xi,m] - 4m^4 + 4m^2 - 1\right)\tau^2 + \frac{1}{24}k^{13} mcn[k\xi,m] \\
&\quad \times \left([384m^{12} - 768m^{10} + 576m^8 - 192m^6 + 24m^4]sn^4[k\xi,m]\right. \\
&\quad - [128m^{12} + 64m^{10} - 448m^8 + 416m^6 - 152m^4 + 20m^2] \\
&\quad \left. \times sn^2[k\xi,m] + 64m^{10} - 112m^8 + 64m^6 - 8m^4 - 4m^2 + 1\right)\tau^4 + \cdots
\end{aligned} \tag{3.251}$$

The solution $\phi(\xi,\tau)$ of (3.246) in the closed form is given by

$$\phi(\xi,\tau) = kmcn[k(\xi - k^2(2m^2-1)\tau),m] \tag{3.252}$$

References

[1] I. N. Sneddon, *Elements of Partial Differential Equations*, Dover Publications, INC, Mineola, New York, 2006.
[2] B. B. Kadomtsev and V. I. Petviashvili, *Sov. Phys. Dokl.* 15, 539–541 (1970).
[3] P. Chatterjee, K. Roy, and U. N. Ghosh, *Waves and Wave Interaction in Plasmas*, World Scientific Publishing Co. Ptc. Ltd, Singapore,, 2023.

[4] G. B. Whitham, *Linear and Nonlinear Waves*, John Wiley & Sons, New York, 1973.
[5] A.-M. Wazwaz, *Partial Differential Equation and Solitary Waves Theory*, Springer, Higher Education Press, Beijing, 2009.
[6] I. Shingareva and C. Lizarraga-Celaya, *Solving Nonlinear Partial Differential Equations with Maple and Mathematica*, SpringerWien, New York, 2011.
[7] L. Debnath, *Nonlinear Partial Differential Equations for Scientists and Engineers*, Birkhauser, Boston, 2011.
[8] G. Baumann, *Mathematica for Theoretical Physics*, Springer, New York, 2005.
[9] G. Adomian, A review of the decomposition method in applied mathematics, *J. Math. Anal. Appl.* 135, 501 (1988).
[10] G. Adomian and R. Rach, Noise terms in decomposition solution series, *Comput. Math. Appl.* 24, 61 (1992).
[11] G. Adomian, *Solving Frontier Problems of Physics: The Decomposition Method*, Kluwer, Boston, 1994.
[12] S. J. Liao, The proposed homotopy analysis technique for the solution of nonlinear problems, Ph.D. thesis, Shanghai Jiao Tong University, 1992.
[13] S. J. Liao, An approximate solution technique not depending on small parameters: A special example, *Int. J. Nonlinear Mech.* 30, 371 (1995).
[14] S. J. Liao, *Beyond Perturbation: Introduction to the Homotopy Analysis Method*, Chapman & Hall, CRC Press, Boca Raton, 2003.
[15] J. H. He, A new approach to nonlinear partial differential equations, *Commun. Nonlinear Sci. Numer. Simul.* 2, 203 (1997).
[16] J. H. He, Homotopy perturbation technique, *Comput. Math. Appl. Mech. Eng.* 178, 257 (1999).
[17] M. Sajid and T. Hayat, Comparison of HAM and HPM methods in nonlinear heat conduction and convection equations, *Nonlinear Anal. Real World Appl.* 9, 2296 (2008).
[18] J. H. He, Variational iteration method — a kind of non-linear analytical technique: Some examples, *Int. J. Non-Linear Mech.* 34(4), 699 (1999).
[19] J. H. He, Variational iteration method — some recent results and new interpretations, *J. Comput. Appl. Math.* 207(1), 3 (2007).
[20] J. H. He, Variational iteration method for autonomous ordinary differential systems, *Appl. Math. Comput.* 118 (2–3), 115 (2000).

Chapter 4

Integrability and Some Modern Methods

4.1 Introduction

Nonlinear partial differential equations (NLPDEs) explain various physical systems. The main governing equations of fluid dynamics, gravitation, and other areas of physics and engineering are modeled by NLPDEs. They are challenging to study since there are no universal methods for solving these equations, and often, each equation needs to be approached as a distinct problem. It is customary to describe the differences between a linear and nonlinear PDE in terms of the characteristics of the operator that defines the PDE. Over the past few decades, there has been a growing interest among mathematicians in issues related to the behavior of partial differential space-time system solutions, where the highest-order terms occur linearly with modest coefficient requirements. These equations are frequently used for studying nonlinear physical phenomena in mathematics, physics, chemistry, biology, and engineering.

4.2 Integrability

Dynamical systems can be broadly categorized into two groups. Systems that display chaotic behavior, or whose solutions are highly dependent on the original data, are found in the first class. These systems are typically not explicitly solvable in terms of "elementary functions" as a result of their chaotic nature.

Equations that are algebraically entirely integrable are found in the other class of systems. It makes sense to describe systems that prevent chaos. When discussing ODEs and PDEs, the idea of integrability refers to the presence of solutions and the potential for locating them in a closed form or by specific techniques. It's crucial to remember that not every PDE and every ODE can be solved in closed form or that they are integrable.

Numerical methods or other techniques may be necessary to obtain answers for many equations in the complicated and specialized field of integrability investigation. The idea of integrability has applications in many fields of physics and engineering and offers a useful foundation for comprehending the behavior of a specific differential equation.

A nonlinear evolution equation's integrability is a crucial component. To verify that every evolution equation is integrable, there is no universal method. The existence of a Lax pair, the Bäcklund transformation (BT), and the Painlevé analysis of an evolution equation guarantee the complete integrability of the evolution equation. The Lax pair of the aforementioned evolution equation is completely necessary for integrability via the Darboux transformation (DT). However, there is no uniform way of building the Lax pair of the evolution equation. In order to solve the nonlinear evolution equation, one important subfield is the study of the existence of Lax pairs and the Bäcklund transformation (BT). Nonetheless, the inverse scattering transform (IST) can be used to address initial value problems for a specific class of NLPDEs. The IST method's cornerstone, Lax pairs, is essential to the discovery of conservation laws. The NLPDE is typically referred to as an evolution equation if it is first-order in time. Many physical problems involve integrable NLPDEs.

It is possible to substitute a system of linear PDEs in an auxiliary function that requires the original NLPDE to hold in order to make sense of a fully integrable NLPDE, like the KdV equation. The term "Lax pair" refers to that system of compatible linear PDEs. A pair of operators that, if they exist, suggest that a specific evolution equation corresponds; this is known as a Lax pair. The study of numerous complex events in various scientific and nonlinear engineering disciplines has greatly benefited from the precise solutions of nonlinear evolution equations (NLEEs) and their generalized versions.

The KdV equation with time-dependent coefficients and other NLEEs are known to accept one- or two-soliton solutions, respectively, but not N-soliton solutions (for $N > 2$) while using Hirota's method. Examples of these equations are the ninth-order KdV equation and the sixth-order Boussinesq equation. While an equation with three soliton solutions also has N-soliton solutions (for $N > 3$), it is said to be completely integrable if it also contains two soliton solutions. An equation does not necessarily indicate that it has three soliton solutions. There are multiple definitions or requirements for evolution equations to be completely integrable, including the presence of higher-order symmetries, an infinite number of conservation laws, and N-soliton solutions of any order.

A technique for locating solitons for a PDE of any order is the IST. The PDE is completely integrable, or completely solved, if it can be achieved. The complete integrability of a PDE can be predicted by the existence of an infinite number of conservation laws, despite the fact that applying the IST to a PDE can be a very difficult task. For instance, the Burgers' equation is a C-integrable equation with infinitely many symmetries and only one conserved density that may be linked to the linear heat equation via the Hopf-Cole transformation.

Note: For an NLPDE, here we mean a PDE that is not linear.

4.2.1 *Integrability of ODEs*

The integrability of an ODE is a clear concept. It is based on the existence of sufficiently many well-behaved first integrals. These are also called constants of motion. The theory can be studied by using (i) Hamiltonian formalism, (ii) the Arnold–Liouville theorem, (iii) action-angle variables, etc.

4.2.2 *Integrability of PDEs*

There is no universally accepted definition of integrability for studying the integrability of PDEs. Few systems are partially integrable, and few systems are completely integrable. It is well known that the phase space of an NLPDE is infinite-dimensional. But the existence of infinitely many first integrals may not be enough for completely integrable NLPDEs. So we have to focus on the solution generation technique. Solitons, a special type of nonlinear wave, are an important feature of an NLPDE. These solitary nonlinear waves preserve their shape in their evolution. As an example, the KdV equation has an infinite conservation property.

4.3 Painlevé Analysis

Painlevé analysis is a powerful and popular mathematical technique for investigating the integrability of nonlinear differential equations. This method, named after the French mathematician Paul Painlevé (1863–1933), focuses on identifying equations with the Painlevé property, which ensures that their solutions do not contain movable crucial singularities other than poles. This characteristic is critical for differential equation integration and analytical solutions.

The origins of Painlevé analysis trace back to the late 19th and early 20th centuries, when Paul Painlevé, along with his contemporaries, sought to classify second-order ODEs that are solvable in terms of known functions. Their work culminated in the identification of six new transcendental functions, known as the *Painlevé transcendents*, which cannot be expressed in terms of elementary functions but have well-defined, predictable singularities.

In modern research, Painlevé analysis serves as a fundamental tool for distinguishing integrable systems from non-integrable ones. By applying this technique, we can reveal the underlying structure of differential equations, identify special solutions, and explore the interplay between symmetry, singularity structure, and integrability. Painlevé analysis involves several steps, including the transformation of the original differential equation into a form suitable for singularity analysis, the examination of the series expansions around potential singular points, and the identification of any constraints necessary for the Painlevé property to hold. The insights gained from this process not only deepen our understanding of the mathematical properties of differential equations but also have practical applications in several fields. As we delve deeper into the specifics of this analysis, we will uncover the rigorous criteria that define the Painlevé property and explore various examples that demonstrate its utility in solving complex differential equations.

4.3.1 *History*

Painlevé analysis traces its origins to the late 19th and early 20th centuries, rooted in the efforts of mathematicians seeking to classify and solve nonlinear differential equations. The story begins with the work of Augustin–Louis Cauchy and others in the 19th century, who developed the theory of complex functions and the study of singularities in differential equations. Building on these early developments, mathematicians such as Émile Picard and Jules Henri Poincaré investigated the nature of differential equations' solutions, particularly focusing on their singularities. Then Paul Painlevé made significant strides in understanding the solutions of second-order ODEs. His work aimed to classify these equations based on the nature of their singularities. Painlevé introduced the concept of movable singularities, which are singularities whose locations depend on the initial conditions of the solution. He focused on identifying those differential equations whose solutions do not exhibit movable critical singularities (branch points or

Figure 4.1. Paul Painlevé.

essential singularities), other than poles. Painlevé systematic study led to the identification of six new classes of transcendental functions, now known as the Painlevé transcendents. These functions cannot be expressed in terms of elementary functions and have well-defined singularity structures. The equations defining these functions are now referred to as the Painlevé equations (PI to PVI).

In the early 20th century, further formalization and development of Painlevé's ideas were made. Researchers such as Richard Fuchs and Lazarus Fuchs made important contributions to the theory of singular points and the criteria for classifying differential equations based on their singularities. The development of methods for transforming differential equations into forms is amenable to Painlevé analysis.

In the mid-20th century, the resurgence of interest in integrable systems, driven by advances in theoretical physics and applied mathematics, brought Painlevé analysis back into the spotlight. The discovery of solitons and the development of the IST method highlighted the importance of integrability in nonlinear systems, further emphasizing the relevance of Painlevé analysis.

Today, Painlevé analysis is a standard tool in the study of nonlinear differential equations. It plays a crucial role in identifying integrable systems, understanding their solution structures, and exploring their applications in various scientific fields. The technique has been extended to PDEs and discrete systems, broadening its scope and impact. Painlevé analysis continues to be a vibrant area of research, with ongoing efforts to classify new integrable systems, explore the properties of Painlevé transcendents, and apply these insights to physical problems ranging from fluid dynamics to quantum mechanics and general relativity. The historical journey of

Painlevé analysis, from its inception to its current status, underscores its significance as a fundamental tool in mathematical analysis and its enduring influence on the study of differential equations.

4.3.2 *Painlevé transcendents*

Painlevé transcendents are special solutions to a set of six nonlinear second-order ODEs known as the *Painlevé equations* and developed in the early 20th century. The Painlevé equations are distinguished by the fact that their solutions are free from movable critical singularities other than poles, making them particularly important in the study of integrable systems and nonlinear dynamics.

4.3.3 *Painlevé equations*

The six Painlevé equations (PIs to PVIs) are given as follows:

Painlevé I (PI):

$$\frac{d^2u}{dx^2} = 6u^2 + x$$

This equation appears in contexts such as the description of special solutions in statistical mechanics and quantum gravity.

Painlevé II (PII):

$$\frac{d^2u}{dx^2} = 2u^3 + xu + \beta$$

PII arises in models of random matrices and certain types of nonlinear waves.

Painlevé III (PIII):

$$u'' = \frac{u'^2}{u} - \frac{u'}{x} + \frac{\alpha u^2 + \gamma u^3}{4x^2} + \frac{\beta}{4x} + \frac{\delta}{4u}$$

This equation is found in studies of self-dual Einstein spaces and in problems related to heat conduction.

Painlevé IV (PIV):

$$u'' = \frac{u'^2}{2u} + \frac{3}{2}u^3 + 4xu^2 + 2x^2u - 2\alpha u + \frac{\beta}{u}$$

PIV appears in various physical models, including the dynamics of certain fluid flows.

Painlevé V (PV):

$$u'' = \left[\frac{1}{2u} + \frac{1}{u-1}\right]u'^2 - \frac{u'}{x} + \frac{(u-1)^2}{x^2}\left[\alpha u + \frac{\beta}{u}\right] + \gamma\frac{u}{x} + \delta\frac{u(u+1)}{u-1}$$

PV is relevant in studies of correlation functions in random matrix theory and statistical mechanics.

Painlevé VI (PVI):

$$u'' = \frac{1}{2}\left[\frac{1}{u} + \frac{1}{u-1} + \frac{1}{u-x}\right]u'^2 - \left[\frac{1}{x} + \frac{1}{x-1} + \frac{1}{u-x}\right]u'$$
$$+\frac{u(u-1)(u-x)}{x^2(x-1)^2}\left[\alpha + \beta\frac{x}{u^2} + \gamma\frac{x-1}{(u-1)^2} + \delta\frac{x(x-1)}{(u-x)^2}\right]$$

In which $\alpha, \beta, \gamma, \delta$ are arbitrary complex parameters. Their only singularities are movable poles with, in addition, three fixed critical singularities for PVI, located at $x = \infty, 0, 1$. PVI is encountered in connection with isomonodromic deformations and in problems of conformal field theory.

4.4 Series Solutions

Before looking into Painlevé analysis, let us get introduced to the basic series solution of linear ODEs.

4.4.1 *Ordinary point and singular point*

Consider the second-order homogeneous linear differential equation

$$a_0(x)\frac{d^2y}{dx^2} + a_1(x)\frac{dy}{dx} + a_2(x)y = 0 \tag{4.1}$$

A point $x = x_0$ is called an *ordinary point* of the differential Eq. (4.1) if both of the functions $P_1(x) = \frac{a_1(x)}{a_0(x)}$ and $P_2(x) = \frac{a_2(x)}{a_0(x)}$ are analytic at x_0. If either or both of these functions is not analytic at x_0, then x_0 is called a *singular point* of the differential Eq. (4.1).

4.4.2 *Regular singular point and irregular singular point*

If the functions defined by the products $(x - x_0)P_1(x)$ and $(x - x_0)^2P_2(x)$ are both analytic at x_0, then x_0 is called a *regular singular point* of the differential Eq. (4.1). If either or both of the functions defined by the products $(x - x_0)P_1(x)$ and $(x - x_0)^2P_2(x)$ is not analytic at x_0, then x_0 is called an *irregular singular point* of (4.1).

Example: Consider the differential equation

$$(x-1)\frac{d^2y}{dx^2}+x\frac{dy}{dx}+\frac{1}{x}y=0$$

Here $P_1(x)=\frac{x}{x-1}$, $P_2(x)=\frac{1}{x(x-1)}$. The function $P_2(x)$ is analytic except at $x=0$, and $x=1$ which are singular points of the differential equation under consideration. All other points are ordinary points.

4.4.3 *Polynomial coefficients*

We will be primarily interested in the case when the coefficients $a_0(x), a_1(x)$, and $a_2(x)$ in (4.1) are polynomial functions with no common factors. A polynomial function is analytic at any value of x, and a rational function is analytic except at points where its denominator is zero. Thus, we get

$$P(x)=\frac{a_1(x)}{a_0(x)} \quad \text{and} \quad Q(x)=\frac{a_2(x)}{a_0(x)}$$

A point $x=x_0$ is an ordinary point of (4.1) if $a_0(x_0)\neq 0$, whereas $x=x_0$ is a singular point of (4.1) if $a_0(x_0)=0$, that means if at least one of the rational functions $P(x)=\frac{a_1(x)}{a_0(x)}$ and $Q(x)=\frac{a_2(x)}{a_0(x)}$ in the standard form fails to be analytic at that point. But since $a_0(x)$ is a polynomial and x_0 is one of its zeros, it follows from the factor theorem of algebra that $x-x_0$ is a factor of $a_0(x)$. This means that after $\frac{a_1(x)}{a_0(x)}$ and $\frac{a_2(x)}{a_0(x)}$ are reduced to lowest terms, the factor $x-x_0$ must remain, to some positive integer power, in one or both denominators. Now suppose that $x=x_0$ is a singular point of (4.1) but both the functions defined by the products $p(x)=(x-x_0)P(x)$ and $q(x)=(x-x_0)^2Q(x)$ are analytic at x_0. We are led to the conclusion that multiplying $P(x)$ by $x-x_0$ and $Q(x)$ by $(x-x_0)^2$ has the effect (through cancellation) that $(x-x_0)$ no longer appears in either denominator. We can now determine whether x_0 is regular by a quick visual check of denominators.

If $(x-x_0)$ appears at most to the first power in the denominator of $P(x)$ and at most to the second power in the denominator of $Q(x)$, then $x=x_0$ is a regular singular point.

Moreover, observe that if $x=x_0$ is a regular singular point and we multiply the standard form by $(x-x_0)^2$, then the original differential equation can be put into the form

$$(x-x_0)^2y''+(x-x_0)p(x)y'+q(x)y=0$$

where p and q are analytic at $x=x_0$.

Example (Ordinary points):

(a) The following is the homogeneous linear second-order differential equation with constant coefficients,

$$y'' + y = 0, \quad \text{and} \quad y'' + 3y' + 2y = 0$$

have no singular points. In other words, every finite value of x is an ordinary point of such equations.

(b) Every finite value of x is an ordinary point of the differential equation

$$y'' + e^x y' + (\sin x)y = 0$$

Specifically $x = 0$ is an ordinary point of the differential equation, because both e^x and $\sin x$ are analytic at this point.

Example (Classification of Singular points): It should be clear that $x = 2$ and $x = -2$ are singular points of

$$(x^2 - 4)^2 y'' + 3(x - 2)y' + 5y = 0$$

After dividing the equation by $\left(x^2 - 4\right)^2 = (x - 2)^2(x + 2)^2$ and reducing the coefficients to lowest terms, we find that

$$P(x) = \frac{3}{(x - 2)(x + 2)^2}, \quad Q(x) = \frac{5}{(x - 2)^2(x + 2)^2}$$

We now test $P(x)$ and $Q(x)$ at each singular point.

For $x = 2$ to be a regular singular point, the factor $x - 2$ can appear at most to the first power in the denominator of $P(x)$ and at most to the second power in the denominator of $Q(x)$. A check of the denominators of $P(x)$ and $Q(x)$ shows that both these conditions are satisfied, so $x = 2$ is a regular singular point. Alternatively, we are led to the same conclusion by noting that both rational functions

$$p(x) = (x - 2)P(x) = \frac{3}{(x + 2)^2}, \quad q(x) = (x - 2)^2 Q(x) = \frac{5}{(x + 2)^2}$$

are analytic at $x = 2$.

Now, since the factor $x - (-2) = x + 2$ appears to be the second power in the denominator of $P(x)$, we can conclude immediately that $x = -2$ is an irregular singular point of the equation. This also follows from the fact that

$$p(x) = (x + 2)P(x) = \frac{3}{(x - 2)(x + 2)}$$

is not analytic at $x = -2$.

In this example, note that since $x = 2$ is a regular singular point, the original equation can be written as

$$(x-2)^2 y'' + (x-2)\frac{3}{(x+2)^2}y' + \frac{5}{(x+2)^2}y = 0$$

As another example, we can see that $x = 0$ is an irregular singular point of $x^3y'' - 2xy' + 8y = 0$ by inspection of the denominators of $P(x) = \frac{-2}{x^2}$ and $Q(x) = \frac{8}{x^3}$. On the other hand, $(x = 0)$ is a regular singular point of $xy'' - 2xy' + 8y = 0$, since $(x = 0)$ and $(x-0)^2$ do not even appear in the respective denominators of $P(x) = -2$ and $Q(x) = \frac{8}{x}$. For a singular point $x = x_0$ any non-negative power of $x - x_0$ less than one (namely, zero) and any non-negative power less than two (namely, zero and one) in the denominators of $P(x)$ and $Q(x)$, respectively, imply that x_0 is a regular singular point. A singular point can be a complex number. You should verify that $x = 3i$ and $x = -3i$ are two regular singular points of $(x^2+9)y'' - 3xy' + (1-x)y = 0$.

Example (Singular points):

(a) The differential equation

$$y'' + xy' + (\ln x)y = 0$$

is already in standard form. The coefficient functions are

$$P(x) = x \quad \text{and} \quad Q(x) = \ln x$$

Now, $P(x) = x$ is analytic at every real number, and $Q(x) = \ln x$ is analytic at every positive real number. However, since $Q(x) = \ln x$ is discontinuous at $x = 0$, so, it cannot be represented by a power series in x, that is, a power series centered at 0. We conclude that $x = 0$ is a singular point of the differential equation.

(b) By putting $xy'' + y' + xy = 0$ in the standard form, we have

$$y'' + \frac{1}{x}y' + y = 0$$

We see that $P(x) = 1/x$ fails to be analytic at $x = 0$. Hence $x = 0$ is a singular point of the equation.

Example (Ordinary and Singular points):

(a) The only singular points of the differential equation

$$(x^2-1)y'' + 2xy' + 6y = 0$$

are the solutions of $x^2 - 1 = 0$ or $x = \pm 1$. All other values of x are ordinary points.

(b) Inspection of the Cauchy–Euler equation

$$x^2y'' + y = 0$$

shows that it has a singular point at $x = 0$. All other values of x are ordinary points.

(c) Singular points need not be real numbers. For example, consider the equation

$$(x^2 + 1)y'' + xy' - y = 0$$

which has singular points at the solutions of $x^2+1 = 0$, namely $x = \pm i$. All other values of x, real or complex, are ordinary points.

Note: To solve the linear ODE with an ordinary point, we use the power series method. To solve the linear ordinary differential equation with a regular singular point, we use the Frobenius method.

4.4.4 *Movable singularity*

More than a century ago, the relationship between the singularities of differential equation solutions and their structures was discovered. Solvabilities of linear ODEs coefficients entirely dictate the singularities of their solutions. We refer to these singularities as fixed singularities since their position is independent of the integration constants. Solutions to nonlinear equations may also have singularities, the location of which is dependent upon the initial/boundary conditions (integration constants). It is known as movable singularities for these singularities.

The movable singularities have significant effects on the equation's integrability, especially for nonlinear differential equations. The procedure to ascertain whether a differential equation has the Painlevé property, that is, that its solutions do not have movable singularities in the complex plane, is known as the Painlevé test. In areas like nonlinear dynamics, soliton theory, and integrable systems, an understanding of movable singularities is essential. When a system has the Painlevé property and no movable singularities, it is sometimes implied that the system may have certain integrable properties and that the system's solutions can be stated in terms of standard functions. Because of this, moveable singularity research is a crucial component of mathematical analysis across a wide range of scientific and academic fields.

Algebraic branch points, logarithmic branch points, and essential singular points are called critical singular points. We are interested in all singularities — aside from crucial singular points.

Examples:

Equations	**Solutions**	**Type of Singularity**
$\frac{dy}{dx} = -y^2$	$y(x) = \frac{1}{\sqrt{x-c}}$	movable pole
$\frac{dy}{dx} = \frac{1}{y}$	$y(x) = 2\sqrt{x-c}$	algebraic branch point
$\frac{dy}{dx} = \mathrm{e}^{-y}$	$y(x) = \ln(x-c)$	logarithmic branch point
$\frac{dy}{dx} = -y(\ln(y))^2$	$y(x) = \mathrm{e}^{\frac{1}{x-c}}$	essential singular point

(i) The function f is analytic with respect to the first argument and rational with respect to the second argument in the first-order nonlinear ODE of the form

$$\frac{dy}{dx} = f(x, y) \tag{4.2}$$

Only when it coincides with the generic Riccati equation $\frac{dy}{dx} = A_0(x) + A_1(x)y + A_2(x)y^2$. Equation (4.2) admits solutions without movable critical points.

(ii) Consider the second-order nonlinear ODE of the form

$$\frac{d^2y}{dx^2} = f\left(x, y, \frac{dy}{dx}\right) \tag{4.3}$$

where the function f is analytic with respect to the first argument x and a rational function of y and $\frac{dy}{dx}$. All equations of this type with solutions that lack moveable critical points (apart from movable poles and fixed singular points) can be categorized into 50 classes, as demonstrated by P. Painlevé and B. Gambier. Canonical form equations for six of these classes are irreducible and referred to as Painlevé equations; their solutions are known as Painlevé transcendents.

4.4.5 *Critical points and types of singularity*

Definition: A critical point is a point where multivaluedness can occur.

Example: (i)

$$W(z) = \frac{1}{z - z_0}$$

Here, z_0 is a singular point which is a pole.

(ii)

$$W(z) = e^{\frac{1}{z-z_0}}$$

Here, the function $W(z)$ has essential singularity at its singular point z_0 which is single-valued.

(iii)

$$W(z) = \sqrt{z - z_0}$$

has an algebraic branch point and is multivalued.

(iv)

$$W(z) = \log(z - z_0)$$

is multivalued logarithmic branch point.

(v)

$$W(z) = \tan(ln(z - z_0))$$

is multivalued essential singularity.

(vi) Consider a linear ODE

$$\frac{d^2w}{dz^2} + \frac{1}{z}\frac{dw}{dz} + \left(1 - \frac{\nu^2}{z^2}\right) w = 0$$

Here, the singularities are at 0 and infinity which are fixed singularities. But for the nonlinear equation

$$\frac{dw}{dz} = w^2 + 1$$

or

$$\frac{dw}{w^2 + 1} = dz$$

or

$$\tan^{-1}(w) = z + c$$

where c is an arbitrary constant. Putting $z = z_0$ when $w = 0$, we get,

$$w = \tan(z - z_0)$$

Therefore, the singularities are

$$z = z_0 + \left(n + \frac{1}{2}\right)\pi$$

These are movable singularities.

4.4.6 *Algorithm for the Painlevé test*

Consider a system of M polynomial differential equations

$$F_i(u(z), u'(z), u''(z), \ldots, u^{(m_i)}(z)) = 0, \quad i = 1, 2, \ldots, M \tag{4.4}$$

Let's discuss the steps for the Painlevé test.

Step 1: To determine the dominant behavior

If we substitute

$$u_i(z) = \chi_i g^{\alpha_i}(z), \quad i = 1, 2, 3, \ldots, M \tag{4.5}$$

where χ_i is a constant, in to (4.4), we can determine the leading exponents α_i. In the resulting polynomial system, equating every two or more possible lowest exponents of $g(z)$ in each equation gives a linear system for α_i. The linear system is then solved for α_i; all the α_i are integers, and at least one is negative. If any of the α_i are non-integer in a given branch, then that branch of the algorithm terminates. A nonlinear α_i implies that some solutions of (4.4) have movable algebraic branch points.

For each solution α_i, we substitute

$$u_i(z) = u_{i,0}(z) g^{\alpha_i}(z), i = 1, 2, 3, \ldots, M \tag{4.6}$$

into (4.4). We then solve the nonlinear equation for $u_{i,0}(z)$, which is found by balancing the leading terms. If any of the solutions contradict the assumption that $u_{i,0}(z) \neq 0$, then that branch of the algorithm fails the Painlevé test.

Step 2: To determine the resonances

For each α_i and $u_{i,0}(z)$, we calculate the $r_1 \leq \cdots \leq r_m$ for which $u_{i,r}(z)$ is an arbitrary function. To do this, we substitute

$$u_i(z) = u_{i,0}(z) g^{\alpha_i}(z) + u_{i,r}(z) g^{\alpha_i + r}(z) \tag{4.7}$$

into (4.4), and keep only the lowest order terms in $g(z)$ that are linear in $u_{i,r}$. This can be done by computing the solutions for r of $det(Q_r) = 0$, where the $M \times M$ matrix Q_r satisfies

$$Q_r u_r = 0, u_r = (u_{1,r}, u_{2,r}, \ldots, u_{M,r})^T \tag{4.8}$$

If r_m is not a positive integer, then the algorithm terminates. If $r_1 = -1$, $r_2 = \cdots = -r_m = 0$ and $m - 1$ of the $u_{i,0}(z)$ found in step 1 are arbitrary, then (4.4) passes the Painlevé test.

There is always a resonance $r = -1$ which corresponds to the arbitrariness of $g(z_0)$; as such, it is often called the universal resonance. When there are negative resonances other than $r = -1$, (or, more than one resonance equals -1), then the Laurent series solution is not the general solution, and further analysis is needed to determine if (4.4) passes the Painlevé test.

Step 3: To find the constant of integration and check compatibility conditions

For the system to possess the Painlevé property, the arbitrariness of $u_{i,r}(z)$ must be verified up to the highest resonance level. This is done by substituting

$$u_i(z) = g^{\alpha_i}(z)\Sigma_{i,k}(z)g^k(z) \tag{4.9}$$

into (4.4), where r_m is the largest positive integer resonance.

When testing ODEs (4.4), it must be replaced by

$$u_i(z) = g^{\alpha_i}(z)\Sigma u_{i,k}g^k(z), i = 1, 2, 3, \ldots, M \tag{4.10}$$

where the coefficients $u_{i,k}$ are constants, $g(z) = z - z_0$, and z_0 is an arbitrary constant. If z explicitly occurs in the ODE, then it is (automatically) replaced by $g(z) + z_0$ prior to step 1 of the test.

Here we will discuss the solutions of the first and second Painlevé equations only.

4.4.7 *Solutions of PI equation and PII equation*

1. The first Painlevé equation has the form

$$\frac{d^2u}{dx^2} = 6u^2 + x \tag{4.11}$$

Solution: Suppose $\phi = x - c$. The general solution of Eq. (4.11) is in the form

$$u(x) = \frac{1}{\phi^\alpha} \sum_{n=0}^{\infty} u_n \phi^n \tag{4.12}$$

Substituting the leading term $u(x) = \frac{u_0}{\phi^\alpha}$ of the expansion (4.12) into Eq. (4.11), we obtain

$$\frac{u_0\alpha(\alpha + 1)}{\phi^{\alpha+2}} = \frac{6u_0^2}{\phi^{2\alpha}} + x$$

Omitting the last term on the right, we find that

$$\alpha + 2 = 2\alpha \implies \alpha = 2$$

and

$$6u_0 = 6u_0^2 \implies u_0 = 1 \quad (\text{since } u_0 \neq 0)$$

Therefore, Eq. (4.12) becomes

$$u(x) = \frac{u_0}{\phi^2} + \frac{u_1}{\phi} + u_2 + u_3\phi + u_4\phi^2 + u_5\phi^3 + u_6\phi^4 + \cdots$$

Now,

$$\frac{du}{dx} = -\frac{2u_0}{\phi^3} - \frac{u_1}{\phi^2} + u_3 + 2u_4\phi + 3u_5\phi^2 + 4u_6\phi^3 + \cdots$$

$$\frac{d^2u}{dx^2} = \frac{6u_0}{\phi^4} + \frac{2u_1}{\phi^3} + 2u_4 + 6u_5\phi + 12u_6\phi^2 + \cdots$$

Substituting these values in Eq. (4.11) and collecting the same powers of ϕ, we obtain

$$\begin{aligned}\frac{d^2u}{dx^2} - 6u^2 - x = & \left(\frac{6u_0}{\phi^4} - \frac{6u_0^2}{\phi^4}\right) + \left(\frac{2u_1}{\phi^3} - \frac{12u_0u_1}{\phi^3}\right) + \left(-\frac{6u_1^2}{\phi^2} - \frac{12u_0u_2}{\phi^2}\right) \\ & + \left(\frac{-12u_0u_3}{\phi} - \frac{12u_1u_2}{\phi}\right) \\ & + \left(2u_4 - 6u_2^2 - 12u_1u_3 - 12u_0u_4 - c\right) \\ & + \left(6u_5 - 12u_1u_4 - 12u_2u_3 - 12u_0u_5 - 1\right)\phi \\ & + \left(12u_6 - 6u_3^2 - 12u_2u_4 - 12u_1u_5 - 12u_0u_6\right)\phi^2 + \cdots \\ = & \, 0\end{aligned}$$

Now, equating the coefficients of the same powers of ϕ to zero, we obtain a system of ODEs as

$$\begin{aligned}& u_0^2 - u_0 = 0 \implies u_0 = 1, \\ & 2u_1 - 12u_0u_1 = 0 \implies u_1 = 0, \\ & -6u_1^2 - 12u_0u_2 = 0 \implies u_2 = 0, \\ & -12u_0u_3 - 12u_1u_2 = 0 \implies u_3 = 0, \\ & 2u_4 - 6u_2^2 - 12u_1u_3 - 12u_0u_4 - c = 0 \implies u_4 = -\frac{c}{10}. \\ & 6u_5 - 12u_5 - 12u_2u_3 - 1 = 0 \implies u_5 = 0, \\ & 12u_6 - 6u_3^2 - 12u_2u_4 - 12u_6 = 0 \implies u_6 = d\end{aligned}$$

So,

$$\begin{aligned} u(x) &= \frac{u_0}{\phi^2} + \frac{u_1}{\phi} + u_2 + u_3\phi + u_4\phi^2 + u_5\phi^3 + u_6\phi^4 + \cdots \\ &= \frac{1}{[x-c]^2} - \frac{c}{10}[x-c]^2 - \frac{1}{6}[x-c]^3 \\ &\quad + d[x-c]^4 + \cdots \end{aligned} \tag{4.13}$$

where c, d are arbitrary constants. Equation (4.13) is the required general solution.

2. The second Painlevé equation has the form

$$\frac{d^2u}{dx^2} = 2u^3 + xu + \beta \tag{4.14}$$

Solution: Suppose $\phi = x - c$. The general solution of Eq. (4.14) is in the form

$$u(x) = \frac{1}{\phi^\alpha}\sum_{n=0}^{\infty} u_n\phi^n \tag{4.15}$$

Substituting the leading term $u(x) = \frac{u_0}{\phi^\alpha}$ of the expansion (4.15) into Eq. (4.14), we obtain

$$\frac{u_0\alpha(\alpha+1)}{\phi^{\alpha+2}} = \frac{2u_0^3}{\phi^{3\alpha}} + \frac{xu_0}{\phi^\alpha} + \beta$$

Omitting the last two terms on the right, we find that

$$\alpha + 2 = 3\alpha \implies \alpha = 1$$

and

$$2u_0 = 2u_0^3 \implies u_0 = \pm 1 = m \ \ (\text{since } u_0 \neq 0)$$

Therefore, Eq. (4.15) becomes

$$u(x) = \frac{m}{\phi} + u_1 + u_2\phi + u_3\phi^2 + u_4\phi^3 + u_5\phi^4 + u_6\phi^5 + \cdots$$

Now,

$$\begin{aligned} \frac{du}{dx} &= -\frac{m}{\phi^2} + u_2 + 2u_3\phi + 3u_4\phi^2 + 4u_5\phi^3 + 5u_6\phi^4 + \cdots \\ \frac{d^2u}{dx^2} &= \frac{2m}{\phi^3} + 2u_3 + 12u_4\phi + 12u_5\phi^2 + 20u_6\phi^3 + \cdots \end{aligned}$$

Substituting these values in Eq. (4.14) and collecting the same powers of ϕ, we obtain

$$\begin{aligned}\frac{d^2u}{dx^2} - 2u^3 - xu - \beta = &\left(\frac{2m}{\phi^3} - \frac{2m^3}{\phi^3}\right) + \left(-\frac{6m^2u_1}{\phi^2} - 0\right)\\ &+ \left(\frac{-6u_1^2m}{\phi} - \frac{cm}{\phi} - \frac{6m^2u_2}{\phi}\right)\\ &+ \left(2u_3 - 2u_1^3 - 6m^2u_3 - m - cu_1 - \beta\right)\\ &+ \left(6u_4 - 6m^2u_4 - 6mu_2^2 - 6u_1^2u_2 - u_1 - cu_2\right)\phi\\ &+ \cdots\\ =&\, 0\end{aligned}$$

Now, equating the coefficients of the same powers of ϕ to zero, we obtain a system of ODEs as

$$\begin{aligned}&2m - 2m^3 = 0 \implies m = \pm 1 = u_0,\\ &-6m^2u_1 = 0 \implies u_1 = 0,\\ &-6u_1^2m - cm - 6m^2u_2 = 0 \implies u_2 = -\frac{c}{6m},\\ &2u_3 - 2u_1^3 - 6m^2u_3 - m - \beta = 0 \implies u_3 = -\frac{m+\beta}{4},\\ &6u_4 - 6m^2u_4 - 6mu_2^2 - 6u_1^2u_2 - u_1 - cu_2 = 0 \implies u_4 = d\end{aligned}$$

So,

$$u(x) = \frac{m}{x-c} - \frac{c}{6m}(x-c) - \frac{m+\beta}{4}(x-c)^2 + d(x-c)^3 + \cdots \tag{4.16}$$

where c, d are arbitrary constants and $m = \pm 1$. Equation (4.16) is the required general solution.

4.4.8 *Weiss, Tabor, Carnevale (WTC) algorithm*

In 1983, J. Weiss and J. Carnevale introduced a new method to study the Painlevé property, which is known as the *WTC method*. Using this method, one can check the integrability of many nonlinear evolution equations. The procedure is as follows:

(i) The determination of the leading order of the Laurent series.
(ii) The identification of powers at which the arbitrary function can enter the Laurent series is called resonances and

(iii) verifying that at the resonance value, a sufficient number of arbitrary functions exist without introducing the movable critical manifold.

In this method, the general solution of a PDE is considered in the form

$$u_{x,t} = (\phi_{x,t})^p \Sigma u_k(x,t)(\phi(x,t))^k \tag{4.17}$$

where p is negative. $\phi(x,t) = 0$ is the equation of a singular manifold, and the function $u_k(x,t)$ has to be determined by substituting the expression (4.17) in the PDE, which will become

$$\Sigma_{k=0}^{\infty} E_k(u_0, u_1, \ldots, u_k, \phi)\phi^{k+q} = 0 \tag{4.18}$$

where q is some negative constant, E_k depends on ϕ only by the derivatives of ϕ.

The successive steps for Painlevé analysis are

(i) To determine the possible leading order p by balancing two or more terms of the PDE and expressing that they dominate other terms.
(ii) Then to solve $E_0 = 0$ for non-zero values of u_0. Several solutions may appear. These are called branches.
(iii) Finding the resonance. The values of k for which u_k cannot be determined from the equation $E_k = 0$. This last equation has usually the form

$$E_k = (k+1)P(k)\phi_n^j \phi_k^{n-j} u_k + Q(u_0, u_1, \ldots, u_{k-1}, \phi) = 0 \tag{4.19}$$

where n is the order of the PDE, $0 \leq j \leq n$ and P is a polynomial of degree $(n-1)$. The values of the resonance are of the zeros of P. At resonance, after substitution in (4.19) of the previously computed u_i, $i \leq k-1$ into given PDE, the function Q is either zero, in which case u_k can be arbitrarily chosen and the resonance is said to be compatible, or non-zero, the expression (4.17) does not exist for arbitrarily Q.

The WTC algorithm considers the singularity structure of the solutions around the non-characteristic manifold of the form $g(z) = 0$ where $g(z)$ is an analytic function of z in the neighborhood of the manifold. If the singularity manifold is determined by $g(z) = 0$ and $u(z)$ is a solution of the NLPDE, then we consider a Laurent series for $u(z)$. Accordingly we consider

$$u_i(z) = g^{\alpha_i}(z)\Sigma_{k=0}^{\infty} u_{i,k}(z) g^k(z) \tag{4.20}$$

where $i = 1, 2, \ldots, M$, where the coefficients $u_{i,k}(z)$ are analytic functions of z with $u_{i,o}(z)$ not equal to 0 in the neighborhood of the manifold. So the

α_i's are integers with at least one exponent $\alpha_i < 0$. If we substitute (4.20) in the PDE and equate coefficients of like powers of $g(z)$, it determines the possible values of α_i. And ultimately defines a relation for $u_{i,k}(z)$. The recurrence relation is of the form

$$Q_k U_k = G_k(u_0, u_1, \ldots, u_{k-1}, g, z), \quad u_k = (u_{1,k}, u_{2,k}, \ldots, u_{M,k})^T \tag{4.21}$$

where Q_k is an $M \times M$ matrix and T denotes transpose. For an NLPDE to pass the Painlevé test, the series (4.20) should have $(M - 1)$ arbitrary functions. Then the Laurent series solution corresponds to the general solution of the NLPDE. The $(M-1)$ arbitrary functions $u_{i,k}(z)$ occur when k is one of the roots of determinant Q_k. These roots $r_1 \leq r_2 \leq \cdots \leq r_M$ are called resonances. Since WTC algorothm is unable to detect essential singularity, it is only a necessary condition for the PDE to have the Painlevé property. The resonances are also equal to the functional indices of the auxiliary equation of DT.

Painlevé test for Burgers' equation

Let Burgers' equation be

$$\frac{\partial u}{\partial t} + u\frac{\partial u}{\partial x} - a\frac{\partial^2 u}{\partial x^2} = 0 \tag{4.22}$$

Let the of (4.273) solution be

$$u(x,t) = \frac{1}{\phi^\alpha}\sum_{n=0}^{\infty} u_n(t)\phi^n(x,t) \tag{4.23}$$

where $\phi = x - x_0(t)$ and α belongs to the set of natural numbers.

Let the leading term be $u(x,t) = \frac{u_0(t)}{\phi^\alpha}$. Now,

$$\frac{\partial u}{\partial t} = u_0(t)(-\alpha)\phi^{-\alpha-1}\phi_t + \frac{1}{\phi^\alpha}u_0'(t) = \frac{\alpha u_0(t)x_0'(t)}{\phi^{\alpha+1}} + \frac{u_0'(t)}{\phi^\alpha} \tag{4.24}$$

and

$$\frac{\partial u}{\partial x} = -\frac{\alpha u_0(t)}{\phi^{\alpha+1}}\phi_x = -\frac{\alpha u_0(t)}{\phi^{\alpha+1}} \tag{4.25}$$

and

$$\frac{\partial^2 u}{\partial x^2} = \frac{\alpha(\alpha+1)u_0(t)}{\phi^{\alpha+2}}\phi_x = \frac{\alpha(\alpha+1)u_0(t)}{\phi^{\alpha+2}} \tag{4.26}$$

Putting (4.274)–(4.276) and (4.280) in (4.273), we get,

$$\frac{\alpha u_0(t)x_0'(t)}{\phi^{\alpha+1}} + \frac{u_0'(t)}{\phi^{\alpha}} - \frac{\alpha u_0^2(t)}{\phi^{2\alpha+1}} - \frac{a\alpha(\alpha+1)u_0(t)}{\phi^{\alpha+2}} = 0 \tag{4.27}$$

Using homogeneous balancing, we get,

$$\begin{aligned} 2\alpha + 1 &= \alpha + 2 \\ \text{i.e., } \alpha &= 1 \end{aligned} \tag{4.28}$$

Neglecting the first two terms in (4.281), we get,

$$-\frac{\alpha u_0^2(t)}{\phi^{2\alpha+1}} - \frac{a\alpha(\alpha+1)u_0(t)}{\phi^{\alpha+2}} = 0 \tag{4.29}$$

Now, putting the value of α in (4.29), results

$$\begin{aligned} &-u_0^2(t) - 2au_0(t) = 0 \\ &or, -u_0(t)(u_0(t) + 2a) = 0 \end{aligned}$$

Since $u_0(t)$ cannot be equal to zero, so

$$u_0(t) = -2a$$

Thus, (4.274) gives

$$u = \frac{u_0}{\phi} + \sum_{n=1}^{\infty} u_n \phi^{n-1}; \quad \phi = x - x_0(t) \tag{4.30}$$

Now,

$$\begin{aligned} \frac{\partial u}{\partial t} &= -\frac{u_0(t)}{\phi^2}\phi_t + \frac{u_0'(t)}{\phi} + \sum_{n=2}^{\infty} u_n(n-1)\phi^{n-2}\phi_t + \sum_{n=1}^{\infty} u_n'(t)\phi^{n-1} \\ &= -\frac{u_0(t)}{\phi^2}\phi_t + \frac{u_0'(t)}{\phi} + u_1'(t) + \sum_{n=2}^{\infty}[u_n(n-1)\phi^{n-2}\phi_t + u_n'(t)\phi^{n-1}] \end{aligned} \tag{4.31}$$

and

$$\begin{aligned} \frac{\partial u}{\partial x} &= -\frac{u_0(t)}{\phi^2}\phi_x + \sum_{n=2}^{\infty} u_n(n-1)\phi^{n-2}\phi_x \\ &= -\frac{u_0(t)}{\phi^2} + \sum_{n=2}^{\infty} u_n(n-1)\phi^{n-2} \end{aligned} \tag{4.32}$$

and

$$\begin{aligned}\frac{\partial^2 u}{\partial x^2} &= \frac{2u_0}{\phi^3}\phi_x + \sum_{n=3}^{\infty} u_n(n-1)(n-2)\phi^{n-3}\phi_x \\ &= \frac{2u_0}{\phi^3} + \sum_{n=3}^{\infty} u_n(n-1)(n-2)\phi^{n-3}\end{aligned} \tag{4.33}$$

Putting in (4.282), (4.32) and (4.33) in (4.273) we get,

$$\begin{aligned}&-\frac{u_0(t)}{\phi^2}\phi_t + \frac{u_0'(t)}{\phi} + u_1'(t) + \sum_{n=2}^{\infty}[u_n(n-1)\phi^{n-2}\phi_t + u_n'(t)\phi^{n-1}] \\ &+ \left[-\frac{u_0}{\phi^2} + \sum_{n=1}^{\infty} u_n\phi^{n-1}\right]\left[\frac{u_0(t)}{\phi} + \sum_{n=2}^{\infty} u_n(n-1)\phi^{n-2}\right] \\ &- a\left[\frac{2u_0}{\phi^3} + \sum_{n=3}^{\infty} u_n(n-1)(n-2)\phi^{n-3}\right] = 0 \\ &\Longrightarrow -\frac{u_0(t)}{\phi^2}\phi_t + \frac{u_0'(t)}{\phi} + u_1'(t) + u_2(t)\phi_t + u_2'(t)\phi + \sum_{n=3}^{\infty}[u_n(n-1)\phi^{n-2}\phi_t \\ &+ u_n'(t)\phi^{n-1}] + \left[\frac{u_0(t)}{\phi} + u_1(t) + u_2(t)\phi + \sum_{n=3}^{\infty} u_n(t)\phi^{n-1}\right] \\ &\times \left[-\frac{u_0(t)}{\phi^2} + u_2(t) + \sum_{n=3}^{\infty}[u_n(n-1)\phi^{n-2}]\right] \\ &- \frac{2au_0(t)}{\phi^3} - a\sum_{n=3}^{\infty}[u_n(n-1)(n-2)\phi^{n-3}] = 0\end{aligned}$$

Then we have,

$$\begin{aligned}&-\sum_{n=3}^{\infty} u_n(t)u_0(t)\phi^{n-3} + \sum_{n=3}^{\infty}(n-1)u_n(t)u_0(t)\phi^{n-3} \\ &= a\sum_{n=3}^{\infty}(n-1)(n-2)u_n(t)\phi^{n-3}\end{aligned}$$

Comparing the coefficients of ϕ^{n-3}, we get

$$\begin{aligned}
&-u_0(t)+(n-1)u_0(t)=a(n-1)(n-2)\\
&\Longrightarrow 2a+(n-1)(-2a)=a(n-1)(n-2)\\
&\Longrightarrow a(n^2-n-2)=0\\
&\Longrightarrow (n+1)(n-2)=0\\
&\Longrightarrow n=-1,2
\end{aligned}$$

Hence, we get the resonance for the positive value of n i.e., $n=2$. Thus, we get the series

$$\begin{aligned}
u(x,t)&=\frac{u_0}{\phi}+u_1+u_2\phi\\
&=\frac{-2a}{\phi}+u_1+u_2\phi
\end{aligned}\tag{4.34}$$

Therefore,

$$u_t=\frac{-2ax_0'(t)}{\phi^2}+u_1'(t)+u_2'(t)\phi-u_2(t)x_0'(t)+u_3'(t)\phi^2-2u_3(t)\phi x_0'(t)+\cdots\cdots$$

and

$$u_x=\frac{2a}{\phi^2}+u_2+2u_3(t)\phi+\cdots$$

and

$$u_{xx}=\frac{-4a}{\phi^3}+2u_3+\cdots$$

Putting in (4.273), we get,

$$\begin{aligned}
&\frac{-2ax_0'(t)}{\phi^2}+u_1'(t)+u_2'(t)\phi-u_2(t)x_0'(t)\\
&\quad+\left[\frac{-2a}{\phi}+u_1(t)+u_2(t)\phi\right]\left[\frac{2a}{\phi^2}+u_2(t)\right]+\frac{-4a^2}{\phi^3}=0
\end{aligned}$$

Compairing the coefficients of like powers of ϕ, we get,

Coefficient of $\frac{1}{\phi^2}$:

$$-2ax_0'(t)+2au_1(t)=0$$
$$\text{i.e., } u_1(t)=x_0'(t)$$

Coefficient of $\frac{1}{\phi}$:

$$-2au_2(t)+2au_2(t)=0$$

Since this is an identity, so $u_2(t)$ is arbitrary, suppose. $u_2(t) = \psi(t)$ Therefore, the solution of (4.273) is given by

$$u(x,t)=\frac{-2a}{x-x_0(t)}+x_0'(t)+\psi(t)(x-x_0(t))$$

Solution of KdV equation:

Let us consider the KdV equation

$$\frac{\partial u}{\partial t}+6u\frac{\partial u}{\partial x}+\frac{\partial^3 u}{\partial x^3}=0 \tag{4.35}$$

For the leading term, let us substitute $u_i = \chi_i g^{\alpha_i}$. So we get

$$\begin{aligned}
(u_i)_t &= \chi_i\alpha_i g^{\alpha_i-1}g_t\\
(u_i)_x &= \chi_i\alpha_i g^{\alpha_i-1}g_x\\
(u_i)_{xx} &= \chi_i\alpha_i(\alpha_i-1)g^{\alpha_i-2}g_x^2+\chi_i\alpha_i g^{\alpha_i-1}g_{xx}\\
(u_i)_{3x} &= \chi_i\alpha_i(\alpha_i-1)(\alpha_i-2)g^{\alpha_i-3}g_x^3+3\chi_i\alpha_i(\alpha_i-1)g^{\alpha_i-2}g_xg_{xx}\\
&\quad+\chi_I\alpha_i g^{\alpha_i-1}g_{xxx}
\end{aligned}$$

So, putting these in (4.35), we get,

$$\begin{aligned}
\chi_i\alpha_i[g^{\alpha_i-1}g_t &+ 6\chi_i g^{2\alpha_i-1}g_x+(\alpha_i-1)(\alpha_i-2)g^{\alpha_i-3}g_x^3\\
&+3(\alpha_i-1)g^{\alpha_i-2}g_xg_{xx}+g^{\alpha_i-1}g_{xxx}]=0
\end{aligned} \tag{4.36}$$

Using homogeneous balancing, we can write $2\alpha_i-1=\alpha_i-3$, i.e., $\alpha_i=-2$. Therefore,

$$u(x,t)=\chi_i g^{-2}$$

Taking the initial value of χ_i as $u_0(x,t)$, we get

$$u(x,t)=u_0(x,t)g^{-2}(x,t)$$

Putting these values in (4.36), we get

$$-2u_0(x,t)[g^{-3}g_t + 6u_0g^{-5}g_x + 12g^{-5}g_x^3 - 9g^{-4}g_xg_{xx} + g^{-3}g_{xxx}] = 0$$

Equating the coefficient of leading order terms with zero, we get $u_0 = -2g_x^2(x,t)$.

To find resonance, we write

$$u_i(z) = u_{i,0}(z)g^{\alpha_i}(z) + u_{i,r}(z)g^{\alpha_i+r}(z)$$

Therefore,

$$u_i(z) = -2g_x^2(x,t)g^{-2}(x,t) + u_{i,r}(x,t)g^{r-2}(x,t)$$

Without any loss of generality, we can write

$$\begin{aligned}
u_t &= 4g^{-3}g_x^2g_t - 4g^{-2}g_xg_{xt} + (r-2)g^{r-3}u_rg_t + (u_r)_tg^{r-2}\\
u_x &= 4g_x^2g^{-3}g_x - 4g_xg_{xx}g^{-2} + (r-2)u_rg^{r-3}g_x + (u_r)_xg^{r-2}\\
&= 4g^{-3}g_x^3 - 4g^{-2}g_xg_{xx} + (r-2)u_rg^{r-3}g_x + (u_r)_xg^{r-2}\\
u_{xx} &= 12g^{-3}g_x^2g_{xx} - 12g^{-4}g_x^4 - 4g^{-2}g_xg_{xxx} - 4g^{-2}g_{xx}^2 + 8g^{-3}g_x^2g_{xx}\\
&\quad + (r-2)u_rg^{r-3}g_{xx} + (r-2)(r-3)u_rg^{r-4}g_x^2 + (r-2)(u_r)_xg^{r-3}g_x\\
&\quad + (u_r)_x(r-2)g^{r-3}g_x + (u_r)_xxg^{r-2}\\
&= 20g^{-3}g_x^2g_{xx} - 12g^{-4}g_x^4 - 4g^{-2}g_xg_{xxx} - 4g^{-2}g_{xx}^2 + (r-2)u_rg^{r-3}g_{xx}\\
&\quad + (r-2)(r-3)u_rg^{r-4}g_x^2 + 2(r-2)g^{r-3}(u_r)_xg_x + (u_r)_{xx}g^{r-2}\\
u_{xxx} &= -108g^{-4}g_x^3g_{xx} + 48g^{-3}g_xg_{xx}^2 + 28g^{-3}g_x^2gxxx + 48g^{-5}g_x^5\\
&\quad -12g^{-2}g_{xx}g_{xxx} - 4g^{-2}g_xg_{4x} + 3(r-2)g^{r-3}(u_r)_xg_{xx}\\
&\quad + 3(r-2)(r-3)u_rg^{r-4}g_xg_{xx}\\
&\quad + (r-2)u_rg^{r-3}g_{xxx} + 3(r-2)(r-3)g^{r-4}g_x^2(u_r)_x\\
&\quad + (r-2)(r-3)(r-4)u_rg^{r-5}g_x^3 + 3(r-2)g^{r-3}(u_r)_{xx}g_x\\
&\quad + (u_r)_{xxx}g^{r-2}
\end{aligned}$$

Putting these in (4.35), we get,

$$\begin{aligned}
&4g^{-3}g_x^2g_t - 4g_xg_{xt}g^{-2} + (r-2)g^{r-3}u_rg_t + (u_r)_tg^{r-2} - 60g^{-4}g_x^3g_{xx}\\
&\quad + [-12r + 24 + 24 + (r-2)(r-3)(r-4)]u_rg^{r-5}g_x^3 - 12g^{r-4}g_x^2(u_r)_x\\
&\quad + [-24 + 3(r-2)(r-3)]u_rg^{r-4}g_xg_{xx} + 6(r-2)u_r^2g^{2r-5}g_x\\
&\quad + 6g^{2r-4}u_r(u_r)_x + 48g^{-3}g_xg_{xx}^2 + 28g^{-3}g_x^2g_{xxx} - 12g^{-2}g_{xx}g_{xxx}\\
&\quad - 4g^{-2}g_xg_{4x} + 3(r-2)g^{r-3}(u_r)_xg_{xx} + (r-2)u_rg^{r-3}g_{xxx}\\
&\quad + 3(r-2)(r-3)g^{r-4}g_x^2(u_r)_x + 3(r-2)g^{r-3}(u_r)_{xx}g_x + (u_r)_{xxx}g^{r-2} = 0
\end{aligned}$$

Comparing the coefficients of g_x^3, which is linear in $u_r(x,t)$, we get $-12r + 24 + 24 + (r-2)(r-3)(r-4) = 0$, i.e., $r = -1, 4, 6$ Therefore,

$$\begin{aligned}
u(x,t) &= \Sigma u_r(x,t)g^{r-2}(x,t)\\
&= -2g_x^2(x,t)g^2 + u_1(x,t)g^{-1}(x,t) + u_2(x,t)\\
&\quad + u_3(x,t)g(x,t) + \cdots + u_6(x,t)g^4(x,t)
\end{aligned}$$

is a solution of KdV equation. Solutions contains finite term.

4.5 Lax Pair

The integrability of an evolution equation is very essential to studying nonlinear phenomena in various fields of science. An evolution equation is integrable, which means the equation is properly solvable. We cannot say that every evolution equation is properly integrable. Many researchers have tried to develop various methods in a scientific way to understand the integrability of the evolution equation. The Lax pair is one such method. It was first introduced by Peter David Lax in 1968. It represents a set of two linear differential operators with the commutativity condition, which is equivalent to the nonlinear evolution equation. The existence of the Lax pair is a criterion of integrability; however, it is insufficient to construct solutions. To build solutions, Lax developed a general principle for describing integrable nonlinear evolution equations that can be solved exactly by IST or DT. In addition, if a suitable Lax pair can be found for a particular NLPDE, then it is possible that they can be used to solve the associated Cauchy problem using a method such as IST or by DT.

The existence of a Lax pair for an evolution equation ensures that the evolution equation is properly integrable in the Lax sense, which is called *Lax integrability*. Lax pairs are a tool for finding conserved quantities

Figure 4.2. Peter David Lax

in some evolutionary differential equations. In 1974, Ablowitz, Kaup, Newell, and Segur published a matrix formalism for Lax pairs where they introduced a construction that avoids the need to consider higher-order Lax operators. This method is referred to as the Ablowitz-Kaup-Newell-Segur (AKNS) method. A system with possibly nonlinear equations of motion will be said to admit a Lax pair if one can find a pair of matrices or operators (L, A) such that the equations of motion are equivalent to the Lax equation $L_t = LA - AL$. However, a Lax pair is not unique. For instance, we may add to 'A' a matrix that commutes with L and add to L a time-independent matrix that commutes with A without altering the zero curvature equation. As we will explain, Lax pairs are based on the idea of expressing (typically) nonlinear evolution equations as 'compatibility' conditions for a pair of auxiliary linear equations to admit simultaneous solutions. The idea only works for certain special systems (integrable systems), which, however, play an important role in our understanding of more general dynamical systems. Unfortunately, there is no recipe for finding a Lax pair for a given system or knowing in advance whether one exists. So some knowledge of the nature of the system and its solutions, coupled with educated guesswork, is involved.

However, as we will see, once a Lax pair is known, it can be very helpful in understanding the system.

The Lax pair of an NLPDE $F(x, t, u, \ldots) = F[u] = 0$ is a system of two linear differential operators $L = L(u, \lambda)$ and $M = M(u, \lambda)$ (depending on a solution u of the PDE and on the spectral parameter λ) such that $[L, M] = 0$ if and only if $f[u] = 0$, where $[L, M] = LM - ML$ is the commutator of the operators L and M.

A Lax pair can be represented in several equivalent forms: the Lax representation, the zero-curvature representation, the scalar representation, and the projective Riccati representation.

The key components of the Lax pair of an NLPDE are as follows:

(i) **NLPDE**: Start with a given NLPDE.
(ii) **Lax Pair**: Construct a pair of linear differential equations, typically involving matrices or operators, that are related to the original nonlinear equation.
(iii) **Compatibility Condition**: Investigate the compatibility between the linear equations derived from the Lax pair. The compatibility condition ensures that the original NLPDE is satisfied.
(iv) **Integrability**: If the compatibility requirement is satisfied, it indicates that the original equation is integrable, which means it may contain extra symmetries, conservation laws, or special solutions.

4.6 Lax Compatibility Equation

The Lax pair method often gives the equations known as the *Lax compatibility equation* or the *zero curvature equation*. These equations play a crucial role in establishing the integrability of certain NLPDEs in the Lax sense.

The Lax pair consists of two linear differential equations, involving matrices or operators, which are written in terms of a spectral parameter λ. For a pair of matrices $M(x, t, \lambda)$ and $N(x, t, \lambda)$, the Lax compatibility condition (zero curvature condition) is given by

$$\frac{\partial M}{\partial t} - \frac{\partial N}{\partial x} + [M, N] = 0 \tag{4.37}$$

The symbol [.,.] denotes the matrix commutator (the Lie bracket), which is the difference between the matrix products MN and NM, i.e., $(MN - NM)$. So if matrices M and N can satisfy this equation, it ensures the integrability of the associated NLPDE through the Lax pair. The forms of M and N are constructed based on the properties of the given NLPDE.

The zero curvature equation is crucial in the theory of integrable systems as it connects the linear equations of the Lax pair. It guarantees that certain nonlinear equations admit solutions that can be explicitly obtained from the solutions of the linear equations, thereby indicating integrability and the existence of conserved quantities.

This condition establishes connections between different mathematical structures and provides a means to find exact solutions for an NLPDE by solving the associated linear differential equation.

4.7 Lax Pair in Operator Form

Consider the following (1+1)-dimensional evolution equation:

$$\frac{\partial u}{\partial t} = F(x, t, u, u_x, u_{xx}, u_{xxx} \cdots) \tag{4.38}$$

where $u = u(x,t)$, $u_x = \frac{\partial u}{\partial x}$, $u_{xx} = \frac{\partial^2 u}{\partial x^2}$, $u_{xxx} = \frac{\partial^3 u}{\partial x^3}, \cdots$ (4.38) is integrable. They represent a pair of differential operators having a characteristic whereby they yield nonlinear evolution equations when they commute. Let us consider a linear differential operator L, corresponding to the evolution equation (4.38) that may depend upon function $u(x,t)$ and its spatial derivatives $u_x, u_{xx}, u_{xxx}, \ldots,$ but not explicitly depend upon the temporal variable t, such that

$$L\psi = \lambda\psi \qquad \text{where} \qquad \psi = \psi(x,t) \tag{4.39}$$

For another operator M, ψ satisfy

$$\psi_t = M\psi \tag{4.40}$$

We take the time derivative of Eq. (4.39), and using Eq. (4.40), we get

$$\begin{aligned}
&L_t\psi + L\psi_t = \lambda_t\psi + \lambda\psi_t, \\
\Longrightarrow\ &L_t\psi + LM\psi = \lambda_t\psi + \lambda M\psi, \quad [As\ \psi_t = M\psi] \\
\Longrightarrow\ &L_t\psi + LM\psi = \lambda_t\psi + M(\lambda\psi), \\
\Longrightarrow\ &L_t\psi + LM\psi = \lambda_t\psi + ML\psi, \\
\Longrightarrow\ &L_t\psi + LM\psi - ML\psi = \lambda_t\psi, \\
\Longrightarrow\ &(L_t + LM - ML)\psi = \lambda_t\psi
\end{aligned}$$

Hence, in order to solve for the non-trivial eigenfunction $\psi(x,t)$ with spectral parameter λ, we must have

$$L_t + LM - ML = 0 \implies L_t + [L, M] = 0 \tag{4.41}$$

where $[L, M] = (LM - ML)$ and $L_t = \frac{\partial L}{\partial t}$, which will be true iff λ_t=0. Equation (4.41) is the Lax representation of the given PDE (4.38), also known as the zero curvature equation or Lax compatibility equation, and $[L, M] = (LM - ML)$ represents the commutator of the operators L and M that form the Lax operators.

Note: We have shown that for a Lax pair, Eqs. (4.39) and (4.40), have the following properties:

(i) The eigenvalues λ are independent of time, i.e., $\lambda_t = 0$.
(ii) The compatibility relationship $L_t + (LM - ML) = 0$ must be true; this, together with property (i), implies that the operator L must be self-adjoint, i.e., L is equal to its complex conjugate.

Another form:

$$L\psi = \lambda\psi$$
$$(L - \lambda I)\psi = 0$$
$$\psi_t - M\psi = 0$$
$$(\partial_t - M)\psi = 0$$
$$L'\psi = 0$$
$$M'\psi = 0$$

By the comparison

$$(L'M' - M'L') = 0$$

which is called the zero curvature condition.

Note: Lax adds the compatibility condition for higher dimensional nonlinear evolution equation with some example.

4.8 How to Construct $[L, M]$

Let us now discuss the ways one may give how to construct L and M, so that $[L, M] = LM - ML$ be constructed. L is a composite operator that may consist of a differential function, and scalars and non-differentials are considered to include an identity operator. As an example, an operator could be defined as

$$L = \frac{\partial^n}{\partial x^n} + \alpha u(x,t)I \tag{4.42}$$

where $\frac{\partial^n}{\partial x^n}$ represents the nth order total differential operator and $I = \frac{\partial^0}{\partial x^0}$, being the identity operator. Here α is a scalar and $u = u(x,t)$ is the function of spatial variable x and temporal variable t. If L operates on an auxiliary function, then the result is a new function. Now, L operates on ψI, then we obtained

$$\begin{aligned} L\psi &= \left(\frac{\partial^n}{\partial x^n} + \alpha u(x,t)I\right)(\psi I), \\ \text{or, } L\psi &= \frac{\partial^n(\psi I)}{\partial x^n} + \alpha u(x,t)\psi I \end{aligned} \tag{4.43}$$

However, if L operates on an operator, then the result is also an operator. Let the new operator $M = \psi I$.

For $n = 2$

$$\begin{aligned} LM &= \left(\frac{\partial^2}{\partial x^2} + \alpha u I\right)(\psi I) \\ &= \frac{\partial^2}{\partial x^2}(\psi I) + (\alpha u I)(\psi I) \\ &= \frac{\partial}{\partial x}\left(\frac{\partial \psi}{\partial x} I + \psi \frac{\partial}{\partial x}\right) + \alpha u I \psi I \\ &= \frac{\partial}{\partial x}\left(\frac{\partial \psi}{\partial x} I\right) + \frac{\partial}{\partial x}\left(\psi \frac{\partial}{\partial x}\right) + \alpha u \psi I \\ &= \frac{\partial^2 \psi}{\partial x^2} I + 2\frac{\partial \psi}{\partial x}\frac{\partial}{\partial x} + \psi \frac{\partial^2}{\partial x^2} + \alpha u \psi I \end{aligned} \tag{4.44}$$

For $n = 3$,

$$\begin{aligned} LM &= \left(\frac{\partial^3}{\partial x^3} + \alpha u I\right)\psi I \\ &= \frac{\partial}{\partial x}\left(\frac{\partial^2}{\partial x^2}(\psi I)\right) + \alpha u \psi I \\ &= \frac{\partial}{\partial x}\left(\frac{\partial^2 \psi}{\partial x^2} I + 2\frac{\partial \psi}{\partial x}\frac{\partial}{\partial x} + \psi \frac{\partial^2}{\partial x^2}\right) + \alpha u \psi I \\ &= \frac{\partial^3 \psi}{\partial x^3} I + 3\frac{\partial^2 \psi}{\partial x^2}\frac{\partial}{\partial x} + 3\frac{\partial \psi}{\partial x}\frac{\partial^2}{\partial x^2} + \psi \frac{\partial^3}{\partial x^3} + \alpha u \psi I \end{aligned} \tag{4.45}$$

So by the principle of induction, for n we may consider

$$\begin{aligned} LM = &\binom{n}{0}\frac{\partial^n\psi}{\partial x^n}I + \binom{n}{1}\frac{\partial^{n-1}\psi}{\partial x^{n-1}}\frac{\partial}{\partial x} + \binom{n}{2}\frac{\partial^{n-2}\psi}{\partial x^{n-2}}\frac{\partial^2}{\partial x^2} \\ &+ \binom{n}{3}\frac{\partial^{n-3}\psi}{\partial x^{n-3}}\frac{\partial^3}{\partial x^3} + \cdots + \binom{n}{n-1}\frac{\partial\psi}{\partial x}\frac{\partial^{n-1}}{\partial x^{n-1}} \\ &+ \binom{n}{n}\psi\frac{\partial^n}{\partial x^n} + \alpha u\psi I \end{aligned} \tag{4.46}$$

where $\binom{n}{k} = \frac{n!}{k!(n-k)!}$, $\quad i = 0, 1, 2, 3, 4, \ldots, n$, all are binomial coefficients. Accordingly, $[L, M] = LM - ML$ can be constructed easily.

4.9 Lax Pair of Some Nonlinear Evolution Equations

4.9.1 *Advection equation*

Consider the one-dimensional advection equation

$$\frac{\partial u}{\partial t} + c\frac{\partial u}{\partial x} = 0, \quad u = u(x,t) \tag{4.47}$$

where $c \in \mathbb{R}$.

Also, we assume the Lax pair of the advection equation as

$$L = \frac{\partial^2}{\partial x^2} - uI, \quad M = \alpha\frac{\partial}{\partial x} + B(x,t)I \tag{4.48}$$

where L and M are the Lax operators, α is a real constant, and $B(x,t)$ in M has to be determined. Inserting L and M into the Lax compatibility Eq. (4.41), thus we obtain

$$-\frac{\partial u}{\partial t}I + \left[\frac{\partial^2}{\partial x^2} - uI, \alpha\frac{\partial}{\partial x} + B(x,t)I\right] = 0 \tag{4.49}$$

Now,

$$\begin{aligned} LM &= \left(\frac{\partial^2}{\partial x^2} - uI)(\alpha\frac{\partial}{\partial x} + BI\right) \\ &= \frac{\partial^2}{\partial x^2}\left(\alpha\frac{\partial}{\partial x} + BI\right) - u\alpha\frac{\partial}{\partial x} - uIBI \\ &= \frac{\partial}{\partial x}\left(\alpha\frac{\partial^2}{\partial x^2} + \frac{\partial B}{\partial x}I + B\frac{\partial}{\partial x}\right) - u\alpha\frac{\partial}{\partial x} - uBI \\ &= \alpha\frac{\partial^3}{\partial x3} + B\frac{\partial^2}{\partial x^2} + \left(2\frac{\partial B}{\partial x} - u\alpha\right)\frac{\partial}{\partial x} + \frac{\partial^2 B}{\partial x^2}I - uBI \end{aligned}$$

and similarly

$$ML = \left(\alpha\frac{\partial}{\partial x} + B\right)\left(\frac{\partial^2}{\partial x^2} - uI\right)$$
$$= \alpha\frac{\partial^3}{\partial x^3} - \alpha\frac{\partial u}{\partial x}I - u\alpha\frac{\partial}{\partial x} + B\frac{\partial^2}{\partial x^2} - uBI$$

Now, the commutator

$$[L, M] = 2\frac{\partial B}{\partial x}\frac{\partial}{\partial x} + \frac{\partial^2 B}{\partial x^2}I + \alpha\frac{\partial u}{\partial x}I$$

Then

$$\frac{\partial L}{\partial t} + [L, M] = -\frac{\partial u}{\partial t}I + 2\frac{\partial B}{\partial x}\frac{\partial}{\partial x} + \left(\frac{\partial^2 B}{\partial x^2} + \alpha\frac{\partial u}{\partial x}\right)I$$

Now, for $[L, M]$ to be a multiplicative operator, the following conditions must hold: $\frac{\partial B}{\partial x} = 0$, and setting $\alpha = -c$, then $B(x,t) = p$, for some real constant. If we assume the value of p is 0, then we get,

$$\frac{\partial L}{\partial t} + [L, M] = -\left(\frac{\partial u}{\partial t} + c\frac{\partial u}{\partial x}\right)I$$

Therefore, we finally arrived at

$$\frac{\partial u}{\partial t} + c\frac{\partial u}{\partial x} = 0$$

which is a familiar form of the advection Eq. (4.47). Thus, we have shown that if L and M satisfy the zero curvature Eq. (4.41), then u satisfies the advection Eq. (4.47). It therefore follows that the advection equation can be thought of as the compatibility condition for the Lax pair.

Hence, the *Lax operators* for the advection equation becomes

$$L = \frac{\partial^2}{\partial x^2} - uI, \quad M = -c\frac{\partial}{\partial x}$$

and corresponding the *Lax pair* of the advection equation is

$$\psi_{xx} - u\psi = \lambda\psi, \quad \psi_t + c\psi_x = 0.$$

4.9.2 *KdV equation*

The well-known (1+1)-dimensional KdV equation is defined as

$$\frac{\partial u}{\partial t} - 6u\frac{\partial u}{\partial x} + \frac{\partial^3 u}{\partial x^3} = 0, \quad u = u(x,t) \tag{4.50}$$

Consider the Lax operators in the following form

$$L = \frac{\partial^2}{\partial x^2} - uI,$$

$$M = \alpha\frac{\partial^3}{\partial x^3} - B(x,t)\frac{\partial}{\partial x} - C(x,t)I$$

where the variables α, B, and C in M are unknown. Inserting L and M into the Lax compatibility Eq. (4.41), we have

$$-\frac{\partial u}{\partial t} + \left[\frac{\partial^2}{\partial x^2} - uI, \alpha\frac{\partial^3}{\partial x^3} - B(x,t)\frac{\partial}{\partial x} - C(x,t)I\right] = 0$$

Now, the commutator

$$\begin{aligned}
LM &= \left(\frac{\partial^2}{\partial x^2} - uI\right)\left(\alpha\frac{\partial^3}{\partial x^3} - B\frac{\partial}{\partial x} - CI\right), \\
&= \frac{\partial^2}{\partial x^2}\left(\alpha\frac{\partial^3}{\partial x^3} - B\frac{\partial}{\partial x} - CI\right) - uI\left(\alpha\frac{\partial^3}{\partial x^3} - B\frac{\partial}{\partial x} - CI\right), \\
&\quad - u\alpha\frac{\partial^3}{\partial x^3} + uB\frac{\partial}{\partial x} + uCI, \\
&= \alpha\frac{\partial^5}{\partial x^5} - (B + u\alpha)\frac{\partial^3}{\partial x^3} - \left(2\frac{\partial B}{\partial x} + C\right)\frac{\partial^2}{\partial x^2} \\
&\quad + \left(uB - \frac{\partial^2 B}{\partial x^2} - 2\frac{\partial C}{\partial x}\right)\frac{\partial}{\partial x} + uCI - \frac{\partial^2 C}{\partial x^2}I
\end{aligned}$$

and

$$\begin{aligned}
ML &= \left(\alpha\frac{\partial^3}{\partial x^3} - B\frac{\partial}{\partial x} - CI\right)\left(\frac{\partial^2}{\partial x^2} - uI\right) \\
&= \left(\alpha\frac{\partial^3}{\partial x^3} - B\frac{\partial}{\partial x} - CI\right)\frac{\partial^2}{\partial x^2} - \left(\alpha\frac{\partial^3}{\partial x^3} - B\frac{\partial}{\partial x} - CI\right)uI \\
&= \alpha\frac{\partial^5}{\partial x^5} - (B + \alpha u)\frac{\partial^3}{\partial x^3} - \left(3\alpha\frac{\partial u}{\partial x} + C\right)\frac{\partial^2}{\partial x^2} \\
&\quad + \left(uB - 3\alpha\frac{\partial^2 u}{\partial x^2}\right)\frac{\partial}{\partial x} + \left(B\frac{\partial u}{\partial x} - \alpha\frac{\partial^3 u}{\partial x^3} + uC\right)I
\end{aligned}$$

So,

$$\begin{aligned}
[L,M] &= \left(3\alpha\frac{\partial u}{\partial x} - 2\frac{\partial B}{\partial x}\right)\frac{\partial^2}{\partial x^2} + \left(3\alpha\frac{\partial^2 u}{\partial x^2} - \frac{\partial^2 B}{\partial x^2} - 2\frac{\partial C}{\partial x}\right)\frac{\partial}{\partial x} \\
&\quad + \left(\alpha\frac{\partial^3 u}{\partial x^3} + B\frac{\partial u}{\partial x} - \frac{\partial^2 C}{\partial x^2}\right)I
\end{aligned}$$

Now, for $[L, M]$ to be a multiplicative operator, the following must be true:

$$3\alpha\frac{\partial u}{\partial x} - 2\frac{\partial B}{\partial x} = 0, \tag{4.51}$$

$$3\alpha\frac{\partial^2 u}{\partial x^2} - \frac{\partial^2 B}{\partial x^2} - 2\frac{\partial C}{\partial x} = 0 \tag{4.52}$$

To solve for B and C, we integrate Eqs. (4.51) and (4.52) with respect to x, ignoring the constant of integration. We get

$$3\alpha u - 2B = 0,$$

$$3\alpha\frac{\partial u}{\partial x} - \frac{\partial B}{\partial x} - 2C = 0,$$

which yields

$$B = \frac{3}{2}\alpha u, \quad C = \frac{3}{4}\alpha\frac{\partial u}{\partial x}$$

Using the value of B and C, the commutator becomes

$$[L, M] = \left(-\frac{3}{2}\alpha u\frac{\partial u}{\partial x} + \frac{1}{4}\alpha\frac{\partial^3 u}{\partial x^3}\right)I$$

Setting $\alpha = -4$, thus we finally arrive at

$$\frac{\partial u}{\partial t} - 6u\frac{\partial u}{\partial x} + \frac{\partial^3 u}{\partial x^3} = 0$$

Which again is the familiar form of the KdV Eq. (4.50), so the Lax operators of the KdV equation becomes

$$L = \frac{\partial^2}{\partial x^2} - uI,$$

$$M = -4\frac{\partial^3}{\partial x^3} + 6u\frac{\partial}{\partial x} + 3\frac{\partial u}{\partial x}I$$

corresponding the Lax pair is

$$\psi_{xx} - u\psi = \lambda\psi,$$

$$\psi_t = -4\psi_{xxx} + 6u\psi_x + 3\frac{\partial u}{\partial x}\psi.$$

4.9.3 *mKdV equation*

The well-known mKdV equation is defined as

$$\frac{\partial v}{\partial t} - 6v^2\frac{\partial v}{\partial x} + \frac{\partial^3 v}{\partial x^3} = 0, \quad v = v(x, t) \tag{4.53}$$

The Miura transformation is given as follows:

$$u = \frac{\partial v}{\partial x} + v^2, \quad u = u(x,t), \quad v = v(x,t) \tag{4.54}$$

converts the KdV equation (4.50) into the mKdV. This suggests that it may be possible to use the Miura transformation to convert the KdV Lax pair into an mKdV Lax pair; in fact, this does turn out to be a good approach. The Miura transformation, when applied to the Lax operators for the KdV equation, yields the following Lax operators for the mKdV equation:

$$L = \frac{\partial^2}{\partial x^2} - \left(\frac{\partial v}{\partial x} + v^2\right) I,$$

$$M = -4\frac{\partial^3}{\partial x^3} + 6\left(\frac{\partial v}{\partial x} + v^2\right)\frac{\partial}{\partial x} + 3\left[\frac{\partial}{\partial x}\left(\frac{\partial v}{\partial x} + v^2\right)\right] I$$

Inserting L and M into the Lax compatibility equation (4.41), we obtain

$$\frac{\partial v}{\partial x \partial t} + 2v\frac{\partial v}{\partial t} + \frac{\partial^4 v}{\partial x^4} + 2v\frac{\partial^3 v}{\partial x^3} - 6v^2\frac{\partial^2 v}{\partial x^2} - 12v\left(\frac{\partial v}{\partial x}\right)^2 - 12v^3\frac{\partial v}{\partial x} = 0$$

which can be simplified to

$$\left(2v + \frac{\partial}{\partial x}\right)\left(\frac{\partial v}{\partial t} - 6v^2\frac{\partial v}{\partial x} + \frac{\partial^3 v}{\partial x^3}\right) = 0 \tag{4.55}$$

Clearly, the term $(\frac{\partial v}{\partial t} - 6v^2\frac{\partial v}{\partial x} + \frac{\partial^3 v}{\partial x^3})$ must be zero, from which we deduce that v must be a solution to the mKdV equation. Thus, we have shown that L and M satisfy the zero curvature Eq. (4.41) and v satisfies the mKdV equation. Therefore, it follows that the mKdV equation can be thought of as the compatibility condition for the Lax operators.

4.9.4 *Gardner equation*

Consider the well-known (1+1)-dimensional Gardner equation defined by

$$\frac{\partial u}{\partial t} + \alpha(u + u^2)\frac{\partial u}{\partial x} + \gamma\frac{\partial^3 u}{\partial x^3} = 0, \quad u = u(x,t) \tag{4.56}$$

where α and γ are real non-zero constants.

We proposed the Lax operators in the following form

$$L = \frac{\partial}{\partial x} - uI,$$

$$M = l\frac{\partial^3}{\partial x^3} + B(x,t)\frac{\partial^2}{\partial x^2} + C(x,t)\frac{\partial}{\partial x} + D(x,t)I$$

where l is a real constant and variables B, C, and D in M have yet to be determined. Inserting L and M into the Lax compatibility Eq. (4.41), thus we obtain

$$-\frac{\partial u}{\partial t}I + \left[\frac{\partial}{\partial x} - uI, l\frac{\partial^3}{\partial x^3} + B(x,t)\frac{\partial^2}{\partial x^2} + C(x,t)\frac{\partial}{\partial x} + D(x,t)I\right] = 0$$

Now,

$$\begin{aligned}
LM &= \left(\frac{\partial}{\partial x} - uI\right)\left(l\frac{\partial^3}{\partial x^3} + B\frac{\partial^2}{\partial x^2} + C\frac{\partial}{\partial x} + DI\right), \\
&= \frac{\partial}{\partial x}\left(l\frac{\partial^3}{\partial x^3} + B\frac{\partial^2}{\partial x^2} + C\frac{\partial}{\partial x} + DI\right) \\
&\quad - uI\left(l\frac{\partial^3}{\partial x^3} + B\frac{\partial^2}{\partial x^2} + C\frac{\partial}{\partial x} + DI\right), \\
&\quad - \left(ul\frac{\partial^3}{\partial x^3} + uB\frac{\partial^2}{\partial x^2} + uC\frac{\partial}{\partial x} + uDI\right), \\
&= l\frac{\partial^4}{\partial x^4} + (B - ul)\frac{\partial^3}{\partial x^3} + \left(\frac{\partial B}{\partial x} - uB + C\right)\frac{\partial^2}{\partial x^2} \\
&\quad + \left(\frac{\partial C}{\partial x} + D - uC\right)\frac{\partial}{\partial x} + \left(\frac{\partial D}{\partial x}I - uDI\right)
\end{aligned}$$

and

$$\begin{aligned}
ML &= \left(l\frac{\partial^3}{\partial x^3} + B\frac{\partial^2}{\partial x^2} + C\frac{\partial}{\partial x} + DI\right)\left(\frac{\partial}{\partial x} - uI\right) \\
&= \left(l\frac{\partial^3}{\partial x^3} + B\frac{\partial^2}{\partial x^2} + C\frac{\partial}{\partial x} + DI\right)\frac{\partial}{\partial x} \\
&\quad - \left(l\frac{\partial^3}{\partial x^3} + B\frac{\partial^2}{\partial x^2} + C\frac{\partial}{\partial x} + DI\right)uI \\
&= l\frac{\partial^4}{\partial x^4} + (B - lu)\frac{\partial^3}{\partial x^3} + \left(C - Bu - 3l\frac{\partial u}{\partial x}\right)\frac{\partial^2}{\partial x^2} \\
&\quad + \left(D - Cu - 3l\frac{\partial^2 u}{\partial x^2} - 2B\frac{\partial u}{\partial x}\right)\frac{\partial}{\partial x} \\
&\quad - \left[\left(B\frac{\partial^2 u}{\partial x^2} + l\frac{\partial^3 u}{\partial x^3} + C\frac{\partial u}{\partial x} + Du\right)I\right]
\end{aligned}$$

So the commutator is

$$[L, M] = \left(\frac{\partial B}{\partial x} + 3l\frac{\partial u}{\partial x}\right)\frac{\partial^2}{\partial x^2} + \left(\frac{\partial C}{\partial x} + 3l\frac{\partial^2 u}{\partial x^2} + 2B\frac{\partial u}{\partial x}\right)\frac{\partial}{\partial x}$$
$$+ \left(B\frac{\partial^2 u}{\partial x^2} + \frac{\partial D}{\partial x} + C\frac{\partial u}{\partial x} + l\frac{\partial^3 u}{\partial x^3}\right)I$$

The commutator $[L, M]$ is to be a multiplicative operator; then the following must be true:

$$\frac{\partial B}{\partial x} + 3l\frac{\partial u}{\partial x} = 0, \tag{4.57}$$

$$\frac{\partial C}{\partial x} + 3l\frac{\partial^2 u}{\partial x^2} + 2B\frac{\partial u}{\partial x} = 0 \tag{4.58}$$

$$B\frac{\partial^2 u}{\partial x^2} + \frac{\partial D}{\partial x} + C\frac{\partial u}{\partial x} + l\frac{\partial^3 u}{\partial x^3} = -\alpha(u + u^2)\frac{\partial u}{\partial x} - \gamma\frac{\partial^3 u}{\partial x^3} \tag{4.59}$$

Then

$$\frac{\partial L}{\partial t} + [L, M] = -\left[\frac{\partial u}{\partial t} + \alpha(u + u^2)\frac{\partial u}{\partial x} + \gamma\frac{\partial^3 u}{\partial x^3}\right]I \tag{4.60}$$

Therefore, we finally arrive at

$$\frac{\partial u}{\partial t} + \alpha(u + u^2)\frac{\partial u}{\partial x} + \gamma\frac{\partial^3 u}{\partial x^3} = 0 \tag{4.61}$$

which is a familiar form of Gardner equation (4.56). Thus, we have shown that L and M satisfy the Lax compatibility Eq. (4.41) and u satisfies the Gardner Eq. (4.61). It therefore follows that the Gardner equation can be thought of as the compatibility condition for the Lax operators.

Now, our job is to find the expression for B, C, D, and l for Eqs. (4.57)–(4.59). In Eq. (4.57), we integrate with respect to x, and ignoring the constant of integration, we have

$$B = -3lu$$

Substituting the value of B in Eq. (4.58), we have

$$\frac{\partial C}{\partial x} + 3l\frac{\partial^2 u}{\partial x^2} - 6lu\frac{\partial u}{\partial x} = 0$$

Now, integrate the above equation with respect to x, and ignoring the constant of integration, we get

$$C = -3l\frac{\partial u}{\partial x} + 3lu^2$$

Substituting the values of B and C with $l = \gamma$ in Eq. (4.59), we get

$$\frac{\partial D}{\partial x} + \left(3\gamma\frac{\partial u}{\partial x} - 3\gamma u^2\right)\frac{\partial u}{\partial x} + 3\gamma u\frac{\partial^2 u}{\partial x^2} = -\alpha(u+u^2)\frac{\partial u}{\partial x}$$

$$\Longrightarrow \frac{\partial D}{\partial x} + 3\gamma\left[u\frac{\partial^2 u}{\partial x^2} + \left(\frac{\partial u}{\partial x}\right)^2\right] - 3\gamma u^2\frac{\partial u}{\partial x} = -\alpha(u+u^2)\frac{\partial u}{\partial x}$$

$$\Longrightarrow \frac{\partial D}{\partial x} + 3\gamma\frac{\partial}{\partial x}\left(u\frac{\partial u}{\partial x}\right) = -[\alpha(u+u^2) - 3\gamma u^2]\frac{\partial u}{\partial x}$$

$$\Longrightarrow D + 3\gamma u\frac{\partial u}{\partial x} = -\alpha\left(\frac{u^2}{2} + \frac{u^3}{3}\right) + \gamma u^3$$

Therefore, we have

$$l = -\gamma, \quad B = 3\gamma u, \quad C = 3\gamma\frac{\partial u}{\partial x} - 3\gamma u^2,$$

$$D = \gamma u^3 - \alpha\left(\frac{u^2}{2} + \frac{u^3}{3}\right) - 3\gamma u\frac{\partial u}{\partial x}$$

Substituting the values of l, B, C, and D in Eq. (4.9.4), we get the Lax operator of the Gardner equation. Therefore, the Lax operator becomes

$$L = \frac{\partial}{\partial x} - uI,$$

$$M = -\gamma\frac{\partial^3}{\partial x^3} + 3\gamma u\frac{\partial^2}{\partial x^2} + \left(3\gamma\frac{\partial u}{\partial x} - 3\gamma u^2\right)\frac{\partial}{\partial x} + \left[\gamma u^3 - \alpha\left(\frac{u^2}{2} + \frac{u^3}{3}\right) - 3\gamma u\frac{\partial u}{\partial x}\right] I$$

corresponding lax pair is

$$\psi_x - u\psi = \lambda\psi$$

$$\psi_t = -\gamma\psi_{xxx} + 3\gamma u\psi_{xx} + \left(3\gamma\frac{\partial u}{\partial x} - 3\gamma u^2\right)\psi_x + \left[\gamma u^3 - \alpha\left(\frac{u^2}{2} + \frac{u^3}{3}\right) - 3\gamma u\frac{\partial u}{\partial x}\right]\psi$$

4.9.5 *Burgers' equation*

The Burgers' equation is defined by

$$\frac{\partial u}{\partial t} + \alpha u\frac{\partial u}{\partial x} + \beta\frac{\partial^2 u}{\partial x^2} = 0, \quad u = u(x,t) \tag{4.62}$$

where α, β are real non-zero constant.

Consider the Lax operators

$$L = l\frac{\partial}{\partial x} + uI, \quad M = m\frac{\partial^2}{\partial x^2} + B(x,t)\frac{\partial}{\partial x} \tag{4.63}$$

where constants l, m, and variables $B(x,t)$ in M have yet to be determined. Inserting L and M into Lax compatibility equation (4.41).

We obtained

$$\frac{\partial u}{\partial t}I + \left[l\frac{\partial}{\partial x} + uI, m\frac{\partial^2}{\partial x^2} + \beta\frac{\partial}{\partial x}\right] = 0 \tag{4.64}$$

Now,

$$\begin{aligned}
LM &= \left(l\frac{\partial}{\partial x} + uI\right)\left(m\frac{\partial^2}{\partial x^2} + B(x,t)\frac{\partial}{\partial x}\right)\\
&= lm\frac{\partial^3}{\partial x^3} + l\frac{\partial B}{\partial x}\frac{\partial}{\partial x} + lB\frac{\partial^2}{\partial x^2} + um\frac{\partial^2}{\partial x^2} + uB\frac{\partial}{\partial x}\\
&= lm\frac{\partial^3}{\partial x^3} + (lB + um)\frac{\partial^2}{\partial x^2} + \left(uB + \frac{\partial B}{\partial x}\right)\frac{\partial}{\partial x}\\
ML &= \left(m\frac{\partial^2}{\partial x^2} + B\frac{\partial}{\partial x}\right)\left(l\frac{\partial}{\partial x} + uI\right)\\
&= ml\frac{\partial^3}{\partial x^3} + (Bl + mu)\frac{\partial^2}{\partial x^2} + \left(2m\frac{\partial u}{\partial x} + Bu\right)\frac{\partial}{\partial x} + \left(m\frac{\partial^2 u}{\partial x^2} + B\frac{\partial u}{\partial x}\right)I
\end{aligned}$$

Therefore, commutator

$$[L, M] = \left(l\frac{\partial B}{\partial x} - 2m\frac{\partial u}{\partial x}\right)\frac{\partial}{\partial x} - \left[m\frac{\partial^2 u}{\partial x^2} + B\frac{\partial u}{\partial x}\right]I \tag{4.65}$$

Now, for $[L, M]$ to be a multiplicative operator, the following must be true:

$$l\frac{\partial B}{\partial x} - 2m\frac{\partial u}{\partial x} = 0$$

Now, we find the expression of m, l, and B. To solve for B, we integrate the above equation with respect to x, ignoring the constant of integration. We get

$$lB - 2mu = 0$$

which yields

$$B = \frac{2mu}{l}$$

Substituting the value of B in Eq. (4.65), we obtained

$$[L, M] = -\left(m\frac{\partial^2 u}{\partial x^2} + \frac{2um}{l}\frac{\partial u}{\partial x}\right)I$$

Setting $m = -\beta$ and $l = \frac{2\beta}{\alpha}$, we have

$$\frac{\partial L}{\partial t} + [L, M] = -\left(\frac{\partial u}{\partial t} + \alpha\frac{\partial u}{\partial x} + \beta\frac{\partial^2 u}{\partial x^2}\right)I$$

Therefore, we finally arrive at

$$\frac{\partial u}{\partial t} + \alpha\frac{\partial u}{\partial x} + \beta\frac{\partial^2 u}{\partial x^2} = 0$$

which is a familiar form of the Burger equation. Thus, we have shown that L and M satisfy the zero curvature Eq. (4.41) and u satisfies the Burger equation. It therefore follows that the Burger equation can be thought of as the compatibility condition for the Lax pair operator.

Substituting the values of m, l, and B in Eq. (4.63), we get the Lax operators of the Burgers' equation

$$L = \frac{2\beta}{\alpha}\frac{\partial}{\partial x} + uI,$$
$$M = -\beta\frac{\partial^2}{\partial x^2} - \alpha u\frac{\partial}{\partial x}$$

corresponding Lax pair of the Burger equation is

$$\frac{2\beta}{\alpha}\psi + u\psi = \lambda\psi,$$
$$\psi_t = -\beta\psi_{xx} - \alpha u\psi_x.$$

4.9.6 *Fifth-order KdV equation*

The generalized Lax equation is written in the form

$$u_t + \frac{3}{10}\gamma^2 u^2 u_x + 2\gamma u_x u_{xx} + \gamma u u_{xxx} + u_{xxxxx} = 0 \tag{4.66}$$

where $u = u(x, t)$ is a sufficiently often differential (or differentiable) function. Equation (4.66) is a special case of the generalized fifth-order KdV equation (fKdV)

$$u_t + \alpha u^2 u_x + \beta u_x u_{xx} + \gamma u u_{xxx} + u_{xxxxx} = 0 \tag{4.67}$$

where α, β, and γ are real non-zero parameters. The general Lax equation is characterized by $\beta = 2\gamma$ and $\alpha = \frac{3}{10}\gamma^2$. Using the transformation

$$u(x,t) = \frac{10}{\gamma} v(x,t) \tag{4.68}$$

in Eq. (4.66), one can obtain the standard Lax equation as

$$v_t + 30v^2 v_x + 20v_x v_{xx} + 10v v_{xxx} + v_{xxxxx} = 0 \tag{4.69}$$

On the other hand, Eq. (4.69) can be obtained from Eq. (4.66) for the choice of the parameter $\gamma = 10$. By choosing the different real values of the parameters, a class of fKdV equations can be obtained from Eq. (4.67). The most celebrated fKdV equation, which was obtained by different real values of the parameters α, β, and γ, is the Caudrey–Dodd–Gibbon (CDG) equation, the Swada–Kotera (SK) equation, the Kaup–Kuperschmidt (KP) equation, and the Ito equation. These forms include the following:

(i) the Caudrey–Dodd–Gibbon equation is

$$u_t + 180u^2 u_x + 30u_x u_{xx} + 30u u_{xxx} + u_{xxxxx} = 0$$

(ii) the Sawada–Kotera equation is

$$u_t + \frac{1}{5}\gamma^2 u^2 u_x + \gamma u_x u_{xx} + \gamma u u_{xxx} + u_{xxxxx} = 0$$

where the value of $\gamma = 15$ is commonly used.

(iii) the Kaup–Kuperschmidt equation is

$$u_t + 20u^2 u_x + 25u_x u_{xx} + 10u u_{xxx} + u_{xxxxx} = 0$$

which is characterized by $\beta = \frac{5}{2}\gamma$, $\alpha = \frac{1}{5}\gamma^2$,

(iv) the Ito equation is

$$u_t + 2u^2 u_x + 6u_x u_{xx} + 3u u_{xxx} + u_{xxxxx} = 0$$

Equation (4.69) plays an important role in studying mathematical models in quantum mechanics, nonlinear optics, solid-state physics, and plasma dynamics. Furthermore, it describes the motions of long waves in shallow water under gravity and a one-dimensional nonlinear lattice. It is well known that Eqs. (4.66) and (4.69) are completely integrable, and they have infinite sets of conservation laws. The Lax operators for Eq. (4.66) are defined by

$$L = \partial_x^2 + \frac{1}{10}\gamma u I$$

$$M = -16\partial_x^5 - 4\gamma u\partial_x^3 - 6\gamma u_x\partial_x^2 - \left(\frac{3}{10}\gamma^2u^2 + 5\gamma u_{xx}\right)\partial_x - \left(\frac{3}{10}\gamma^2uu_x + \frac{3}{2}\gamma u_{xxx}\right)I$$

Thus, we have shown that L and M satisfy the zero curvature Eq. (4.41) and u satisfies the Lax equation. It follows that the Lax equation can be thought of as the compatibility condition for the Lax pair operators. Therefore, the Lax pair corresponding to L and M is

$$\psi_{xx} = \lambda\psi - \frac{1}{10}\gamma u\psi,$$

$$\psi_t = -16\psi_{xxxxx} - 4\gamma u\psi_{xxx} - 6\gamma u_x\psi_{xx} - \left(\frac{3}{10}\gamma^2u^2 + 5\gamma u_{xx}\right)\psi_x - \left(\frac{3}{10}\gamma^2uu_x + \frac{3}{2}\gamma u_{xxx}\right)\psi$$

4.9.7 *Boussinesq equation*

The Boussinesq equation is

$$u_{tt} - u_{xt} - u_{xx} + 3(u^2)_{xx} - u_{xxxx} = 0 \tag{4.70}$$

A new variable that is defined as

$$u = -\sigma_{xx} \tag{4.71}$$

Substituting (4.71) into Eq. (4.70) and integrating twice with respect to the variable x and taking the constant of integration to be zero, therefore, we have

$$\sigma_{2t} - \sigma_{2x} - (3\sigma_{2x}^2 + \sigma_{4x}) = 0$$

The Lax operators of the Boussinesq equation are as follows:

$$L = 4\partial_x^3 + (1 + 6\sigma_{2x})\partial_x + (3\sigma_{3x} - 2w_0a\sigma_{2x} + a\sigma_{x,t} - \lambda),$$

$$M = \partial_t + a\partial_x^2 + a\sigma_{2x}$$

where $a^2 = -3$ and λ is the spectral parameter. Thus, the compatibility condition $[L, M] \equiv LM - ML = 0$ is equivalent to the following results:

$$\partial_x(\sigma_{2t} - \sigma_{2x} - (3\sigma_{2x}^2 + \sigma_{4x})) = 0$$

4.9.8 *Geophysical Boussinesq equation*

The geophysical Boussinesq (GB) equation is

$$u_{tt} - 2w_0 u_{xt} - u_{xx} + 3(u^2)_{xx} - u_{xxxx} = 0 \tag{4.72}$$

A new variable that is defined as

$$u = -\sigma_{xx} \tag{4.73}$$

Substituting (4.73) into Eq. (4.72) and integrating twice with respect to the variable x and taking the constant of integration is zero, therefore we have

$$\sigma_{2t} - 2w_0\sigma_{x,t} - \sigma_{2x} - (3\sigma_{2x}^2 + \sigma_{4x}) = 0$$

The Lax operators of geophysical Boussinesq equation are

$$\begin{aligned} L &= 4\partial_x^3 - 2w_0 a\partial_x^2 + (1 + 6\sigma_{2x})\partial_x + (3\sigma_{3x} - 2w_0 a\sigma_{2x} + a\sigma_{x,t} - \lambda), \\ M &= \partial_t + a\partial_x^2 + a\sigma_{2x} \end{aligned}$$

where $a^2 = -3$ and λ is the spectral parameter. Thus, the compatibility condition $[L, M] \equiv LM - ML = 0$, is equivalent to the following results:

$$\partial_x(\sigma_{2t} - 2w_0\sigma_{x,t} - \sigma_{2x} - (3\sigma_{2x}^2 + \sigma_{4x})) = 0$$

4.10 Lax Pairs in Matrix Form

In 1974, Ablowitz, Kaup, Newell, and Segur published a matrix formalism for Lax pairs where they introduced a construction that avoids the need to consider higher-order Lax operators. This method is also referred to as the *AKNS method.* In their analysis, they introduced the following system:

$$D_x\psi = U\psi \tag{4.74}$$

$$D_t\psi = V\psi \tag{4.75}$$

where U and V correspond to operators L and M, respectively, and ψ is an auxiliary vector function. Matrices U and V will, in general, both depend on the time-independent eigenvalue λ, and the size of ψ will depend upon the order of L. Thus, if L is of order 2, then the vector ψ will have two elements, and U and V will each be a 2×2 matrix. The compatibility condition for Eqs. (4.74) and (4.75) is

$$\begin{aligned} [D_t, D_x]\psi &= D_t(U\psi) - D_x(V\psi) \\ &= (D_tU)\psi - UD_t\psi - (D_xV)\psi - VD_x\psi = 0 \end{aligned}$$

This can be written as

$$(D_tU - D_xV + [U,V])\psi = 0 \tag{4.76}$$

which is known as the *matrix Lax* equation, and where

$$[U,V] = UV - VU$$

is defined as the matrix commutator. Eq. (4.76) is also known as the *zero-curvature* equation.

4.10.1 *ODEs and PDEs*

Simple harmonic oscillator (SHO): The simple harmonic oscillator is a system that describes oscillations of a particle of mass m due to a restoring force $-kx$, where x is the displacement from equilibrium and $k(>0)$ is the force constant. The equation of motion is

$$m\ddot{x} = -kx \tag{4.77}$$

Now, we find the Lax pair of (4.77). The general solution of (4.77) can be written as $x(t) = A\cos(\omega t + \phi)$, where A and ϕ are constants of integration with $\omega = \sqrt{k/m}$.

The momentum $p = m\dot{x}$, so (4.77) can be written as a pair of first-order equations $\dot{x} = p/m$ and $\dot{p} = -m\omega^2 x$. Then

$$\begin{aligned}\frac{d(\omega x)}{dt} &= \omega(\frac{p}{m}),\\ \frac{d(p/m)}{dt} &= -\omega(\omega x)\end{aligned} \tag{4.78}$$

Now, we need to find two matrices U and V such that $\dot{U} = [U,V]$. These equations are equivalent to the Lax equation $\dot{U} = [U,V]$ for the pair of 2×2 matrices

$$U = \begin{pmatrix} p/m & \omega x \\ \omega x & -p/m \end{pmatrix}, \quad V = \begin{pmatrix} 0 & \omega/2 \\ -\omega/2 & 0 \end{pmatrix} \tag{4.79}$$

The entries rely on ωx and p/m. Indeed, (4.78) is linear in $\omega\dot{x}$ and $\dot{p}/m$. To reproduce (4.78) for $\dot{U} = [U,V]$, we set U to be linear in ωx and p/m. Consider U as a 2×2 real matrix. This produces four equations of motion. To ensure there are only 2 independent equations as in (4.78), we treat U as a traceless symmetric matrix with entries linear in p/m and ωx, as in (4.79). So, $\dot{U}$ is also symmetric. Because the commutator of symmetric and antisymmetric matrices is symmetric, we choose V as antisymmetric. Since the RHS of (4.78) is linear in ωx and p/m, we assume V is independent of these variables, implying that $[U, V]$ is similarly linear. The RHS of (4.78) is independent of m and linear in ω, implying that the elements of the antisymmetric matrix V can only depend on ω and must be linear inside it. This indicates that V is present in (4.79). Then, it is verified that (4.79) corresponds to (4.78).

$$\dot{U} = \begin{pmatrix} \dot{p}/m & \omega\dot{x} \\ \omega\dot{x} & -\dot{p}/m \end{pmatrix}, \quad [U, V] = \begin{pmatrix} -\omega^2 x & \omega p/m \\ \omega p/m & \omega^2 x \end{pmatrix} \tag{4.80}$$

Hence, $\dot{U} = [U, V]$ gives Eq. (4.78).

mKdV equation:

For the mKdV equation

$$u_t + 6u^2 u_x + u_{xxx} = 0, \quad u = u(x, t) \tag{4.81}$$

If we choose

$$U = \begin{pmatrix} \lambda & u \\ -u & -\lambda \end{pmatrix},$$

$$V = \begin{pmatrix} -4\lambda^3 - 2u^2\lambda & -4u\lambda^2 - 2u_x\lambda - 2u^3 - u_{xx} \\ 4u\lambda^2 - 2u_x\lambda + 2u^3 + u_{xx} & 4\lambda^3 + 2u^2\lambda \end{pmatrix}$$

then

$$U_t = \begin{pmatrix} 0 & u_t \\ -u_t & -\lambda \end{pmatrix}$$

$$V_x = \begin{pmatrix} -4\lambda u u_x & -4\lambda^2 u_x - 2\lambda u_{xx} - 6u^2 u_x - u_{xxx} \\ 4\lambda^2 u_x - 2\lambda u_{xx} + 6u^2 u_x + u_{xxx} & 4\lambda u u_x \end{pmatrix}$$

Therefore,

$$[U, V] = UV - VU = \begin{pmatrix} -4\lambda u u_x & -2\lambda(2\lambda u_x + u_{xx}) \\ 2\lambda(2\lambda u_x - u_{xx}) & 4\lambda u u_{xx} \end{pmatrix}$$

So,

$$U_t - V_x + [U, V] = \begin{pmatrix} 0 & u_t + 6u^2 u_x + u_{xxx} \\ -u_t - 6u^2 u_x - u_{xxx} & 0 \end{pmatrix}$$

Therefore, $U_t - V_x + [U, V] = O \implies u_t + 6u^2 u_x + u_{xxx} = 0$.

Here, U and V are Lax pair of the mKdV equation.

Another matrix Lax pair may be given by

$$U = \begin{pmatrix} -i\lambda & u \\ u & i\lambda \end{pmatrix},$$

$$V = \begin{pmatrix} -4i\lambda^3 - 2i\lambda u^2 & 4\lambda^2 u + 2i\lambda u_x + 2u^3 - u_{xx} \\ 4\lambda^2 u - 2i\lambda u_x + 2u^3 - u_{xx} & 4i\lambda^3 + 2i\lambda u^2 \end{pmatrix}$$

Similarly, one can obtain the matrix Lax pair for the KdV equation also.

Note: Matrix Lax pair is not unique.

KdV equation:

For the KdV equation

$$u_t + auu_x + u_{xxx} = 0 \tag{4.82}$$

The matrix Lax pair for (4.82) is given by

$$U = \begin{pmatrix} 0 & 1 \\ \lambda - \frac{a}{6}u & 0 \end{pmatrix},$$

$$V = \begin{pmatrix} \frac{a}{6}u_x & -4\lambda - \frac{a}{3}u \\ -4\lambda^2 + \frac{\lambda a}{3}u + \frac{a^2}{18}u^2 + \frac{a}{6}u_{xx} & -\frac{a}{6}u_x \end{pmatrix}$$

and for the KdV equation is

$$u_t + muu_x + nu_{xxx} = 0, \tag{4.83}$$

where m, n are arbitrary constants. The matrix Lax pair is given by

$$U = \begin{pmatrix} 0 & 1 \\ \lambda - \frac{1}{6}\frac{m}{n}u & 0 \end{pmatrix}, \quad V = \begin{pmatrix} A & B \\ C & -A \end{pmatrix}$$

where $A = \frac{1}{6}mu_{xx}$, $B = -4n\lambda - \frac{1}{3}mu$, $C = (-4n\lambda - \frac{1}{3}mu)(\lambda - \frac{1}{6}\frac{m}{n}u) + \frac{1}{6}mu_{xx}$.

Burgers' equation: Consider the Burgers' equation

$$u_t - 2uu_x - u_{xx} = 0, \quad u = u(x,t) \tag{4.84}$$

A matrix Lax pair is given by

$$U = \begin{pmatrix} \frac{1}{2}(u+\lambda) & 0 \\ 1 & -\frac{1}{2}(u+\lambda) \end{pmatrix},$$

$$V = \begin{pmatrix} \frac{1}{2}(u_x + u^2 - \lambda^2) & 0 \\ u - \lambda & -\frac{1}{2}(u_x + u^2 - \lambda^2) \end{pmatrix}$$

Gardner equation:

Consider the Gardner equation

$$u_t + auu_x + bu^2u_x + cu_{xxx} = 0 \tag{4.85}$$

where $u(x,t)$ is an independent variable that presents the amplitude of the wave model and evaluates the time of the virtual displacement over the isopycnal surface, x signifies the scaled space variable along the direction of wave mode, and t determines the scaled time. a and b are the coefficients of the quadratic and cubic nonlinear terms, respectively, whereas c presents a dispersive effect. Considering the constants a, b, c are taken as $a = k_1c$ and $b = k_2c$ (k_1, k_2 are non-zero constants). The Lax pair is written as

$$\psi_x = U\psi = \begin{pmatrix} \lambda & u \\ -\frac{1}{6}(k_2u + k_1) & -\lambda \end{pmatrix} \psi$$

$$\psi_t = V\psi = \begin{pmatrix} A & B \\ C & -A \end{pmatrix} \psi$$

where $A = -4c\lambda^3 - \frac{1}{3}c(k_2u^2 + k_1u)\lambda - \frac{1}{6}k_1cu_x$, $B = -(4cu\lambda^2 + 2cu_x\lambda + \frac{1}{3}k_2cu^3 + \frac{1}{3}k_1cu^2 + cu_{xx})$, $C = \frac{2}{3}c(k_1 + k_2u)\lambda^2 - \frac{1}{3}k_2cu_x\lambda + \frac{1}{18}k_2^2cu^3 + \frac{1}{9}k_1k_2cu^2 + \frac{1}{18}k_1^2cu + \frac{1}{6}k_2cu_{xx}$.

Variable-coefficient mKdV equation:

Consider the variable-coefficient mKdV (vc-mKdV) equation

$$u_t + \{f(t)u^2 + q(t)\}u_x + g(t)u_{xxx} + l(t)u = 0 \tag{4.86}$$

It has been shown that when $f(t), g(t), l(t)$ satisfy the relation

$$f(t) = \frac{6}{C^2} g(t) e^{2\int l(t)\,dt} \tag{4.87}$$

where C is an arbitrary nonzero constant, Eq. (4.86) becomes integrable. According to the AKNS procedure, the Lax pair of Eq.(4.86) is given by

$$\Phi_x = U\Phi, \tag{4.88}$$

where $U = \lambda \begin{pmatrix} 1 & 0 \\ 0 & -1 \end{pmatrix} + \begin{pmatrix} 0 & p \\ q & 0 \end{pmatrix} = \begin{pmatrix} \lambda & e^{\int l(t)\,dt} u \\ -\frac{1}{C^2} e^{\int l(t)\,dt} u & -\lambda \end{pmatrix}$,

$$\Phi_t = V\Phi \tag{4.89}$$

where $V = \begin{pmatrix} A & B \\ C & -A \end{pmatrix}$, $A = \sum_{j=0}^{3} a_j(x,t)\lambda^{3-j}$, $B = \sum_{j=0}^{3} b_j(x,t)\lambda^{3-j}$, $C = \sum_{j=0}^{3} c_j(x,t)\lambda^{3-j}$
where $a_0 = \alpha_0(t)$, $b_0 = c_0 = 0$, $a_1 = \alpha_1(t)$, $b_1 = \alpha_0(t)p$, $c_1 = \alpha_0(t)q$, $a_2 = -\frac{1}{2}\alpha_0(t)pq + \alpha_2(t)$, $b_2 = \frac{1}{2}\alpha_0(t)p_x + \alpha_1(t)p$, $c_2 = -\frac{1}{2}\alpha_0(t)q_x + \alpha_1(t)q$, $a_3 = \frac{1}{4}\alpha_0(t)(pq_x - qp_x) - \frac{1}{2}\alpha_1(t)pq + \alpha_3(t)$, $b_3 = \frac{1}{4}\alpha_0(t)(p_{xx} - 2p^2q) + \frac{1}{2}\alpha_1(t)p_x + \alpha_2(t)p$, $c_3 = \frac{1}{4}\alpha_0(t)(q_{xx} - 2pq^2) - \frac{1}{2}\alpha_1(t)q_x + \alpha_2(t)q$. $\alpha_0(t)$, $\alpha_1(t)$, $\alpha_2(t)$, $\alpha_3(t)$ are arbitrary functions of t.

In the above example, $\alpha_0 = (-4)$(coefficient of u_{xxx}), $\alpha_1(t) = 0$, $\alpha_2(t) = (-1)$(coefficient of u_x), $\alpha_3(t) = 0$.

Remark: (i) To find the Lax pair in matrix form it is enough to find the values of p, q, $\alpha_0(t)$, $\alpha_1(t)$, $\alpha_2(t)$, and $\alpha_3(t)$.
(ii) trace$(U) = 0 =$ trace(V).

Variable coefficient Gardner equation:

Let us consider the Gardner equation with the time-dependent coefficients,

$$u_t + \{b(t)u^2 + d(t) + a(t)u\}u_x + c(t)u_{xxx} + f(t)u = 0 \tag{4.90}$$

Eq.(4.90) will be integrable if $b(t)$, $a(t)$, $c(t)$, $f(t)$ satisfy the relations

$$b(t) = k_2 c(t) e^{2\int f(t)\,dt}, \quad a(t) = k_1 c(t) e^{\int f(t)\,dt} \tag{4.91}$$

Letting

$$p = e^{\int f(t)\,dt} u, \quad q = -\frac{1}{6}\left(k_2 e^{\int f(t)\,dt} u + k_1\right) \tag{4.92}$$

where k_1, k_2 are arbitrary constants, the Lax pair of Eq. (4.90) is given by

$$\Phi_x = U\Phi$$

where $U = \lambda \begin{pmatrix} 1 & 0 \\ 0 & -1 \end{pmatrix} + \begin{pmatrix} 0 & p \\ q & 0 \end{pmatrix} = \begin{pmatrix} \lambda & e^{\int f(t)\,dt}u \\ -\frac{1}{6}(k_2 e^{\int f(t)\,dt}u + k_1) & -\lambda \end{pmatrix}$,

$$\Phi_t = V\Phi$$

where $V = \begin{pmatrix} A & B \\ C & -A \end{pmatrix}$, $A = \sum_{j=0}^{3} a_j(x,t)\lambda^{3-j}$, $B = \sum_{j=0}^{3} b_j(x,t)\lambda^{3-j}$, $C = \sum_{j=0}^{3} c_j(x,t)\lambda^{3-j}$.

4.11 Inverse Scattering Method

Let us first discuss the Fourier transformation (FT) before we plunge into the inverse scattering method. Let us assume the following conditions on $f(x)$:

(i) $f(x)$ and $f'(x)$ are piecewise continous in every finite integral.
(ii) $\int_{-\infty}^{\infty} |f(x)|dx$ converges, i.e., $f(x)$ is absolutely integrable in $(-\infty, \infty)$.

Then *Fourier's integral theorem* states that

$$f(x) = \int_0^{\infty} (A(\alpha)\cos(\alpha x) + B(\alpha)\sin(\alpha x))d\alpha \tag{4.93}$$

where

$$A(\alpha) = \frac{1}{\pi}\int_{-\infty}^{\infty} f(x)\cos(\alpha x)dx \tag{4.94}$$

$$B(\alpha) = \frac{1}{\pi}\int_{-\infty}^{\infty} f(x)\sin(\alpha x)dx \tag{4.95}$$

The result (4.93) holds $f(x)$ is a point of continuity of $f(x)$. The right-hand side of (4.93) is sometimes called a *Fourier integral expansion* of $f(x)$.

(4.93) can also be written as

$$f(x) = \frac{1}{\pi}\int_{\alpha=0}^{\infty}\int_{u=-\infty}^{\infty} f(u)\cos\alpha(x-u)dud\alpha \tag{4.96}$$

$$f(x) = \frac{1}{2\pi}\int_{-\infty}^{\infty}\int_{-\infty}^{\infty} f(u)e^{i\alpha(x-u)}dud\alpha \tag{4.97}$$

$$f(x) = \frac{1}{2\pi}\int_{-\infty}^{\infty} e^{i\alpha x}d\alpha\int_{-\infty}^{\infty} f(u)e^{-i\alpha u}du \tag{4.98}$$

From (4.98) it follows that if

$$F(\alpha) = \int_{-\infty}^{\infty} f(u)e^{-i\alpha u} du \tag{4.99}$$

then

$$f(x) = \frac{1}{2\pi} \int_{-\infty}^{\infty} f(\alpha)e^{i\alpha x} d\alpha \tag{4.100}$$

The function $F(\alpha)$ is called the *Fourier transform* of $f(x)$ and is sometimes written as $F(\alpha) = \mathscr{F}\{f(x)\}$. The function $f(x)$ is the *inverse Fourier transform* of $F(\alpha)$ and is written $f(x) = \mathscr{F}^{-1}\{F(\alpha)\}$. By using the Fourier transformation method, we can solve linear PDEs only. So a new method is developed to handle nonlinear PDEs, which is called the inverse scattering method.

Now, we will discuss in detail the inverse scattering method. The inverse scattering method is important because it uses the linear technique to solve the initial value problem for a wide variety of nonlinear wave equations of physical interest and to obtain N-soliton solutions. The inverse scattering method was first discovered and developed by Gardner, Geene, Kruskal, and Miura for the KdV equation. A general formulation of the method by Peter Lax soon followed. The inverse scattering method derives its name from its close mathematical connection for the KdV case to the quantum mechanical scattering of a particle by a localized potential or tunneling through a barrier.

The method of inverse scattering, or the inverse scattering transform, will be used to solve the initial-value problem. To solve the initial-value problem, we shall introduce an auxiliary or transformed problem, namely a linear one-dimensional scattering problem in which the time t of the KdV equation appears only as a parameter. The initial solution of the KdV equation is the scattering "potential", which gives the scattering "data", i.e., the discrete eigenvalues and the reflection coefficient of the scattering problem, at $t = 0$. With the aid of an ingeniously chosen evolution equation, we shall then deduce quite simply the scattering data for $t > 0$ in terms of the scattering data for $t = 0$. Finally, we shall invert the scattering data to find the "potential" for $t > 0$ and identify the potential with the solution $u(x, t)$ of the KdV equation. The method takes us a long way to solve the nonlinear equation and seems to be tractable because the scattering problem is not only an ordinary differential system but also a linear one. The final solution, when all is done, is not easy to untangle, but it does give qualitative as well as quantitative results.

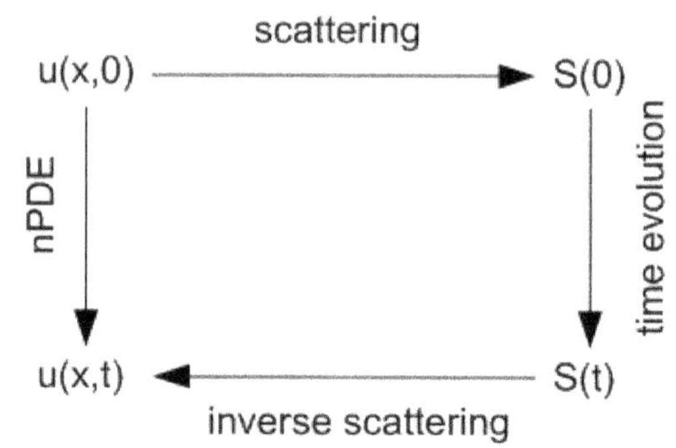

Figure 4.3. Diagram of the Inverse Scattering Method

The solution steps for the inverse scattering transform are summarized as follows:

(i) The starting point is to consider an NLPDEs for certain initial conditions $u(x, 0)$.
(ii) By a scattering process, the scattering data $S(0)$ is to be obtained at the initial time $t = 0$ from the initial data.
(iii) Since the characteristic data of the scattering process is related to a linear problem, the time evolution of the scattering data for the asymptotic behavior $|x| \to \infty$, can be determined easily.
(iv) Finally the solution $u(x, t)$ be obtained by using the inverse scattering process. The inverse scattering process is closely related to a linear integro-differential equation, the Marchenko equation.

The most prominent solutions contained in this process are for solitons and multisolitions. The solution process discussed so far is not only applicable to the KdV equation but also delivers solutions for more complicated equations. The IST procedure is not only applicable to all nonlinear initial value problems. There exists, however, a set of equations for which the IST procedure works very well. One of these equations is the KdV equation, which is the completely integrable equation. Other types of nonlinear equations can be solved by Lie's symmetry analysis. The IST method is closely related to its linear counterpart, the FT. Both IST and FT are also logically related to the Strum–Liouville problem. The main difference between IST and FT is that the FT is only capable of solving linear problems, whereas the IST can also be applied to nonlinear differential equations.

4.12 Solution of KdV Equation

Now, we derive the analytical solutions of the KdV equations using initial conditions. The KdV equation is given by

$$u_t + 6uu_x + u_{xxx} = 0, \quad t > 0, \quad -\infty < x < \infty \tag{4.101}$$

and the initial condition $u(x, t = 0) = u_0(x)$. We assume natural boundary conditions, i.e., the solution of the KdV Eq. (4.101) vanishes sufficiently fast at $|x| \to \infty$ ($u \to 0$ as $x \to \pm\infty$). We use the inverse scattering theory.

A Lax pair for the KdV equation is considered as

$$v_{xx} + uv = \lambda v \tag{4.102}$$

$$v_t = (u_x + \gamma)v - (4\lambda + 2u)v_x \tag{4.103}$$

where γ is constant depending on the normalization of u and λ is the spectrum parameter. The direct problem is at time $t = 0$, given $u(x, 0) = u_0$. The spectrum of these equations consists of a finite number of discrete eigenvalues, $\lambda = K_n^2$, $n = 1, 2, \ldots, N$, for $\lambda > 0$, and a continuum, $\lambda = -k^2$, for $\lambda < 0$.

When $\lambda = -k^2$, the Lax pair for the KdV equation is

$$v_{xx} + (u + k^2)v = 0 \tag{4.104}$$

$$v_t = (u_x + \gamma)v + (4k^2 - 2u)v_x \tag{4.105}$$

Since λ and γ are independent of time t. So, $\gamma_t = k_t = 0$. The consistency $v_{txx} = v_{xxt}$ of the system of u, it satisfies the KdV equation, we get

$$v_{txx} - v_{xxt} = (u_t + 6uu_x + u_{xxx})v \tag{4.106}$$

Equation (4.104) defines the forward scattering problem, while Eq. (4.105) defines the time evolution of the scattering problem. Let's assume that there is some real-valued initial condition $u(x, 0)$ that satisfies

$$\int_{-\infty}^{\infty} |u(x, 0)|(1 + |x|))dx < \infty \tag{4.107}$$

that is related to the integrability condition. This initial condition is called the *Faddeev condition.*

The forward scattering problem determines $v(x, 0; \lambda)$ to

$$v_{xx} + (u(x, 0) + k^2)v = 0 \tag{4.108}$$

This equation is a second-order linear ODE and is called a *Sturm-Louiville equation.* In order to solve this equation, we consider the limit $|x| \to \infty$, then $u(x,0) \to 0$, in which Eq. (4.104) becomes

$$v_{xx} + k^2 v = 0 \tag{4.109}$$

and the solution for $v(x,0;\lambda)$ becomes

$$v(x,0;\lambda) \sim Ae^{ikx} + Be^{-ikx} \quad \text{as } |x| \to \infty \tag{4.110}$$

where A and B are constants. After this, introducing two unique Jost solutions based on the boundary conditions

$$\begin{cases} \phi(x,0;k) \sim e^{-ikx} \\ \bar{\phi}(x,0;k) \sim e^{ikx} \end{cases} \quad \text{as } |x| \to -\infty \tag{4.111}$$

$$\begin{cases} \psi(x,0;k) \sim e^{ikx} \\ \bar{\psi}(x,0;k) \sim e^{-ikx} \end{cases} \quad \text{as } |x| \to \infty \tag{4.112}$$

Since Eq. (4.110) is invariant under the transformation $k \to -k$, it follows that $\phi(x,0;k) = \bar{\phi}(x,0;-k)$ and $\psi(x,0;k) = \bar{\psi}(x,0;-k)$. Furthermore, since the solution of Eq. (4.110) is a two-dimensional space, we get

$$\phi(x,0;k) = A(0;k)\bar{\psi}(x,0;k) + B(0;k)\psi(x,0;k) \tag{4.113}$$

$$\phi(x,\bar{0};k) = -\bar{A}(0;k)\psi(x,0;k) + \bar{B}(0;k)\bar{\psi}(x,0;k) \tag{4.114}$$

The functions $A(0;k)$, $B(0;k)$, $\bar{A}(0;k)$, and $\bar{B}(0;k)$ are independent of x, thus

$$\bar{A}(0;k) = -A(0;-k) = -A^*(0;k^*) \tag{4.115}$$

$$\bar{B}(0;k) = B(0;-k) = B^*(0;k^*) \tag{4.116}$$

where $A^*(0;k^*)$, $B^*(0;k^*)$ are the complex conjugates of $A(0;k)$ and $B(0;k)$, respectively. Again, the functions $A(0;k), A(0;k), \bar{A}(0;k)$, and $\bar{B}(0;k)$ satisfy the condition

$$A(0;k)\bar{A}(0;k) + B(0;k)\bar{B}(0;k) = -1 \tag{4.117}$$

and this satisfies

$$|A(0;k)|^2 - |B(0;k)|^2 = 1 \tag{4.118}$$

Let us consider the Wronskian equation

$$W(\phi,\bar{\phi}) = \phi\bar{\phi}_x - \phi_x\bar{\phi} \tag{4.119}$$

on $Im(k) = 0$. Note that this implies that A does not vanish in $Im(k) = 0$.

Let us consider a few properties of the Jost solution and the functions A and B and theorems.

Proposition-1. The Jost solution and the spectral functions A and B have the following analytic properties:

(i) $\phi(x,0;k)$ and $\psi(x,0;k)$ exist, and they are continuous at the eigenvalue k in the closed half-plane $Im(k) \geq 0$ and are analytic in k in the open half-plane $Im(k) > 0$.
(ii) $\bar{\phi}(x,0;k)$ and $\bar{\psi}(x,0;k)$ exists and they are continuous at the eigenvalue k in the closed half-plane $Im(k) \leq 0$ and are analytic in k in the open half-plane $Im(k) < 0$.
(iii) $A(0;k)$ exists and is continuous in k in the closed half-plane $Im(k) \geq 0$, and is analytic in k in the open half-plane $Im(k) > 0$.
(iv) $B(0;k)$ exists and is continuous in k on $Im(k) = 0$.

Proposition-2. The Jost solution and the spectral function A and B have the following asymptotic properties:

$$\begin{cases} \phi(x,0;k) = e^{-ikx} + O\left(\dfrac{1}{k}\right) \\ \psi(x,0;k) = e^{ikx} + +O\left(\dfrac{1}{k}\right) \end{cases} \quad \text{as } |k| \to \infty \; in \; Im(k) \geq 0 \tag{4.120}$$

$$\begin{cases} \dot{\phi}(x,0;k) = e^{ikx} + O\left(\dfrac{1}{k}\right) \\ \dot{\psi}(x,0;k) = e^{-ikx} + +O\left(\dfrac{1}{k}\right) \end{cases} \quad \text{as } |k| \to \infty \; in \; Im(k) \leq 0 \tag{4.121}$$

$$A(0;k) = 1 + O\left(\frac{1}{k}\right) \quad \text{as } |k| \to \infty \; in \; Im(k) \leq 0 \tag{4.122}$$

$$B(0;k) = O\left(\frac{1}{k}\right) \quad \text{as } |k| \to \infty \; in \; Im(k) = 0 \tag{4.123}$$

Theorem: The function $A(0;k)$ has a finite number of zeroes $k_1, k_2, \ldots, K_N$ in the open half-plane $Im(k) > 0$ and does not vanish on $Im(k) = 0$. Moreover, all of these zeroes are simple and lie on the imaginary axis $Re(k_n) = 0$. At each k_n, we have $\phi(x,0;k_n) = c_n(0)\psi(x,0;k_n)$ for some constant $c_n(0)$.

Let's now look into the evolution of the normalization constant $c_n(t)$ for the eigenfunction $\phi_n(x)$. By definition, the normalization constant is defined as:

$$c_n(t) = \left\{ \int_{-\infty}^{\infty} \phi_n^2(x,t)dx \right\}^{-1} \tag{4.124}$$

To study the normalization constant, we differentiate both sides with respect to time and use the second Lax equation $\phi_t = (u_x + \gamma)\phi - (4\lambda + 2u)\phi_x$.

$$\begin{aligned}
\frac{dc^{-1}(t)}{dt} &= 2\int_{-\infty}^{\infty} \phi\phi_t dx \\
&= 2\int_{-\infty}^{\infty} \phi(u - x\phi - 2(u+2\lambda)\phi_x + 4k^3\phi)dx \\
&= 2\int_{-\infty}^{\infty} u_x\phi^2 - 2(u+2\lambda)\phi_x\phi + 4k^3\phi^2 dx \\
&= 2\int_{-\infty}^{\infty} (4(\phi_{xx} - \lambda\phi)\phi_x + 4\lambda\phi_x\phi - (-u\phi^2)_x)dx + 8k^3\int_{-\infty}^{\infty} \phi^2 dx \\
&= 2[2\phi_x^2 + 2\lambda\phi^2 + 2\lambda\phi^2 - u\phi^2]_{-\infty}^{\infty} + 8k^3c^{-1}(t) \\
&= 0 + 8k^3c^{-1}(t) \\
&= 8k^3c^{-1}(t)
\end{aligned}$$

the solution to the above equation gives us the equation of evolution for the normalizing constant

$$c_n(t)Ae^{8k_n^3 t} \implies c_n(0) = A$$

$$c_n(t) = c_n(0)e^{8k_n^3 t} \tag{4.125}$$

Then the transmission and reflection coefficients, since $\lambda > 0$ gives rise to an unbound state, then define the following:

$$\begin{aligned}
\frac{\psi}{\phi} &= \text{Function of time only} \\
&= \left[\frac{\phi_t + u_x\phi - 2(u+2\lambda)\phi_x}{\phi} \right] \\
\lim_{x\to\infty}\left(\frac{\psi}{\phi}\right) &= \left[\frac{b_t e^{ikx} + 0 - 2(0+2\lambda)ik(-1e^{-ikx} + be^{ikx})}{e^{-ikx} + be^{ikx}} \right]_{x=\infty}
\end{aligned} \tag{4.126}$$

Therefore, to eliminate these exponential functions, we must equate the coefficients of e^{ikx} and e^{-ikx}. This can be rewritten as

$$= \left[\frac{(b_t - 4\lambda ikb)e^{ikx} + (4\lambda ik)e^{-ikx}}{e^{-ikx} + be^{ikx}}\right]_{x=\infty}$$

$$= \left[\frac{Ae^{ikx} + Be^{-ikx}}{Ce^{-ikx} + De^{ikx}}\right] = \alpha$$

$$(A - \alpha C)e^{ikx} + (B - \alpha D)e^{-ikx} = 0 \tag{4.127}$$

Since e^{ikx} and e^{-ikx} are linearly independent then

$$(B - \alpha D) = 0 \implies 4ik\lambda - \alpha = 0 \implies \alpha = 4ik\lambda$$

To vanish the second term, the coefficient must be equal to zero

$$(A - \alpha C) = 0 \implies (b_t - 4\lambda ikb) - 4ik\lambda)(b) = 0 \implies b_t = 8ik^3 b$$

$$b(k, t) = b(k, 0)e^{8ik^3 t}, \quad \text{for all } t > 0 \tag{4.128}$$

Similarly,

$$\frac{\psi}{\phi} = \text{Function of time only}$$

$$= \left[\frac{\phi_t + u_x\phi - 2(u + 2\lambda)\phi_x}{w}\right]$$

$$= \left[\frac{a_t e^{-ikx} - 4\lambda(-ika)e^{-kx}}{ae^{-kx}}\right]$$

$$a(k, t) = a(k, 0), \quad \text{for all } t > 0 \tag{4.129}$$

where the reflection and transmission coefficients (R and T) satisfy

$$|T|^2 + |R|^2 = 1. \tag{4.130}$$

Finally, the scattering data:

If $C_m(0)$, $R(k, 0)$, $T(k, 0)$ are given as above, then

$$\begin{cases} c_n = c_n(0)e^{8k_n^3 t} \\ R(k, t) = R(k, 0)e^{8ik^3 t} \\ T(k, t) = T(k, 0) \end{cases} \tag{4.131}$$

where $c_m = \lim_{x\to\infty} \phi_m(x,0)e^{4k_m x}$, $R(k,0)$ and $T(k,0)$ are obtained from the initial data for the KdV equation $u(x,0) = u(x)$.

The n zeroes of A gives a set of discrete eigenvalues, that associated with the N solitons of $u(x,t)$ of the KdV equation.

The inverse transform is reconstructed for the time-dependent solution $u(x,t)$. The heat equation sets up a Riemann–Hilbert problem. Using Propositions 1 and 2 and the theorem, we get the relation

$$\frac{\phi(x,t;k)}{A(0;k)} - \bar{\psi}(x,t;k) = \left(\frac{B(t;k)}{A(0;k)}\right)\psi(x,0;k)$$

This gives a condition between the two functions $\frac{\phi}{A}$ and $\bar{\psi}$ along the real k-axis. For $Im(k) > 0$, the solution is given by the singular integral.

$$\psi(x,t;k)e^{-ikx} = 1 - \sum_{n=1}^{N}\left(\frac{c_n(0)\psi(x,t;k_n)}{A_k(0;k_n)(k+k_n)}\right)e^{ik_n x+8ik_n^3 t} + \frac{1}{2i\pi}\int_{-\infty}^{\infty}\left(\frac{B(0;\sigma)\psi(x,0;\sigma)}{A(0;\sigma)(k+\sigma)}\right)e^{i\sigma x+8i\sigma^3 t}d\sigma \tag{4.132}$$

This is a closed-form singular integral equation for $\psi(x,t;k)$, where all time dependence is known from $B(t;k)$.

The dependence of ψ on the k is isolated taking the form

$$\psi(x,t;k) = e^{ikx} + \int_x^{\infty} K(x,y,t)e^{iky}dy \tag{4.133}$$

To find K, we (4.132) in (4.120) and get a Goursat problem for K. This solution exists and is unique. One of the boundary conditions for K gives a relation between it and the solution $u(x,t)$ of the KdV equation by

$$u(x,t) = -2\frac{\partial}{\partial x}[K(x,y,t)] \tag{4.134}$$

To obtain an equation for $K(x,y,t)$, we substitute Eq. (4.133) into the singular integral equation for ψ. K is related to the FT of ψ, and by taking the inverse FT, we get the following Gel'fand-Levitan equation, valid for $y \geq x$:

$$K(x,y,t) + L(x+y,t) + \int_x^{\infty} K(x,s,t)L(s+y,t)ds = 0 \tag{4.135}$$

where the quantity L is given by

$$L(x,t) = -i\sum_{n=1}^{N}\left(\frac{c_n(0)}{A_{\zeta k}(0;k_n)}\right)e^{ik_nx+8ik_n^3t} + \frac{1}{2i\pi}\int_{-\infty}^{\infty}\left(\frac{B(0;\sigma)}{A(0;\sigma)}\right)e^{i\sigma x+8i\sigma^3 t}d\sigma \tag{4.136}$$

Moreover, by considering the Wronskian $W(k,k_n)$, one can also show that

$$\frac{-ic_n(0)}{A_k(0;k_n)} = \left(\int_{-\infty}^{\infty}\psi(x,0;k_n)^2dx\right)^{-1} > 0 \tag{4.137}$$

which follows from the fact that the Jost solutions are real whenever $Re(k_n) = 0$. Thus, from the knowledge of $\{A(0;k), B(0;k), c_n(0), \{k_n : n = 1,\ldots,N\}\}$. The full time-dependent solution $u(x,t)$ through the linear Volterra-type integral equation (4.132) can be constructed. Where $T = \frac{1}{A}$ is the transmission coefficient, and $R = \frac{B}{A}$ is the reflection coefficient. Initial conditions for which the reflection coefficient is identically zero on $Im(\zeta) = 0$ are known as reflectionless potentials.

4.12.1 *Soliton solutions*

As an initial condition, we choose the potential in the Strum-Liouville problem to be $u_0(x) = -V_0 sech^2(x)$. The reflection index $b(k)$ vanishes if the amplitude of the potential is $V_0 = N(N+1)$, with N an integer.

One-soliton solution:

For $N = 1$, the initial condition is reduced to $u(x,0) = u_0(x) = -2sech^2(x)$.

Consider the IVP

$$u_t - 6uu_x + u_{xxx} = 0 \tag{4.138}$$

with $u(x,0) = u_0(x) = -2sech^2(x)$, where $-u(x,t)$ physically represent to the vertical displacement of the water (or other medium where the wave is formed) from the equilibrium at the location x at time t. Replacing u by $-u$ is nothing but replacing -6 by $+6$ in 4.139. By scaling x, t, and u, i.e., by so the constants in (4.139) are changable

Solution:

First, we find the eigenvalues and eigenfunctions of

$$\theta_{xx} + (\lambda + 2sech^2(x))\theta = 0 \tag{4.139}$$

where $\theta(\pm\infty) = 0$.

Let $s = \tanh(x)$ this transformation maps from $(-\infty, \infty)$ for x to $[-1, 1]$ for s. So, we have

$$\theta_x = \theta_s(1 - s^2), \quad \theta_{xx} = \theta_{ss}(1 - s^2)^2 - 2s(1 - s^2)\theta_s \tag{4.140}$$

Therefore, the Strum-Louiville problem becomes

$$\frac{d}{ds}\left[(1 - s^2)\frac{d\theta}{ds}\right] + \left[2 + \frac{\lambda}{1 - s^2}\right]\theta = 0, \quad -1 < s < 1 \tag{4.141}$$

Comparing this equation with the associated Legendre equation

$$\frac{d}{d\xi}\left[(1 - \xi^2)\frac{d\xi}{ds}\right] + \left[\alpha(\alpha + 1) - \frac{m^2}{1 - \xi^2}\right]\theta = 0, \quad -1 < s < 1 \tag{4.142}$$

We get

$$\alpha(\alpha + 1) = 2, \quad \lambda = -m^2, \quad \alpha \geq 0, \quad 0 < |m| \leq \alpha$$

Therefore, $\alpha = 1 \implies \lambda_1 = -1$. It is clear that this is the only eigenvalue, and the corresponding eigenfunction can be found in the Lgender polynomials

$$P_N^n(s) = (-1)^n(1 - s^2)^{n/2}\frac{d^n}{ds^n}p_N(s) \tag{4.143}$$

where $n = 0, 1, 2, 3, d, N$ and $p_N(s) = \frac{1}{n!2^N}\frac{d^N}{ds^N}(s^2 - 1)^N$.

Let $N = 1$ then we have

$$\begin{aligned}
\theta(x) &= -i(s^2 - 1)^{1/2}\frac{d}{ds}p_1(s) \\
&= -i(s^2 - 1)^{1/2}\frac{d}{ds}(s) \\
&= -i(s^2 - 1)^{1/2} \\
&= = -isech(x)
\end{aligned}$$

So, the eigenvalue $\lambda_1 = -1$ and the eigenfunction $\theta(x) = -isech(x)$.

Normalize the eigenfunction

$$\int_{-\infty}^{\infty} |\theta(x)|^2 dx = \int_{-\infty}^{\infty} |-isech(x)|^2 dx = 2$$

We get, the normalized eigenfunction is $\bar{\theta}(x) = \frac{\sqrt{2}}{2} sech(x)$.

Now we determine $c(0)$ and $c(t)$ by using the definition

$$c_1(0) = \lim_{x\to\infty} \bar{\theta}(x)e^{k_1 x} = \lim_{x\to\infty} \frac{\sqrt{2}}{2} sech(x)e^{x} = \sqrt{2}$$

Therefore, the evolution equation for the normalization constant is given by

$$c_1(t) = c_1(0)e^{4k_1^2 t} = \sqrt{2}e^{4t}$$

So the integration kernel is determined as

$$\begin{aligned} B(\xi, t) &= \sum_{m=1}^{N} c_m^2(0)e^{8k_m^2 t - k_m \xi} + \frac{1}{2\pi}\int_{-\infty}^{\infty} R(k,0)e^{i(8k^3 t + k\xi)} dk \\ &= \sum_{m=1}^{1} c_m^2(0)e^{8k_m^2 t - k_m \xi} + 0 \\ &= 2e^{8t-\xi} \end{aligned}$$

The Gel'fand and Levitan equation is

$$K(x, y, t) + B(x + y, t) + \int_{x}^{\infty} B(y + z, t)K(x, z, t)dz = 0$$

$$K(x, y, t) + 2e^{8t}e^{-x-y} + \int_{x}^{\infty} 2e^{8t}e^{-z-y}K(x, z, t)dz = 0 \quad (4.144)$$

We solve the above equation by the separation of variables by assuming $K(x, z, t) = f(x, t)e^{-y}$. From (4.144), we get

$$f(x, t)e^{-y} + 2e^{8t}e^{-x-y} + \int_{x}^{\infty} 2e^{8t}e^{-z-y}f(x, t)e^{-z}dz = 0 \quad (4.145)$$

Comparing the coefficients of e^{-y} in Eq. (4.145) gives

$$f(x, t) + 2e^{8t}e^{-x} + \int_{x}^{\infty} 2e^{8t}e^{-z}f(x, t)e^{-z}dz = 0$$

$$f(x, t) + 2e^{8t-x} + 2e^{8t}f(x, t)\int_{x}^{\infty} e^{-2z}dz = 0$$

$$f(x, t) + 2e^{8t-x} + f(x, t)e^{8t-2x} = 0$$

So, we have

$$f(x,t) = \frac{-2e^{8t-x}}{1+e^{8t-2x}}$$

Multiply by $\frac{e^{x-4t}}{e^{x-4t}}$, so we get

$$f(x,t) = -e^{4t} sech(x-4t)$$

so the kernel becomes

$$K(x,x,t) = -e^{4t} sech(x-4t)$$

and the potential $u(x,t)$ can be solved as

$$\begin{aligned} u(x,t) &= -2\frac{\partial}{\partial x} K(x,x,t) \\ &= -2\frac{\partial}{\partial x}(-e^{4t} sech(x-4t)) \\ &= -2\frac{\partial}{\partial x}\left(- e^{4t-x}\frac{-2e^{4t-x}}{e^{x-4t}+e^{4t-x}}\right) \\ &= -2\left(\frac{2}{e^{x-4t}+e^{4t-x}}\right)^2 \end{aligned}$$

Thus, the solution becomes

$$u(x,t) = -2sech^2(x-4t) \tag{4.146}$$

Two-soliton solution

Consider the IVP

$$u_t - 6uu_x + u_{xxx} = 0 \tag{4.147}$$

with $u(x,0) = -n(n+1)sech^2(x)$. Let $n = 2$

$$u(x,0) = u_0(x) = -6sech^2(x)$$

Let $s = \tanh(x)$. So, we have

$$\theta_x = \theta_s(1-s^2), \quad \theta_{xx} = \theta_{ss}(1-s^2)^2 - 2s(1-s^2)\theta_s \tag{4.148}$$

Therefore, the Strum–Louiville problem becomes

$$\frac{d}{ds}\left[(1-s^2)\frac{d\theta}{ds}\right] + \left[6 + \frac{\lambda}{1-s^2}\right]\theta = 0, \quad -1 < s < 1 \tag{4.149}$$

Comparing this equation with the generalized Legender equation

$$\frac{d}{d\xi}\left[(1-\xi^2)\frac{d\xi}{ds}\right]+\left[\alpha(\alpha+1)-\frac{m^2}{1-\xi^2}\right]\theta=0, \quad -1<s<1 \qquad (4.150)$$

We get

$$\alpha(\alpha+1)=6, \quad \lambda=-m^2, \quad \alpha\geq 0, \quad 0<|m|\leq\alpha$$

Therefore, $\alpha = 2 \implies \lambda_1 = -1,\ \lambda_2 = -4$. These are two eigenvalues for the Sturm–Louiville problem, and in order to find the corresponding eigenfunctions, we use the associated Legender polynomials with $N = 2$,

$$\begin{aligned}
\theta_2^1(x) &= -(1-s^2)^{1/2}\frac{d}{ds}p_2(s)\\
&= -(1-s^2)^{1/2}\frac{d}{ds}\left(\frac{1}{2}(3s^2-1)\right)\\
&= -3s(1-s^2)^{1/2}\\
&= = -3(1-s^2)^{1/2}s\\
&= -3sech(x)\tanh(x)
\end{aligned}$$

and

$$\begin{aligned}
\theta_2^2(x) &= (1-s^2)\frac{d^2}{ds^2}p_2(s)\\
&= (1-s^2)\frac{d^2}{ds^2}\left(\frac{3}{2}s^2-\frac{1}{2}\right)\\
&= 3(1-s^2)\\
&= 3sech^2(x)
\end{aligned}$$

So the eigenvalues are $\lambda_1 = -1,\ \lambda_2 = -4$ and the eigenfunctions $\theta_2^1(x) = -3sech(x)\tanh(x),\ \theta_2^2(x) = 3sech^2(x)$.

After that, normalization of the eigenfunction

$$\int_{-\infty}^{\infty}|\theta_2^1(x)|^2dx = \int_{-\infty}^{\infty}|3isech(x)\tanh(x)|^2dx = 6$$

$$\int_{-\infty}^{\infty}|\theta_2^2(x)|^2dx = \int_{-\infty}^{\infty}|-3sech^2(x)|^2dx = 12$$

Therefore, the normalized eigenfunctions are $\bar{\theta_2^1}(x) = \frac{\sqrt{3}}{2}sech(x)\tanh(x)$, $\bar{\theta_2^2}(x) = \frac{\sqrt{3}}{2}sech^2(x)$.

Then, we determine $c(0)$ and $c(t)$ by using the definition

$$c_1(0) = \lim_{x\to\infty} \bar{\theta_1}(x)e^{k_1 x} = \lim_{x\to\infty} \frac{\sqrt{3}}{2} sech(x)\tanh(x)e^{x} = \sqrt{6}$$

$$c_2(0) = \lim_{x\to\infty} \bar{\theta_2}(x)e^{k_2 x} = \lim_{x\to\infty} \frac{\sqrt{3}}{2} sech^2(x)e^{2x} = 2\sqrt{3}$$

Therefore, the evolution equation is given by

$$c_1(t) = c_1(0)e^{4k_1^3 t} = \sqrt{6}e^{4t}$$
$$c_2(t) = c_2(0)e^{4k_2^3 t} = 2\sqrt{3}e^{32t}$$

We determine the integration kernel as

$$\begin{aligned}B(\xi,t) &= \sum_{m=1}^{N} c_m^2(0)e^{8k_m^2 t - k_m\xi} + \frac{1}{2\pi}\int_{-\infty}^{\infty} R(k,0)e^{i(8k^3t+k\xi)}dk \\ &= \sum_{m=1}^{2} c_m^2(0)e^{8k_m^2 t - k_m\xi} + 0 \\ &= 6e^{8t-\xi} + 12e^{64t-2\xi}\end{aligned}$$

Then the Gel'fand and Levitan equation is

$$K(x,y,t) + B(x+y,t) + \int_x^{\infty} B(y+z,t)K(x,z,t)dz = 0$$

$$K(x,y,t) + 6e^{8t}e^{-x-y} + 12e^{64t}e^{-2x-2y} + \int_x^{\infty} [6e^{8t}e^{-z-y} + 12e^{64t}e^{-2z-2y}]K(x,z,t)dz = 0 \qquad (4.151)$$

The above equation is solved by the separation of variables by assuming $K(x,z,t) = f_1(x,t)e^{-y} + f_2(x,t)e^{-2y}$. From (4.151), we get

$$f_1(x,t)e^{-y} + f_2(x,t)e^{-2y} + 6e^{8t}e^{-x-y} + 12e^{64t}e^{-2x-2y} + \int_x^{\infty} [6e^{8t}e^{-z-y} + 12e^{64t}e^{-2z-2y}][f_1(x,t)e^{-z} + f_2(x,t)e^{-2z}dz = 0 \qquad (4.152)$$

Comparing the coefficients of e^{-y} and e^{-2y} in Eq. (4.152) we get

$$f_1(x,t) + 6e^{8t}e^{-x} + \int_x^{\infty} 6e^{8t}e^{-z}[f_1(x,t)e^{-z} + f_2(x,t)e^{-2z}]dz = 0$$

$$f_1(x,t) + 6e^{8t-x} + 3e^{8t+2x}f_1(x,t) + 2e^{8t+3x}f_2(x,t) = 0$$

$$(1 + 3e^{8t-2x})f_1(x,t) + (2e^{8t-3x})f_2(x,t) = -6e^{8t-x}$$

The second equation is

$$f_2(x,t) + 12e^{64t}e^{-2x} + \int_x^\infty 12e^{64t}e^{-2z}[f_1(x,t)e^{-z} + f_2(x,t)e^{-2z}]dz = 0$$

$$f_2(x,t) + 12e^{64t-2x} + 12e^{64t}\left[f_1(x,t)\int_x^\infty e^{-3z}dz + f_2(x,t)\int_x^\infty e^{-4z}dz\right] = 0$$

$$(4e^{64t-3x})f_1(x,t) + (1+3e^{64t-4x})f_2(x,t) = -12e^{64t-2x}$$

which gives the following system:

$$\begin{aligned}
(1+3e^{8t-2x})f_1(x,t) + (2e^{8t-3x})f_2(x,t) &= -6e^{8t-x}\\
(4e^{64t-3x})f_1(x,t) + (1+3e^{64t-4x})f_2(x,t) &= -12e^{64t-2x}
\end{aligned}$$

The above system of algebraic equations can be solved by using Cramer's rule

$$f_1(x,t) = \frac{D_{f_1}}{D}, \quad f_2(x,t) = \frac{D_{f_2}}{D}$$

where

$$D = \begin{vmatrix} 1+3e^{8t-2x} & 2e^{8t-3x} \\ 4e^{64t-3x} & 1+3e^{64t-4x} \end{vmatrix} = 1 + 3e^{64t-4x} + 3e^{8t-2x} + 3e^{72t-6x}$$

$$D_{f_1} = \begin{vmatrix} -6e^{8t-x} & 2e^{8t-3x} \\ -12e^{64t-2x} & 1+3e^{64t-4x} \end{vmatrix} = -6e^{8t-x} + 6e^{72t-5x}$$

$$D_{f_2} = \begin{vmatrix} 1+3e^{8t-2x} & -6e^{8t-x} \\ 4e^{64t-3x} & -12e^{64t-2x} \end{vmatrix} = -12e^{64t-2x} - 12e^{72t-4x}$$

Therefore,

$$\begin{aligned}
K(x,y,t) &= f_1(x,t)e^{-y} + f_2(x,t)e^{-2y}\\
&= \frac{-6(e^{8t-x} + e^{72t-5x} + 2e^{64t-4x})}{1 + 3e^{64t-4x} + 3e^{8t-2x} + e^{72t-6x}}
\end{aligned} \tag{4.153}$$

To derive the solution

$$\begin{aligned}
u(x,t) &= -2\frac{\partial}{\partial x}K(x,x,t)\\
&= -2\frac{\partial}{\partial x}\left(\frac{-6(e^{8t-x} + e^{72t-5x} + 2e^{64t-4x})}{1 + 3e^{64t-4x} + 3e^{8t-2x} + e^{72t-6x}}\right)
\end{aligned}$$

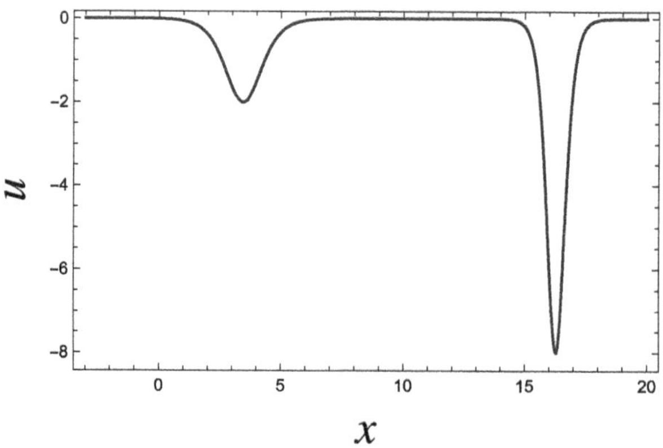

Figure 4.4. Soliton solution of Eq. (4.154).

Thus, the solution becomes

$$u(x,t) = -12\frac{3+4\cosh(2x-8t)+\cosh(4x-64t)}{(3\cosh(x-28t)+\cosh(3x-36t))^2} \tag{4.154}$$

This type of solution is called a *bisoliton solution* in the theory of inverse scattering.

4.13 Hirota Direct Method

In 1971, Ryogo Hirota (1932–2015), a Japanese mathematician, proposed a new method, called the Hirota direct method or the Hirota bilinear formalism, for constructing exact multiple soliton solutions of an NLPDE. Using this method, we can find solutions directly without employing the inverse scattering method. Define a new differential operator called Hirota operator and the method is based on a transformation into new variables so that in these new variables, integrable equations can be written in a Hirota bilinear form. All derivatives appear as Hirota bilinear derivatives, and multiple soliton solutions appear in a particularly simple form. It was shown that the Hirota method gives N-soliton solutions to the KdV, mKdV, sine-Gordon, and nonlinear Schrödinger equations. For nonintegrable equations, only two-soliton solutions can be obtained. Moreover, the Hirota method, reformulated in terms of τ functions, can be applied for further mathematical developments.

This method is one of the most useful techniques for testing and constructing exact soliton solutions to various NLPDEs. The drawback of

Hirota's method is that the construction of a bilinear form is a must. The technique is applicable only if that can be written in bilinear form, either as a single bilinear equation or as a system of coupled bilinear equations. Once the bilinear form is obtained, the method becomes a simple algorithm. Nothing is needed beyond elementary calculus and algebra. Sometimes the calculations are lengthy but straightforward. The complexity of the calculations also increases with the type of soliton solution (two soliton, three soliton, four soliton, etc.) one needs to obtain. Single-soliton solutions are easy to calculate, and two-soliton and three-soliton solutions are also manageable to obtain by hand. But to obtain four solitons or more, a symbolic computation is a must. Once the two- and three-soliton solutions are known, their structure reveals the form of higher soliton solutions, and using mathematical induction, one can obtain the general N-soliton solution of the equation. Other than multisolitons, lump solitons, breather solitons, and hybrid-type solitons, where a combination of solitons, lump, and breather-type solitons may be obtained by this method.

In essence, Hirota's method requires

(i) *an appropriate change in the dependent variable.*
(ii) *introduction of Hirota differential operator D.*
(iii) *Using a perturbation technique to solve the bilinear equation.*

Advantage of Hirota's method: Hirot's method has been used for a large class of NLPDEs, including difference-differential and integrodifferential equations. Other than completely integrable systems, this method is applied to partially integrable equations. It is also shown that an integrable nonlinear evolution equation can be transformed into a bilinear form. Even a bilinear form can be constructed for many nonintegrable equations. Although IST for solving the Initial Value Problem is more general than Hirota's approach, the IST gives several types of solutions, while the latter gives only soliton-type solutions. However, Hirota's method has several advantages over IST. Firstly, the IST is much more complex, mathematically complicated, and time-consuming, but the Hirota method is mainly algebraic, simple, and almost straightforward. Secondly, the existence of two-soliton solutions is a weaker requirement than integrability by IST. Moreover, the Hirota Method can handle a larger class of equations, including nonintegrable PDEs, but only to obtain an N-soliton solution. Initially, a straightforward algorithm is needed to test whether the equation is bilinear or not.

Steps of Hirota's Method: Hirota's method consists of three stages.

Step I: The first step of this method is to make a suitable transformation of the dependent variable by a rational expression(s) of unknown functions such that the NLPDEs will be converted into a quadratic form in dependent variables. This new form is called the "bilinear form." Finding such an unknown function for transformation is typical.

Step II: In the second stage, the nonlinearity will be abolished. A special differential operator called the Hirota D-operator is introduced, and this will convert the bilinear form of the equation into a polynomial of the D-operator. This form is called the Hirota bilinear form. No systematic ways are available to construct the Hirota bilinear form for a given NLPDE. But we assume that all completely integrable NLPDEs can be put into the Hirota bilinear form, and there exist some equations that are not integrable but have Hirota bilinear form, e.g., the extended Kadomtsev–Petviashvili (eKP) equation $(u_t - 6uu_x + u_{xxx})_x + 3u_{yy} + \gamma u_{tt} + \beta u_{ty} = 0$ is an example of it.

Step III: In the final step, we use the perturbation technique in the Hirota bilinear form. The coefficients of the perturbation parameter and its power will be analyzed separately. From the previous information, a multisoliton solution to the equation is obtained.

4.14 Some Important Information

4.14.1 *Bilinear form*

A bilinear form is a type of mathematical function that takes two vector inputs and produces a scalar as the result. It is linear in each of its variables, meaning that it exhibits linearity when each variable is held constant while the other varies. Bilinear forms commonly arise in linear algebra and other mathematical fields.

More formally, let V be a vector space over a field F. A bilinear form on V is a function $B : V \times V \to F$ such that, for all vectors u, v, w in V and scalars α, β in F, the following properties hold:

(i) Linearity in the First Variable:

$$B(\alpha u + \beta v, w) = \alpha B(u, w) + \beta B(v, w)$$

(ii) Linearity in the Second Variable:

$$B(u, \alpha v + \beta w) = \alpha B(u, v) + \beta B(u, w)$$

Bilinear forms are often represented using inner product notation, especially in the context of Euclidean spaces. In such cases, the bilinear form B is denoted by $\langle u, v\rangle$ or $u.v$, and the linearity properties can be expressed as:

(i) $\langle \alpha u + \beta v, w\rangle = \alpha\langle u, w\rangle + \beta\langle v, w\rangle$
(ii) $\langle u, \alpha v + \beta w\rangle = \alpha\langle u, v\rangle + \beta\langle u, w\rangle$

Bilinear forms have applications in various areas of mathematics, physics, and engineering. They are used, for example, in the study of quadratic forms and symmetric matrices and in defining concepts like orthogonality and duality in vector spaces.

4.14.2 *D-operator*

The Hirota D-operator is a specific differential operator that arises in the context of soliton theory, integrable systems, and the study of integrable PDEs. It is named after the Japanese mathematician Ryogo Hirota, who made significant contributions to soliton theory. In the context of soliton equations, the Hirota D-operator is commonly used to formulate the bilinear form of integrable equations. The bilinear form is a convenient way to study soliton solutions and integrability properties of certain PDEs. The Hirota D-operator plays a crucial role in the Hirota bilinear method, which is a powerful technique for finding soliton solutions and studying the integrability of PDEs. The method involves expressing the solutions of certain integrable equations in terms of bilinear forms, which are manipulated using the Hirota D-operator.

Let $S : C^n \to C$ be a space of differentiable functions. Then the Hirota D-operator $D : S \times S \to S$ is defined as

$$[D_x^{m_1} D_t^{m_2} \cdots]\{f.g\} = [(\partial_x - \partial_{x'})^{m_1}(\partial_t - \partial_{t'})^{m_2} \cdots]f(x, t, \cdots) \times g(x', t', \cdots)|_{x'=x,t'=t} \tag{4.155}$$

where $m_i, i = 1, 2, \ldots$ are positive integers and $x, t, \ldots$ are independent variables.

We try to write the bilinear form of the given NLPDE $F[u] = 0$ as a polynomial of the D-operator. We call this polynomial $P(D)$.

NLPDEs and difference equations can be written in Hirota bilinear form if they are equivalent to

$$\sum_{\alpha,\beta=1}^{m} P^{\eta}_{\alpha,\beta}(D) f^{\alpha} f^{\beta} = 0, \quad \eta = 1, \ldots, r$$

For some m, r, and linear operator $P^{\eta}_{\alpha,\beta}(D)$, f^{i}'s are new dependent variables.

Hirota D-operator D_x operates on ordered pairs of functions $f(x)$ and $g(x)$ as

$$D_x(f.g) \equiv \left(\frac{\partial}{\partial x} - \frac{\partial}{\partial x'}\right) f(x)g(x')x' = x$$

For higher-order derivatives, the definition is

$$D_x^m D_t^n(f.g) \equiv \left(\frac{\partial}{\partial x} - \frac{\partial}{\partial x'}\right)^m \left(\frac{\partial}{\partial t} - \frac{\partial}{\partial t'}\right)^n f(x,t)g(x',t')x' = x, t' = t$$

where m, n are non-negative integers, and the dot between the functions f and g is used as a separator. One can also define

$$D_x(f.g) = f_x g - f g_x, \quad D_x D_t(f.g) = f_{xt} g - f_x g_t - f_t g_x + f g_{xt}$$

It is to be noted that the Hirota derivative is the same as the Leibniz product rule with a sign difference. It is worth noting that Hirota's bilinear operator acts on direct spaces, while the ordinary derivative operator acts on single spaces.

4.14.3 *Properties of D-operator*

We now see a few properties, as follows:

(i) The Hirota D-Operator satisfies the following properties of bilinearity:

(a)

$$D_x(cf + g.h) = cD_x(f.h) + D_x(g.h)$$

(b)

$$D_x(f.dg + h) = dD_x(f.g) + D_x(f.h)$$

That's why the Hirota D-Operator is called bilinear.

Proof:

(a)

$$\begin{aligned} D_x(cf+g.h) &= \left(\frac{\partial}{\partial x}-\frac{\partial}{\partial x'}\right)(cf+g)(x)h(x')|_{x'=x} \\ &= \left(\frac{\partial}{\partial x}-\frac{\partial}{\partial x'}\right)(cf(x)+g(x))h(x')|_{x'=x} \\ &= [cf_x(x)+g_x(x)]h(x')-[cf(x)+g(x)]h_x(x')|_{x'=x} \\ &= c[f_xh-fh_x]+[g_xh-gh_x] \\ &= cD_x(f.h)+D_x(g.h) \end{aligned}$$

(b)

$$\begin{aligned} &D_x(f.dg+h) \\ &= \left(\frac{\partial}{\partial x}-\frac{\partial}{\partial x'}\right)f(x)(dg+h)(x')|_{x'=x} \\ &= \left(\frac{\partial}{\partial x}-\frac{\partial}{\partial x'}\right)f(x)[dg(x')+h(x')|_{x'=x} \\ &= df_xg+f_xh-dfg_x-fh_x \\ &= d(f_xg-fg_x)+(f_xh-fh_x) \\ &= dD_x(f.g)+D_x(f.h) \end{aligned}$$

(ii)

$$P(D)(f.1)=P(\partial)f$$

(iii)

$$P(D)(1.f)=P(-\partial)f$$

(iv)

$$D_x^n(f.g)=(-1)^nD_x^n(g.f)\implies P(D)(f.g)=P(-D)(g.f)$$

(v) when

$$f=g,\quad D_x^{2m-1}(f.f)=0,\quad m\geq 1$$

and

$$D_x^{2m}(f.f) = \sum_{i=1}^{2m}(-1)^{2m-i}\binom{2m}{i}(\partial_x^i f)(\partial_x^{2m-i} f), \quad m \geq 1$$

(vi) Take an even polynomial $P(x_1, x_2, \ldots, x_m)$ in m variables and assume that no constant term exists, i.e., $P(0) = P(0, 0, \ldots, 0)$. Then

$$P(D_{x_1}, D_{x_2}, \ldots, D_{x_m})(f.f) = 0$$

(vii)

$$P(D)(e^{px}.e^{qx}) = P(p-q)e^{(p+q)x}$$

(viii)

$$P(D)(e^{px}F.e^{px}G) = e^{2px}P(D)(F.G)$$

(ix) Let $P(D)$ act on two exponential functions e^{θ_1} and e^{θ_2} where $\theta_i = k_i x + \cdots + r_i z + l_i y + \alpha_i$ and $k_i, \ldots, r_i, l_i, \alpha_i$ are constants for $i = 1, 2$. Then

$$P(D)(e^{\theta_1 . \theta_2}) = P(k_1 - k_2, \ldots, r_1 - r_2, l_1 - l_2)e^{(\theta_1 + \theta_2)}$$

(x)

$$D_x^m D_t^n (e^{(k_1 x - \omega_1 t)}.e^{(k_2 x - \omega_2 t)})$$
$$= (k_1 - k_2)^m (-\omega_1 - \omega_2)^n e^{(k_1+k_2)x - (\omega_1+\omega_2)t}$$

(xi)

$$D_x^m D_t^n (c.c) = 0$$

where c is a constant.

4.14.4 *Auxiliary function (f) and Series expansions of different soliton solutions*

Hirota's bilinear method is used to find soliton-type solutions. Soliton solutions are obtained due to the prominent balance between nonlinearity and higher-order linearity that appears in the NLPDE $F(u) = 0$, where u is the wave amplitude. To employ Hirota's bilinear method, In general, we use the transformation $u(x,t) = 2(lnf)_{xx}$ to reduce the given PDE $F(u) = 0$ into the bilinear form then we get various kind of nonlinear waves depending on the choice of auxiliary function f where f is the newly introduced dependent variable. It is known as an *auxiliary function.*

Multiple regular soliton solutions: To obtain multiple regular soliton solutions, we take the following series expansion

$$f = 1 + \sum_{n=1}^{\infty} \epsilon^n f_n$$

for the unknown functions $f_1(x,t), f_2(x,t) \ldots f_n(x,t)$ are in exponential form.

Multiple singular-soliton solutions: The mechanism to derive multiple singular-soliton solutions is just the same as that of multiple regular soliton solutions, along with a different perturbation series expansion

$$f = 1 + \sum_{n=1}^{\infty} (-1)^n \epsilon^n f_n$$

$f_n(x,t)$ is an exponential function in independent variables.

Lump soliton solutions: For lump soliton solutions, the perturbation of f is taken as same form for regular soliton but the auxiliary function 'f' is taken as

$$f(x,y,t) = (a_1 x + a_2 y + a_3 t + a_4)^2 + (a_5 x + a_6 y + a_7 t + a_8)^2 + a_9$$

i.e., the sum of two positive quadratic functions in independent variables.

Breather soliton solutions: Breather soliton solutions are obtained via two steps. Part is based on the derivation of multisoliton solution. In the last part, complex conjugate wave numbers are introduced. in the multisoliton solution of a given PDE.

4.14.5 *Multisoliton*

Multisolitons refer to the phenomenon of multiple solitons existing simultaneously in a physical system. A soliton is a self-reinforcing solitary wave that maintains its shape and velocity even after interacting with other waves or objects. It is a localized, stable solution to a nonlinear wave equation. Multisoliton solutions arise when multiple solitons coexist and propagate independently in a system. These solitons can interact with each other, leading to complex dynamics and sometimes forming bound states called soliton molecules. Multisoliton solutions have been observed in various systems, including optical fibers, plasmas, and nonlinear crystals. The study of multisolitons is important in understanding the behavior and interactions of solitons in nonlinear systems. It has applications in areas such as telecommunications, nonlinear optics, and fluid dynamics. The first multisoliton solution of the KdV equation was found by the Inverse Scattering Method in 1967 by Clifford S. Gardner, John M. Greene, and Martin D. Kruskal *et al.* In 1971, Hirota developed a method for obtaining the multisoliton solutions of the KdV equation that is elegant but easy to handle.

4.14.6 *Hirota integrable*

A set of equations written in the Hirota bilinear form is Hirota integrable if one can combine any number N of one-soliton solution into an N-soliton solution, and the combination is a finite polynomial in the e^{θ}'s involved.

4.15 Hirota's Condition for the Existence of Soliton Solutions

One soliton: Let $\phi = W[p(\xi, \tau, \ldots, \eta)]$ be a bilinearizing transformation of an NLPDE $M[\phi] = 0$, which can be written in the Hirota bilinear form $P(D)(p.p) = 0$.

If $P(D)$ satisfies $P(k_1, \omega_1, \ldots, l_1) = P(s_1) = 0$, then one-soliton solution of this equation exists, which is $\phi = W[p(\xi, \tau, \ldots, \eta)] = W[1 + e^{\theta_1}]$ where $\theta_1 = k_1\xi - \omega_1\tau + \cdots + l_1\eta + \alpha_1$ with the constants $k_1, \omega_1, \ldots, l_1, \alpha_1$. i.e., one-soliton solution is restricted only by dispersion relation.

Two soliton: Let $\phi = W[p(\xi, \tau, \ldots, \eta)]$ be bilinearizing transformation of an NLPDE $M[\phi] = 0$, which can be written in the Hirota bilinear form $P(D)(p.p) = 0$, if this equation satisfies $P(k_r, \omega_r, \ldots, l_r) = P(s_r) = 0$, $r = 1, 2$ and $E_{12} = -\frac{P(s_1 - s_2)}{P(s_1 + s_2)}$, then a two-soliton solution of this equation exists, which is $\phi = W[p(\xi, \tau, \ldots, \eta)] = W[1 + e^{\theta_1} + e^{\theta_2} + E_{12}e^{\theta_1+\theta_2}]$ where $\theta_r = k_r\xi - \omega_r\tau + \cdots + l_r\eta + \alpha_r$, with constants $k_r, \omega_r, \ldots, l_r, \alpha_r$. i.e., two-soliton solutions are restricted by the dispersion relation, together with the phase factor E_{12}, which should be of the form

$$E_{12} = -\frac{P(s_1 - s_2)}{P(s_1 + s_2)} = -\frac{P(k_1 - k_2, -\omega_1 + \omega_2)}{P(k_1 + k_2, -\omega_1 - \omega_2)}$$

Three soliton: Let $\phi = W[p(\xi, \tau, \dots, \eta)]$ be a bilinearizing transformation of a nonlinear PDE $M[\phi] = 0$, which can be written in the Hirota bilinear form $P(D)(p.p) = 0$. Then if $M[\phi] = 0$ satisfies the three soliton condition, which is

$$\sum_{\gamma_r=\pm 1} P(\gamma_1 s_1 + \gamma_2 s_2 + \gamma_3 s_3)P(\gamma_1 s_1 - \gamma_2 s_2)P(\gamma_2 s_2 - \gamma_3 s_3)P(\gamma_3 s_3 - \gamma_1 s_1) = 0 \tag{4.156}$$

with $P(s_r) = 0, r = 1, 2, 3$ then its three-soliton solution is $\phi = W[p(\xi, \tau, \dots, \eta)] = W[1+e^{\theta_1}+e^{\theta_2}+e^{\theta_3}+E_{12}e^{\theta_1+\theta_2}+E_{13}e^{\theta_1+\theta_3}+E_{23}e^{\theta_2+\theta_3}+Be^{\theta_1+\theta_2+\theta_3}]$ where $\theta_r = k_r\xi - \omega_r\tau + \cdots + l_r\eta + \alpha_r, r = 1, 2, 3$. Here, $E_{rj} = -\frac{P(s_r - s_j)}{P(s_r + s_j)}$ for $r, j = 1, 2, 3, r < j$ and $B = E_{12}E_{13}E_{23}$ i.e., the three-soliton solution is restricted by the dispersion relation, the phase factors E_{rj}, in an additional restriction given by (4.156).

This condition can be extended for higher-order soliton solutions. At higher orders, Hirota obtained the conditions $S[X, m] = \sum_{\gamma_r=\pm 1} [X(\sum_{r=1}^{m} \gamma_r k_r - \sum_{r=1}^{m} \gamma_r \omega_r) \prod_{r<j}^{(m)} X(\gamma_r k_r - \gamma_j k_j, -\gamma_r \omega_r + \gamma_j \omega_j)\gamma_r \gamma_j] = 0$.

As a consequence, there is no freedom left in p; the parameters are restricted by the dispersion relation, the phase factors E_{rj}, and also an additional restriction. Thus, the existence of a three-soliton solution or more solitons is not automatic, but rather it imposes severe requirements on the polynomial X.

4.15.1 *Multisoliton of the KdV equation*

We construct one-soliton, two-soliton, and three-soliton solutions of the KdV equation. The KdV equation is given by

$$u_t + 6uu_x + u_{xxx} = 0 \tag{4.157}$$

Step 1. Bilinearization: The KdV equation is bilinearized using the transformation $u(x, t) = 2\frac{\partial^2}{\partial x^2}(\log f)$.

$$u = 2\frac{\partial^2}{\partial x^2}(\log f) = 2(\log f)_{xx} = 2\frac{\partial}{\partial x}\left(\frac{f_x}{f}\right) = 2\left(\frac{f_{xx}f - f_x^2}{f^2}\right)$$

Then

$$\begin{aligned}
u_t &= 2\frac{\partial}{\partial t}\left(\frac{f_{xx}f - f_x^2}{f^2}\right)\\
&= 2\left(\frac{f_{xxt}f^2 + f_{xx}f_tf - 2ff_xf_{xt} - 2ff_{xx}f_t + 2f_tf_x^2}{f^3}\right)\\
&= 2\left(\frac{f_{xxt}f^2 - f_{xx}f_tf - 2ff_xf_{xt} + 2f_tf_x^2}{f^3}\right)\\
u_x &= 2\frac{\partial}{\partial x}\left(\frac{f_{xx}f - f_x^2}{f^2}\right)\\
&= 2\left(\frac{f(f_{xxx}f + f_{xx}f_x - 2f_xf_{xx}) - 2ff_xf_{xx} + 2f_x^3}{f^3}\right)\\
&= 2\left(\frac{f_{xxx}f^2 - 3f_{xx}f_xf + 2f_x^3}{f^3}\right)\\
u_{xx} &= 2\frac{\partial}{\partial x}\left(\frac{f_{xxx}f^2 - 3f_{xx}f_xf + 2f_x^3}{f^3}\right)\\
&= 2\left(\frac{f_{xxxx}f^3 - 4f_{xxx}f_xf^2 + 12f_{xx}f_x^2f - 3f_{xx}^2f^2 - 6f_x^4}{f^4}\right)
\end{aligned}$$

Now, we can write

$$2\frac{\partial}{\partial x}\left(\frac{ff_{xt} - f_xf_t}{f^2}\right) = 2\left(\frac{f_{xxt}f^2 - f_{xx}f_tf - 2ff_xf_{xt} + 2f_tf_x^2}{f^3}\right) = u_t$$

So, Eq. (4.157) can be written as

$$2\frac{\partial}{\partial x}\left(\frac{ff_{xt} - f_xf_t}{f^2}\right) + 3\frac{\partial}{\partial x}(u^2) + u_{xxx} = 0$$

Integrating w.r.t x, we get

$$\begin{aligned}
&2\left(\frac{ff_{xt} - f_xf_t}{f^2}\right) + 3u^2 + u_{xx} = C\\
\Rightarrow\ &2\left(\frac{ff_{xt} - f_xf_t}{f^2}\right) + 12\left(\frac{f_{xx}f - f_x^2}{f^2}\right)^2\\
&+ 2\left(\frac{f_{xxxx}f^3 - 4f_{xxx}f_xf^2 + 12f_{xx}f_x^2f - 3f_{xx}^2f^2 - 6f_x^4}{f^4}\right) = C
\end{aligned}$$

$$\Rightarrow \frac{2}{f^2}(ff_{xt} - f_x f_t) + \frac{2}{f^4}(f_{xxxx}f^3 + 3f_{xx}^2 f^2 - 4f_{xxx}f_x f^2) = C$$

$$\Rightarrow \frac{2}{f^2}(ff_{xt} - f_x f_t) + \frac{2}{f^2}(f_{xxxx}f + 3f_{xx}^2 - 4f_{xxx}f_x) = C$$

$$\Rightarrow ff_{xt} - f_x f_t + 3f_{xx}^2 - 4f_{xxx}f_x + f_{xxxx}f = \frac{Cf^2}{2}$$

Thus, Eq. (4.157) in bilinear form is

$$ff_{xt} - f_x f_t + ff_{xxxx} - 4f_x f_{xxx} + 3f_{xx}^2 = 0 \tag{4.158}$$

Step 2. Transformation to the Hirota bilinear form: We attempt to express the KdV equation (4.157) in the Hirota bilinear form by means of the Hirota-D operator. Let us consider $D_t D_x$ applied on the product $f.f$,

$$\begin{aligned} D_t D_x\{f.f\} &= \left(\frac{\partial}{\partial t} - \frac{\partial}{\partial \acute{t}}\right)\left(\frac{\partial}{\partial x} - \frac{\partial}{\partial \acute{x}}\right)\{f(x,t).f(\acute{x},\acute{t})\}|_{\acute{x}=x,\acute{t}=t} \\ &= f_{xt}f + ff_{xt} - f_t f_x - f_x f_t \\ &= 2(f_{xt}f - f_t f_x) \end{aligned} \tag{4.159}$$

It is to be noted that these terms are the first two terms of Eq. (4.158) multiplied by two. Now, we consider D_x^4.

$$\begin{aligned} D_x^4\{f.f\} &= \left(\frac{\partial}{\partial x} - \frac{\partial}{\partial \acute{x}}\right)^4 \{f(x,t).f(\acute{x},\acute{t})\}|_{\acute{x}=x,\acute{t}=t} \\ &= f_{xxxx}f - 4f_{xxx}f_x + 6f_{xx}f_{xx} - 4f_x f_{xxx} + ff_{xxxx} \\ &= 2(f_{xxxx}f - 4f_{xxx}f_x + 3f_{xx}^2) \end{aligned} \tag{4.160}$$

Note that these terms are the last three terms of Eq. (4.158) multiplied by two. Thus, Eq. (4.158) can be expressed in the Hirota bilinear form as follows.

$$P(D)\{f.f\} = (D_x D_t + D_x^4)\{f.f\} = 0 \tag{4.161}$$

Step 3. Application of the Hirota perturbation: Insert $f = 1 + \epsilon f_1 + \epsilon^2 f_2 + \cdots$ into Eq. (4.161), so we have

$$\begin{aligned} P(D)\{f.f\} = P(D)\{1.1\} + \epsilon P(D)\{f_1.1 + 1.f_1\} + \epsilon^2 P(D)\{f_2.1 + f_1.f_1 \\ + 1.f_2\} + \epsilon^3 P(D)\{f_3.1 + f_2.f_1 + f_1.f_2 + 1.f_3\} + \cdots = 0 \end{aligned} \tag{4.162}$$

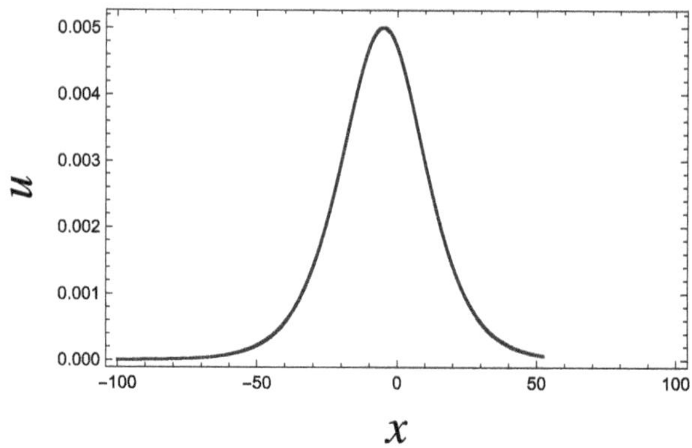

Figure 4.5. One-soliton solution of Eq. (4.163).

One-soliton solution (N = 1):

We take $f = 1 + \epsilon f_1$ where $f_1 = e^{\theta_1}$, and $\theta_1 = k_1 x + \omega_1 t + \alpha_1$ to construct the one-soliton solution of the KdV equation (4.157). Note that $f_j = 0$ for all $j \geq 2$. We insert f into Eq. (4.162) and make the coefficients of ϵ^m, $m = 0, 1, 2$ vanish. The coefficient of ϵ^0 is $P(D)\{1.1\} = 0$ since $P(0,0)\{1\} = 0$.

By the coefficient of ϵ^1

$$P(D)\{1.f_1 + f_1.1\} = P(\partial)e^{\theta_1} + P(-\partial)e^{\theta_1} = 2P(p_1)e^{\theta_1} = 0$$

The dispersion relation $P(p_1) = 0$ gives $\omega_1 = -k_1^3$.

The coefficient of ϵ^2 vanishes trivially since

$$P(D)\{f_1.f_1\} = P(D)\{e^{\theta_1}.e^{\theta_1}\} = P(p_1 - p_1)e^{2\theta_1} = 0$$

Finally, we can set $\epsilon = 1$ without losing generality, resulting in $f = 1 + e^{\theta_1}$. The KdV equation (4.157) has a one-soliton solution, which is

$$u(x,t) = \frac{k_1^2}{2} sech^2\left(\frac{\theta_1}{2}\right) \tag{4.163}$$

where $\theta_1 = k_1 x - k_1^3 t + \alpha_1$.

Two-soliton solution (N = 2):

We take $f = 1 + \epsilon f_1 + \epsilon^2 f_2$, where $f_1 = e^{\theta_1} + e^{\theta_2}$ and $\theta_i = k_i x + \omega_i t + \alpha_i$ for $i = 1, 2$, to construct the two-soliton solution of the KdV equation (4.157). Note that $f_j = 0$ for all $j \geq 3$. We insert f into Eq. (4.162) and make the coefficients of ϵ^m, $m = 0,\ 1,\ 2,\ 3,\ 4$ to vanish.

The coefficient of ϵ^0 is $P(D)\{1.1\} = P(0,0)\{1\} = 0$. By the coefficient of ϵ^1, we have

$$P(D)\{1.f_1 + f_1.1\} = 2P(\partial)\{e^{\theta_1} + e^{\theta_2}\} = 2[P(\partial)e^{\theta_1} + P(\partial)e^{\theta_2}] = 0$$

which implies $P(p_i) = k_i^4 + k_i\omega_i = 0$, i.e., $\omega_i = -k_i^3$ for $i = 1, 2$.

The coefficient of ϵ^2 becomes

$$\begin{aligned}
&P(D)\{1.f_2 + f_2.1\} + P(D)\{f_1.f_1\} \\
&= 2P(\partial)f_2 + P(D)\{(e^{\theta_1} + e^{\theta_2}).(e^{\theta_1} + e^{\theta_2})\} \\
&= 2[P(\partial)f_2 + P(D)\{e^{\theta_1}e^{\theta_2}\}] \\
&= 2[P(\partial)f_2 + P(p_1 - p_2)e^{\theta_1+\theta_2}] \\
&= 0
\end{aligned}$$

This makes f_2 to have the form $f_2 = A(1,2)e^{\theta_1+\theta_2}$. If we put f_2 in the above equation we obtain $A(1,2)$ as

$$A(1,2) = -\frac{P(p_1 - p_2)}{P(p_1 + p_2)} = \frac{(k_1 - k_2)^2}{(k_1 + k_2)^2} \tag{4.164}$$

Since $f_3 = 0$, the coefficient of ϵ^3 turns out to be

$$\begin{aligned}
&P(D)\{f_1.f_2 + f_2.f_1\} \\
&= 2A(1,2)[P(D)\{(e^{\theta_1})(e^{\theta_1+\theta_2})\} + P(D)\{(e^{\theta_2})(e^{\theta_1+\theta_2})\}] \\
&= 2A(1,2)[P(p_2)e^{2\theta_1+\theta_2} + P(p_1)e^{\theta_1+2\theta_2}]
\end{aligned}$$

and this is already zero since $P(p_i) = 0, i = 1, 2$. The coefficient of ϵ^4 also vanishes trivially.

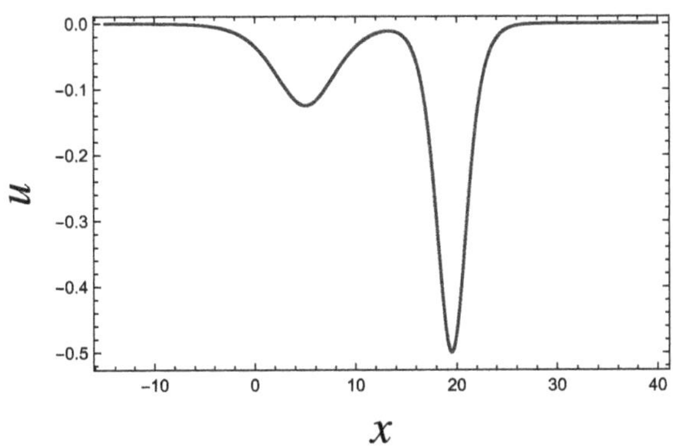

Figure 4.6. Two-soliton solution of Eq. (4.154).

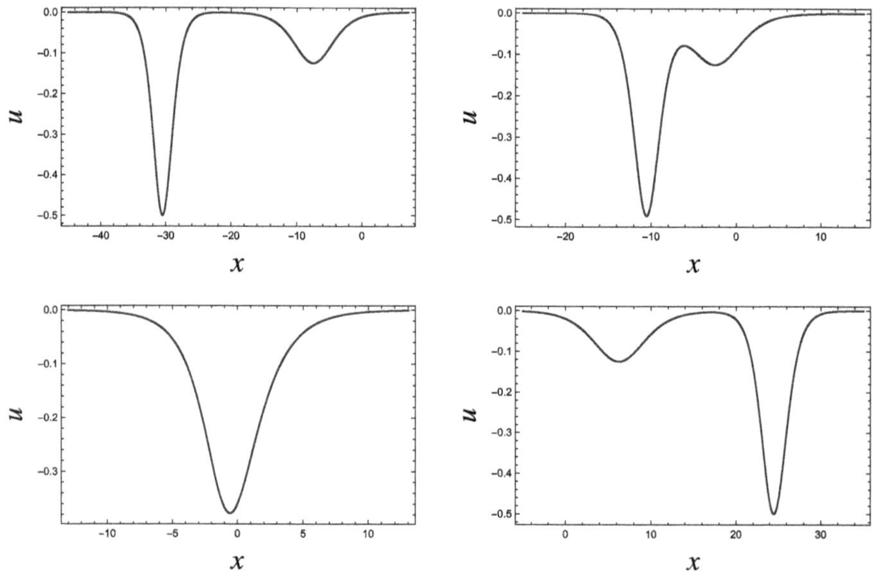

Figure 4.7. Soliton interaction of Eq. (4.165).

Finally, $\epsilon = 1$ can be set, so $f = 1 + e^{\theta_1} + e^{\theta_2} + A(1,2)e^{\theta_1+\theta_2}$ and the two-soliton solution of the KdV equation (4.157) are obtained.

$$u(x,t) = -2\frac{M}{N} \tag{4.165}$$

where $M = \{k_1^2 e^{\theta_1} + k_2^2 e^{\theta_2} + A(1,2)(k_2^2 e^{\theta_1} + k_1^2 e^{\theta_2})e^{\theta_1+\theta_2} + 2(k_1-k_2)^2 e^{\theta_1+\theta_2}\}$, $N = (1 + e^{\theta_1} + e^{\theta_2} + A(1,2)e^{\theta_1+\theta_2})^2$, and $\theta_i = k_i x - k_i^3 t + \alpha_i, i = 1,2$ and $A(1,2) = \frac{(k_1-k_2)^2}{(k_1+k_2)^2}$.

In a limiting sense, we get a two-soliton solution as a combination of two one-soliton solutions.

Now divide the numerator and denominator by $e^{2\theta_1+2\theta_2}$ and then neglecting the terms, we get

$$\begin{aligned} u(x,t) &= -2\left[\frac{A(1,2)k_2^2 e^{-\theta_2}}{(e^{-\theta_2} + A(1,2))^2} + \frac{A(1,2)k_1^2 e^{-\theta_1}}{(e^{-\theta_1} + A(1,2))^2}\right], \\ &= -2\left[\frac{A(1,2)k_2^2}{e^{\theta_2}(e^{-\theta_2} + A(1,2))^2} + \frac{A(1,2)k_1^2}{e^{\theta_1}(e^{-\theta_1} + A(1,2))^2}\right], \\ &= -2\left[\frac{k_1^2}{\{A_{12}e^{\theta_1/2} + \frac{1}{A_{12}}e^{-\theta_1/2}\}^2} + \frac{k_2^2}{\{A_{12}e^{\theta_2/2} + \frac{1}{A_{12}}e^{-\theta_2/2}\}^2}\right], \end{aligned}$$

$$= \left[\frac{A_1}{\left(\frac{e^{\left(\frac{\theta_1}{2}+\Delta_1\right)} + e^{-\left(\frac{\theta_1}{2}+\Delta_1\right)}}{2} \right)^2} + \frac{A_2}{\left(\frac{e^{\left(\frac{\theta_2}{2}+\Delta_2\right)} + e^{-\left(\frac{\theta_2}{2}+\Delta_2\right)}}{2} \right)^2} \right],$$

$$= \left[A_1 sech^2 \left(\frac{\theta_1}{2} + \Delta_1 \right) + A_2 sech^2 \left(\frac{\theta_2}{2} + \Delta_2 \right) \right]$$

Finally,

$$u(x,t) = \left[A_1 sech^2 \left(\frac{k_1 x - \omega_1 t + \alpha_1}{2} + \Delta_1 \right) + A_2 sech^2 \left(\frac{k_2 x - \omega_2 t + \alpha_2}{2} + \Delta_2 \right) \right] \tag{4.166}$$

where $A_i = -2k_i^2$, $\Delta_i = \pm \log |A_{12}|$, $A(1,2) = A_{12}^2$, and $A_{12} = \frac{|k_1 - k_2|}{k_1 + k_2}$. Here, A_{12} is the phase shift. So, Eq. (4.166) may be reproduced from Eq. (4.165) i.e., two distinct solitons can be obtained from the two-soliton solution of the KdV equation. That was the observation of Zabusky and Karuskal. This result is true not only for KdV soliton but also for any multisoliton solution obtained from any nonlinear evolution equation.

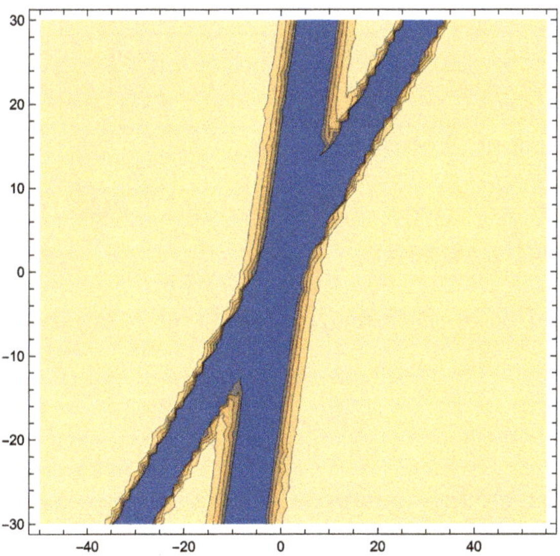

Figure 4.8. Contour plot of the solution of Eq. (4.165).

Three-soliton solution ($N = 3$):

In order to develop a three-soliton solution for the KdV equation (4.157), we take $f = 1 + \epsilon f_1 + \epsilon^2 f_2 + \epsilon^3 f_3$. For $i = 1, 2, 3$, $\theta_i = k_i x + \omega_i t + \alpha_i$, and $f_1 = e^{\theta_1} + e^{\theta_2} + e^{\theta_3}$. Note that $f_j = 0$ for all $j \geq 4$. We insert f into Eq. (4.162) and make the coefficients of ϵ^m, $m = 0,\ 1,\ 2,\ 3,\ 4,\ 5,\ 6$ to vanish. The coefficient of ϵ^0, is identically zero.

By the coefficient of ϵ^1 we have

$$P(D)\{1.f_1 + f_1.1\} = 2P(\partial)\{e^{\theta_1} + e^{\theta_2} + e^{\theta_3}\} = 0$$

which implies $P(p_i)0$, i.e., $\omega_i = -k_i^3$ for $i = 1,\ 2,\ 3$.

From the coefficient of ϵ^2, we get

$$\begin{aligned}-P(\partial)f_2 &= [(k_1 - k_2)(\omega_1 - \omega_2) + (k_1 - k_2)^4]e^{\theta_1+\theta_2}\\ &\quad + [(k_1 - k_3)(\omega_1 - \omega_3) + (k_1 - k_3)^4]e^{\theta_1+\theta_3}\\ &\quad + [(k_2 - k_3)(\omega_2 - \omega_3) + (k_2 - k_3)^4]e^{\theta_2+\theta_3}\end{aligned}$$

where $A_i = \frac{1}{2}k_i^2$, $\Delta_i = \pm \log|A_{12}|$, $A_{12} = \frac{|k_1 - k_2|}{k_1 + k_2}$, $i = 1,\ 2$. This makes f_2 to have the form $f_2 = A(1,2)e^{\theta_1+\theta_2} + A(1,3)e^{\theta_1+\theta_3} + A(2,3)e^{\theta_2+\theta_3}$. If we put f_2 in the above equation, we obtain $A(i,j)$ as

$$A(i,j) = -\frac{P(p_i - p_j)}{P(p_i + p_j)} = \frac{(k_i - k_j)^2}{(k_i + k_j)^2}$$

for $i, j = 1, 2, 3, i < j$.

The coefficient of ϵ^3 becomes

$$\begin{aligned}-P(\partial)f_3 = e^{\theta_1+\theta_2+\theta_3}\{&A(1,2)P(p_3 - p_2 - p_1) + A(1,3)P(p_2 - p_1 - p_3)\\ &+ A(2,3)P(p_1 - p_2 - p_3)\}\end{aligned}$$

Hence f_3 should be of the form $f_3 = Be^{\theta_1+\theta_2+\theta_3}$. So, the above equation gives

$$B = -\frac{M_1}{N_1}$$

where $M_1 = A(1,2)P(p_3 - p_2 - p_1) + A(1,3)P(p_2 - p_1 - p_3) + A(2,3)P(p_1 - p_2 - p_3)$, and $N_1 = P(p_1 + p_2 + p_3)$.

If we make all the simplifications by using $\omega_i = -k_i^3$ for $i = 1, 2, 3$, we see that the above expression is equivalent to $B = A(1,2)A(1,3)A(2,3)$.

Since $f_4 = 0$, from the coefficient of ϵ^4, we have

$$P(D)\{f_1.f_3 + f_3.f_1 + f_2.f_2\} = 0$$

After some calculations the above equation becomes

$$\begin{aligned} &e^{2\theta_1+\theta_2+\theta_3}[BP(p_2+p_3) + A(1,2)A(1,3)P(p_2-p_3)] \\ &\quad + e^{\theta_1+2\theta_2+\theta_3}[BP(p_1+p_3) + A(1,2)A(2,3)P(p_1-p_3)] \\ &\quad + e^{\theta_1+\theta_2+2\theta_3}[BP(p_1+p_2) + A(1,3)A(2,3)P(p_1-p_2)] \\ &= 0 \end{aligned}$$

This is satisfied by $B = A(1,2)A(1,3)A(2,3)$. Finally the coefficient of ϵ^5 and ϵ^6 also vanish automatically. We set $\epsilon = 1$, therefore $f = 1 + e^{\theta_1} + e^{\theta_2} + e^{\theta_3} + A(1,2)e^{\theta_1+\theta_2} + A(1,3)e^{\theta_1+\theta_3} + A(2,3)e^{\theta_2+\theta_3} + Be^{\theta_1+\theta_2+\theta_3}$, so the three-soliton solutions of the KdV equation (4.157) is

$$u(x,t) = -2\frac{L(x,t)}{M(x,t)} \tag{4.167}$$

where

$$\begin{aligned} L(x,t) = {} & e^{\theta_1+\theta_2}[2(k_1-k_2)^2 + 2(k_1-k_2)^2A(1,3)A(2,3)e^{2\theta_3} \\ & + A(1,2)k_1^2e^{\theta_2} + A(1,2)k_2^2e^{\theta_1}] \\ & + e^{\theta_1+\theta_3}[2(k_1-k_3)^2 + 2(k_1-k_3)^2A(1,2)A(2,3)e^{2\theta_2} \\ & + A(1,3)k_1^2e^{\theta_3} + A(1,3)k_3^2e^{\theta_1}] \\ & + e^{\theta_2+\theta_3}[2(k_2-k_3)^2 + 2(k_2-k_3)^2A(1,2)A(1,3)e^{2\theta_1} \\ & + A(2,3)k_2^2e^{\theta_3} + A(2,3)k_3^2e^{\theta_2}] \\ & + k_1^2e^{\theta_1} + k_2^2e^{\theta_2} + k_3^2e^{\theta_3} \\ & + Be^{\theta_1+\theta_2+\theta_3}[A(1,2)k_3^2e^{\theta_1+\theta_2} + A(1,3)k_2^2e^{\theta_1+\theta_3} + A(2,3)k_1^2e^{\theta_2+\theta_3}] \\ & + e^{\theta_1+\theta_2+\theta_3}[A(1,2)(k_1^2+k_2^2+k_3^2+2k_1k_2-2k_1k_3-2k_2k_3) \\ & + A(1,3)(k_1^2+k_2^2+k_3^2+2k_1k_3-2k_1k_2-2k_2k_3) \\ & + A(2,3)(k_1^2+k_2^2+k_3^2+2k_2k_3-2k_1k_2-2k_1k_3) \\ & + B(k_1^2+k_2^2+k_3^2+2k_1k_2+2k_1k_3+2k_2k_3) \end{aligned}$$

and

$$\begin{aligned} M(x,t) = {} & [1 + e^{\theta_1} + e^{\theta_2} + e^{\theta_3} + A(1,2)e^{\theta_1+\theta_2} + A(1,3)e^{\theta_1+\theta_3} \\ & + A(2,3)e^{\theta_2+\theta_3} + Be^{\theta_1+\theta_2+\theta_3}]^2 \end{aligned}$$

for $\theta_i = k_i x - k_i^3 t + \alpha_i$, $A(i,j) = \frac{(k_i - k_j)^2}{(k_i + k_j)^2}$ for $i, j = 1, 2, 3, i < j$ and $B = A(1,2)A(1,3)A(2,3)$.

4.15.2 *Multisoliton of the KP equation*

The KP equation can have one-, two-, or three-soliton solutions. We can determine these solutions by using the Hirota direct method. The KP equation is given by

$$(u_t - 6uu_x + u_{xxx})_x + 3u_{yy} = 0 \tag{4.168}$$

Step 1. Bilinearization: The KP equation is bilinearized using the transformation $u(x,y,t) = -2\frac{\partial^2}{\partial x^2}(\log f)$. Thus, the KP Eq. (4.168) in quadratic form is

$$ff_{xt} - f_x f_t + ff_{xxxx} - 4f_x f_{xxx} + 3f_{xx}^2 + 3f_{yy} - 3f_y^2 = 0 \tag{4.169}$$

Step 2. Transformation to the Hirota bilinear form: We try to express the KP equation's bilinear form in the Hirota bilinear form by utilizing the Hirota-D operator. The KP equation's Hirota bilinear form is

$$P(D)\{f.f\} = (D_x D_t + D_x^4 + 3D_y^2)\{f.f\} = 0 \tag{4.170}$$

Step 3. Application of the Hirota perturbation: Insert $f = 1 + \epsilon f_1 + \epsilon^2 f_2 + \ldots$ into Eq. (4.170), we get

$$\begin{aligned} P(D)\{f.f\} = P(D)\{1.1\} + \epsilon P(D)\{f_1.1 + 1.f_1\} + \epsilon^2 P(D)\{f_2.1 + f_1.f_1 \\ + 1.f_2\} + \epsilon^3 P(D)\{f_3.1 + f_2.f_1 + f_1.f_2 + 1.f_3\} + \cdots = 0 \end{aligned} \tag{4.171}$$

One-soliton solution:

To construct a one-soliton solution of the KP equation, we take $f = 1 + \epsilon f_1$, where $f_1 = e^{\theta_1}$ and $\theta_1 = k_1 x + l_1 y + \omega_1 t + \alpha_1$. Note that $f_j = 0$ for all $j \geq 2$. We insert f into Eq. (4.171) and make the coefficients of ϵ^m, $m = 0, 1, 2$ vanish. The coefficient of ϵ^0 is $P(D)\{1.1\} = 0$ since $P(0,0)\{1\} = 0$.

By the coefficient of ϵ^1

$$P(D)\{1.f_1 + f_1.1\} = P(\partial)e^{\theta_1} + P(-\partial)e^{\theta_1} = 2P(p_1)e^{\theta_1} = 0$$

We have the dispersion relation $P(p_1) = 0$, which implies $\omega_1 = -\frac{k_1^4 + 3l_1^2}{k_1}$. Finally, without loss of generality, we may set $\epsilon = 1$, so $f = 1 + e^{\theta_1}$, and one soliton solution of the KP Eq. (4.168) is

$$u(x,y,t) = -\frac{k_1^2}{2} sech^2\left(\frac{\theta_1}{2}\right) \tag{4.172}$$

where $\theta_1 = k_1 x - \left(\frac{k_1^4+3l_1^2}{k_1}\right)t + l_1 y + \alpha_1$.

Two-soliton solution:

To construct a two-soliton solution of the KP equation, we take $f = 1 + \epsilon f_1 + \epsilon^2 f_2$ where $f_1 = e^{\theta_1} + e^{\theta_2}$ and $\theta_i = k_i x + \omega_i t + l_i y + \alpha_i$ for $i = 1, 2$. Note that $f_j = 0$ for all $j \geq 3$. We insert f into Eq. (4.171) and make the coefficients of ϵ^m, $m = 0,\ 1,\ 2,\ 3,\ 4$ to vanish. We shall only examine the non-trivial ones, which are the coefficients of ϵ^1 and ϵ^2. From the coefficient of ϵ^1, we have

$$P(D)\{1.f_1 + f_1.1\} = 2P(\partial)\{e^{\theta_1} + e^{\theta_2}\} = 0$$

which implies $P(p_i) = k_i^4 + k_i\omega_i + 3l_i^2 = 0$, i.e., $\omega_i = -\frac{k_i^4+3l_i^2}{k_i}$ for $i = 1, 2$. The coefficient of ϵ^2 becomes

$$\begin{aligned}
&P(D)\{1.f_2 + f_2.1\} + P(D)\{f_1.f_1\}\\
&\quad = 2P(\partial)f_2 + P(D)\{(e^{\theta_1}.e^{\theta_2}) + (e^{\theta_2}.e^{\theta_1})\}\\
&\quad = 2[P(\partial)f_2 + P(p_1 - p_2)e^{\theta_1+\theta_2}] = 0
\end{aligned}$$

This makes f_2 to have the form $f_2 = A(1,2)e^{\theta_1+\theta_2}$. If we put f_2 in the above equation, we obtain

$$\begin{aligned}
A(1,2) &= -\frac{P(p_1 - p_2)}{P(p_1 + p_2)}\\
&= \frac{k_1\omega_2 + k_2\omega_1 + 4k_1^3k_2 - 6k_1^2k_2^2 + 4k_1k_2^3 + 6l_1l_2}{k_1\omega_2 + k_2\omega_1 + 4k_1^3k_2 + 6k_1^2k_2^2 + 4k_1k_2^3 + 6l_1l_2}.
\end{aligned} \tag{4.173}$$

At last we may set $\epsilon = 1$, thus $f = 1 + e^{\theta_1} + e^{\theta_2} + A(1,2)e^{\theta_1+\theta_2}$, and the two-soliton solution of the KP equation is given by

$$u = -2\frac{M_2}{N_2} \tag{4.174}$$

where $M_2 = \{k_1^2e^{\theta_1} + k_2^2e^{\theta_2} + [(k_1 - k_2)^2 + A(1,2)((k_1 + k_2)^2 + k_2^2e^{\theta_1} + k_1^2e^{\theta_2})]e^{\theta_1+\theta_2}\}$, $N_2 = (1 + e^{\theta_1} + e^{\theta_2} + A(1,2)e^{\theta_1+\theta_2})^2$ and $\theta_i = k_i x - (\frac{k_i^4+3l_i^2}{k_i})t + l_i y + \alpha_i, i = 1, 2$ and $A(1,2)$ is as given in (4.173).

Three-soliton solution:

To construct a three-soliton solution of the KP equation, we take $f = 1+\epsilon f_1+\epsilon^2 f_2+\epsilon^3 f_3$, where $f_1 = e^{\theta_1}+e^{\theta_2}+e^{\theta_3}$ and $\theta_i = k_i x+\omega_i t+l_i y+\alpha_i$ for $i = 1,2,3$. We note that $f_j = 0$ for all $j \geq 4$. We insert f into Eq. (4.171). Let us consider only the coefficients of ϵ^m, $m = 1,2,3,4$ since others vanish automatically. From the coefficient of ϵ^1, we have

$$P(D)\{1.f_1 + f_1.1\} = 2P(\partial)\{e^{\theta_1} + e^{\theta_2} + e^{\theta_3}\} = 0$$

which implies $P(p_i) = 0$, i.e., $\omega_i = -\frac{k_i^4+3l_i^2}{k_i}$ for $i = 1,2,3$. From the coefficient of ϵ^2 we get

$$\begin{aligned}-P(\partial)f_2 &= [(k_1-k_2)^4 + (k_1-k_2)(\omega_1-\omega_2) + 3(l_1-l_2)^2]e^{\theta_1+\theta_2}\\ &\quad + [(k_1-k_3)^4 + (k_1-k_3)(\omega_1-\omega_3) + 3(l_1-l_3)^2]e^{\theta_1+\theta_3}\\ &\quad + [(k_2-k_3)^4 + (k_2-k_3)(\omega_2-\omega_3) + 3(l_2-l_3)^2]e^{\theta_2+\theta_3}\end{aligned}$$

This makes f_2 to have the form $f_2 = A(1,2)e^{\theta_1+\theta_2} + A(1,3)e^{\theta_1+\theta_3} + A(2,3)e^{\theta_2+\theta_3}$. If we put f_2 in the above equation, we obtain $A(i,j)$ as

$$A(i,j) = -\frac{P(p_i-p_j)}{P(p_i+p_j)} = \frac{k_i\omega_j + k_j\omega_i + 4k_i^3k_j - 6k_i^2k_j^2 + 4k_ik_j^3 + 6l_il_j}{k_i\omega_j + k_j\omega_i + 4k_i^3k_j + 6k_i^2k_j^2 + 4k_ik_j^3 + 6l_il_j}$$

for $i,j = 1,2,3, i < j$. The coefficient of ϵ^3 becomes

$$\begin{aligned}-P(\partial)f_3 = e^{\theta_1+\theta_2+\theta_3}\{&A(1,2)P(p_3-p_2-p_1) + A(1,3)P(p_2-p_1-p_3)\\ &+ A(2,3)P(p_1-p_2-p_3)\}\end{aligned}$$

Hence, f_3 should be of the form $f_3 = Be^{\theta_1+\theta_2+\theta_3}$. If we insert f_3 into the above equation, we find that

$$B = -\frac{M_3}{N_3}$$

where $M_3 = A(1,2)P(p_3-p_2-p_1)+A(1,3)P(p_2-p_1-p_3)+A(2,3)P(p_1-p_2-p_3)$ and $N_3 = P(p_1+p_2+p_3)$.

If we make all the simplifications by using $\omega_i = -\frac{k_i^4+3l_i^2}{k_i}$ for $i = 1,2,3$, we see that the above expression is equivalent to $B = A(1,2)A(1,3)A(2,3)$.

Since $f_4 = 0$, from the coefficient of ϵ^4, we have

$$P(D)\{f_1.f_3 + f_3.f_1 + f_2.f_2\} = 0$$

After some calculations the above equation becomes

$$\begin{aligned} & e^{2\theta_1+\theta_2+\theta_3}[BP(p_2+p_3) + A(1,2)A(1,3)P(p_2-p_3)] \\ & \quad + e^{\theta_1+2\theta_2+\theta_3}[BP(p_1+p_3) + A(1,2)A(2,3)P(p_1-p_3)] \\ & \quad + e^{\theta_1+\theta_2+2\theta_3}[BP(p_1+p_2) + A(1,3)A(2,3)P(p_1-p_2)] \\ & = 0 \end{aligned}$$

This is satisfied by $B = A(1,2)A(1,3)A(2,3)$. We set $\epsilon = 1$, therefore $f = 1+e^{\theta_1}+e^{\theta_2}+e^{\theta_3}+A(1,2)e^{\theta_1+\theta_2}+A(1,3)e^{\theta_1+\theta_3}+A(2,3)e^{\theta_2+\theta_3}+Be^{\theta_1+\theta_2+\theta_3}$, so the three-soliton solution of the KP equation is

$$u(x,t) = -2\frac{L(x,t,y)}{M(x,t,y)} \tag{4.175}$$

where

$$\begin{aligned} & L(x,t,y) \\ & = k_1^2 e^{\theta_1} + k_2^2 e^{\theta_2} + k_3^2 e^{\theta_3} \\ & \quad + e^{2\theta_1+\theta_2+\theta_3}[A(1,2)A(1,3)(k_2-k_3)^2 + B(k_2+k_3)^2] \\ & \quad + e^{\theta_1+2\theta_2+\theta_3}[A(1,2)A(2,3)(k_1-k_3)^2 + B(k_1+k_3)^2] \\ & \quad + e^{\theta_1+\theta_2+2\theta_3}[A(1,3)A(2,3)(k_1-k_2)^2 + B(k_1+k_2)^2] \\ & \quad + e^{\theta_1+\theta_2}[(k_1-k_2)^2 + A(1,2)(k_1^2 e^{\theta_2} + k_2^2 e^{\theta_1} + (k_1+k_2)^2)] \\ & \quad + e^{\theta_1+\theta_3}[(k_1-k_3)^2 + A(1,3)(k_1^2 e^{\theta_3} + k_3^2 e^{\theta_1} + (k_1+k_3)^2)] \\ & \quad + e^{\theta_2+\theta_3}[(k_2-k_3)^2 + A(2,3)(k_2^2 e^{\theta_3} + k_3^2 e^{\theta_2} + (k_2+k_3)^2)] \\ & \quad + Be^{\theta_1+\theta_2+\theta_3}[A(1,2)k_3^2 e^{\theta_1+\theta_2} + A(1,3)k_2^2 e^{\theta_1+\theta_3} + A(2,3)k_1^2 e^{\theta_2+\theta_3}] \\ & \quad + e^{\theta_1+\theta_2+\theta_3}[A(1,2)(k_1^2+k_2^2+k_3^2+2k_1k_2-2k_1k_3-2k_2k_3) \\ & \quad + A(1,3)(k_1^2+k_2^2+k_3^2+2k_1k_3-2k_1k_2-2k_2k_3) \\ & \quad + A(2,3)(k_1^2+k_2^2+k_3^2+2k_2k_3-2k_1k_2-2k_1k_3) \\ & \quad + B(k_1^2+k_2^2+k_3^2+2k_1k_2+2k_1k_3+2k_2k_3) \end{aligned}$$

and

$$\begin{aligned} M(x,t,y) = {} & [1 + e^{\theta_1} + e^{\theta_2} + e^{\theta_3} + A(1,2)e^{\theta_1+\theta_2} + A(1,3)e^{\theta_1+\theta_3} \\ & + A(2,3)e^{\theta_2+\theta_3} + Be^{\theta_1+\theta_2+\theta_3}]^2 \end{aligned}$$

for $\theta_i = k_i x - (\frac{k_i^4+3l_i^2}{k_i})t + \alpha_i$, $A(i,j)$ for $i,j = 1,2,3, i < j$ and $B = A(1,2)A(1,3)A(2,3)$.

4.15.3 *Multiple singular-soliton solution*

The soliton solution carrying singularities gives rise to a singular soliton. The Hirota bilinear method yields similar results for the singularity-containing solutions. It has been noted that regular solitons manifest as $sech^2$, $sech$, etc. Hirota's method will be employed to derive the mKdV equation's singular-soliton solutions.

$$\phi_\tau - 6\phi^2\phi_\xi + \phi_{\xi\xi\xi} = 0 \tag{4.176}$$

Using the transformation

$$\phi(\xi,\tau) = \frac{\partial}{\partial\xi}[lnF(\xi,\tau)] = \frac{\partial}{\partial\xi}\left[ln\frac{p(\xi,\tau)}{q(\xi,\tau)}\right] = \frac{qp_\xi - pq_\xi}{qp}$$

For the mKdV equation, we obtain a system of Hirota bilinear equations, which are given by

$$S_1(D)(p.q) = 0; \quad (D_\tau + D_\xi^3)(p.q) = 0 \tag{4.177}$$

$$S_2(D)(p.p + q.q) = 0; \quad D_\xi^2(p.p + q.q) = 0 \tag{4.178}$$

where S_1 is an odd polynomial in D_ξ, D_τ and S_2 is an even polynomial in D_ξ, but without a constant term.

To determine the singular two-soliton solution, we set

$$p(\xi,\tau) = 1 + e^{\theta_1} + e^{\theta_2} + E_{12}e^{\theta_1+\theta_2}$$
$$q(\xi,\tau) = 1 - e^{\theta_1} - e^{\theta_2} + E'_{12}e^{\theta_1+\theta_2}$$

where $\theta_r = k_r\xi - \omega_r\tau + \alpha_r, r = 1,2$

The coefficients parameter satisfying the relation

$$S_1(k_r, -\omega_r) = 0 \implies \omega_r = k_r^3 \quad \text{for } r = 1,2$$

which is called a dispersion relation.

Substituting the values of the auxiliary functions into the coupled bilinear equations and using the dispersion relation, we found that

$$\phi(\xi,\tau) = \frac{\partial}{\partial\xi}\left[ln\frac{p(\xi,\tau)}{q(\xi,\tau)}\right]$$

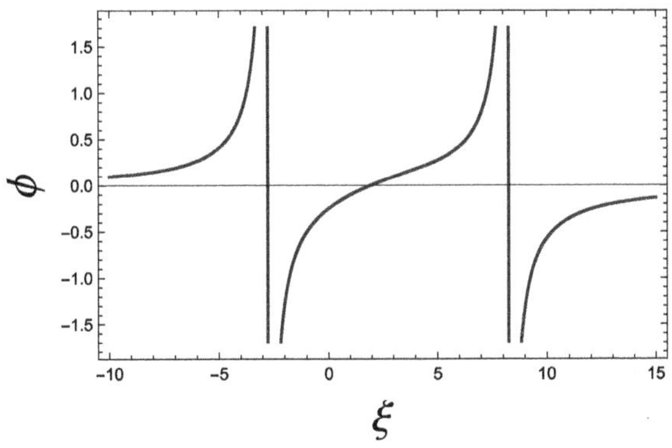

Figure 4.9. Singular two-soliton solution of Eq. (4.179).

is a solution of the mKdV equation if the phase factors E_{12} and E'_{12} are given by

$$E_{12} = E'_{12} = \frac{S_2(k_1 - k_2, -\omega_1 + \omega_2)}{S_2(k_1 + k_2, -\omega_1 - \omega_2)} = \frac{(k_1 - k_2)^2}{(k_1 + k_2)^2}$$

Therefore, the singular two-soliton solution is given by

$$\phi(\xi, \tau) = \frac{\partial}{\partial \xi}\left[ln\left(\frac{P_1}{P_2}\right)\right] \tag{4.179}$$

where $P_1 = 1 + exp(k_1\xi - k_1^3\tau + \alpha_1) + exp(k_2\xi - k_2^3\tau + \alpha_2) + \frac{(k_1-k_2)^2}{(k_1+k_2)^2} exp[(k_1 + k_2)\xi - (\omega_1 + \omega_2)\tau + (\alpha_1 + \alpha_2)]$;

$P_2 = 1 - exp(k_1\xi - k_1^3\tau + \alpha_1) - exp(k_2\xi - k_2^3\tau + \alpha_2) + \frac{(k_1-k_2)^2}{(k_1+k_2)^2} exp[(k_1 + k_2)\xi - (\omega_1 + \omega_2)\tau + (\alpha_1 + \alpha_2)]$

If we use the Cole-Hopf transformation, then the singular-soliton solution of the mKdV equation will be

$$\phi(\xi, \tau) = (lnp)_\xi$$

where the general form of p, for the singular N-soliton solution, is given by

$$p(\xi, \tau) = \sum_{\mu_r \in \{0,1\}} (-1)^{\sum_{r=1}^{N} \mu_r} \left(exp\left[\sum_{r<s}^{(N)} a_{rs}\mu_r\mu_s + \sum_{r=1}^{N} \mu_r\theta_r\right]\right)$$

where

$$\theta_r = k_r\xi - \omega_r\tau + \alpha_r \quad \text{for } r = 1, 2, \ldots, N$$

The dispersion relation will be the same as previous $\omega_r = k_r^3$ for $r = 1, 2, \ldots, N$, and

$$E_{rs} = \exp(a_{rs}) = \frac{(k_r - k_s)^2}{(k_r + k_s)^2}, \quad 1 \leq r < s \leq N.$$

4.16 Lump Soliton

The KP equation can be applied to physics as a way to model water waves of long wavelengths with weakly nonlinear restoring forces and frequency dispersion. It is a two-dimensional generalization of the one-dimensional KdV equation. Like the KdV equation, the KP equation is completely integrable. It is known that all integrable equations possess soliton solutions — exponentially localized solutions in certain directions. Hirota bilinear forms play an important role in presenting soliton solutions. In contrast to soliton solutions, lump solutions are a kind of rational function solution, localized in all directions in the space. A lump soliton is one kind of rough wave that asserts the presence of a large-amplitude wave, generally more than two times, compared to surrounding waves "average amplitude," holding its unpredicted appearance features and its disappearance without any trace. These amazing nonlinear characteristics of lump solitons attracted many researchers. Mathematically, lump solutions can be represented by localized, smooth, and symmetric functions that exist for all time. These functions tend to have fast decay away from a central region. Lump solutions can be classified based on their amplitude, width, skewness, and other properties. Lump solutions are not limited to wave equations but can also appear in other nonlinear systems. Understanding lump solutions can lead to advancements in various applications, including studies of soliton-like behavior, information transmission in nonlinear communication systems, and the characterization of localized phenomena in complex systems.

Lump solutions of KP Equation:

The KP equation's lump solutions will be revealed to us here. Consider the KP equation

$$\frac{\partial}{\partial \bar{\zeta}}\left[\frac{\partial \bar{\phi}_1}{\partial \bar{\tau}} + 6\bar{\phi}_1 \frac{\partial \bar{\phi}_1}{\partial \bar{\zeta}} + \frac{\partial^3 \bar{\phi}_1}{\partial \bar{\zeta}^3}\right] - \sigma \frac{\partial^2 \bar{\phi}_1}{\partial \bar{\chi}^2} = 0 \tag{4.180}$$

If $\sigma = 1$, then Eq. (4.180) is called the KPI equation, and when $\sigma = -1$, Eq. (4.180) is called the KPII equation.

We now consider the transformation

$$\bar{\phi_1}(\bar{\zeta}, \bar{\chi}, \bar{\tau}) = 2\,(lnf)_{\bar{\zeta}\bar{\zeta}} = 2\left[\frac{f_{\bar{\zeta}\bar{\zeta}}f - f_{\bar{\zeta}}^2}{f^2}\right] \tag{4.181}$$

But it is not possible to provide the explanation of Eq. (4.181) because this expression is coming due to the proper guess of the transformation of the dependent variables when applying the Hirota bilinear method. According to the Hirota bilinear method, a suitable transformation has to be set so that it can change the NLPDEs that are in quadratic form in dependent variables into two linear equations, i.e., "Bilinear form", and this is why the method is named as the Hirota bilinear method. This is the most challenging part: setting the transformation to get the soliton solution of a nonlinear evolution equation using the Hirota bilinear method. To obtain the lump soliton solution of the derived KP equation (4.180), rather than solving the standard KP equation (4.180) using the Hirota bilinear method, it could be done by constructing the auxiliary function separately. It becomes the following bilinear equation:

$$\left(f_{\bar{\zeta}\bar{\tau}}f - f_{\bar{\zeta}}f_{\bar{\tau}}\right) - \sigma\left(f_{\bar{\chi}\bar{\chi}}f - f_{\bar{\chi}}^2\right) + f_{\bar{\zeta}\bar{\zeta}\bar{\zeta}\bar{\zeta}}f - 4f_{\bar{\zeta}\bar{\zeta}\bar{\zeta}}f_{\bar{\zeta}} + 3f_{\bar{\zeta}\bar{\zeta}}^2 = 0 \tag{4.182}$$

The Hirota bilinear form of Eq. (4.182) is

$$\left(D_{\bar{\zeta}}D_{\bar{\tau}} + D_{\bar{\zeta}}^4\right)(f.f) - \sigma D_{\bar{\chi}}^2(f.f) = 0, \qquad \sigma = \pm 1 \tag{4.183}$$

Putting $\sigma = 1$ in Eq. (4.182), we then get the (2+1)-dimensional bilinear KPI equation as

$$(f_{\bar{\zeta}\bar{\tau}}f - f_{\bar{\zeta}}f_{\bar{\tau}}) - (f_{\bar{\chi}\bar{\chi}}f - f_{\bar{\chi}}^2) + f_{\bar{\zeta}\bar{\zeta}\bar{\zeta}\bar{\zeta}}f - 4f_{\bar{\zeta}\bar{\zeta}\bar{\zeta}}f_{\bar{\zeta}} + 3f_{\bar{\zeta}\bar{\zeta}}^2 = 0 \tag{4.184}$$

To search for quadratic function solutions to the (4.184), we begin with

$$\begin{cases} f = g^2 + h^2 + A_9 \\ g = A_1\bar{\zeta} + A_2\bar{\chi} + A_3\bar{\tau} + A_4, \\ h = A_5\bar{\zeta} + A_6\bar{\chi} + A_7\bar{\tau} + A_8 \end{cases} \tag{4.185}$$

where $A_i, 1 \le i \le 9$, are real parameters to be determined.

Putting the value of f obtained from (4.185) and its partial derivatives in (4.184), we get the set of constraining equations for the parameters:

$$\begin{cases} A_3 = \dfrac{A_1A_2^2 - A_1A_6^2 + 2A_2A_5A_6}{A_1^2 + A_5^2}, \\ A_7 = \dfrac{2A_1A_2A_6 - A_2^2A_5 + A_5A_6^2}{A_1^2 + A_5^2}, \\ A_9 = \dfrac{3(A_1^2 + A_5^2)^3}{(A_1A_6 - A_2A_5)^2} \end{cases}$$

which needs to satisfy a determinant condition

$$\Delta := A_1A_6 - A_2A_5 = \begin{vmatrix} A_1 & A_2 \\ A_5 & A_6 \end{vmatrix} \neq 0 \tag{4.186}$$

This set leads to a class of positive quadratic function solutions to the bilinear KPI equation (4.184):

$$\begin{aligned} f = {} & \left[A_1\bar{\zeta} + A_2\bar{\chi} + \frac{A_1A_2^2 - A_1A_6^2 + 2A_2A_5A_6}{A_1^2 + A_5^2}\bar{\tau} + A_4\right]^2 \\ & + \left[A_5\bar{\zeta} + A_6\bar{\chi} + \frac{2A_1A_2A_6 - A_2^2A_5 + A_5A_6^2}{A_1^2 + A_5^2}\bar{\tau} + A_8\right]^2 \\ & + \frac{3(A_1^2 + A_5^2)^3}{(A_1A_6 - A_2A_5)^2} \end{aligned} \tag{4.187}$$

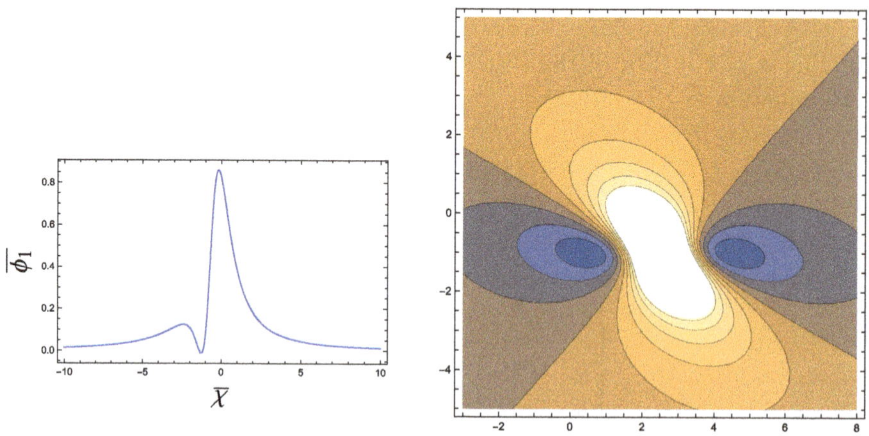

Figure 4.10. Lump soliton solution of Eq. (4.188).

and the resulting class of quadratic function solutions, in turn, yields a class of lump solutions to the (2+1)-dimensional KPI equation (4.184) through the transformation (4.181)

$$\bar{\phi}_1(\bar{\zeta},\bar{\chi},\bar{\tau}) = \frac{4(A_1^2+A_5^2)f - 8(A_1 g + A_5 h)^2}{f^2} \tag{4.188}$$

where the function f is defined by Eq. (4.187), and the functions of g and h are given as follows:

$$g = A_1\bar{\zeta} + A_2\bar{\chi} + \frac{A_1A_2^2 - A_1A_6^2 + 2A_2A_5A_6}{A_1^2+A_5^2}\bar{\tau} + A_4, \tag{4.189}$$

$$h = A_5\bar{\zeta} + A_6\bar{\chi} + \frac{2A_1A_2A_6 - A_2^2A_5 + A_5A_6^2}{A_1^2+A_5^2}\bar{\tau} + A_8 \tag{4.190}$$

In this class of lump solutions, all six involved parameters of A_1, A_2, A_4, A_5, A_6, A_8 are arbitrary, provided that the solutions are well defined, i.e., if the determinant condition equation (4.186) is satisfied. That determinant condition precisely means that two directions (A_1, A_2) and (A_5, A_6) in the $\bar{\zeta}\bar{\chi}$-plane are not parallel.

4.17 Breather Soliton

Breather solitons are exact stationary wave solutions that are specially localized and also time-periodic, which means their profile is not steady and changes periodically in time. They describe an oscillatory wave packet that is determined by the envelope's amplitude, which also determines its width and the frequency of the carrier wave.

Breathers arise both for kink–anti-kink interaction and for multisoliton interaction. During the interaction between two solitons, if there is sufficient energy in the medium, then they pass through without any damage in shape or speed; otherwise, they can't be separated from each other, and as a result, an oscillatory structure occurs over the soliton profile. This new structure, which has become localized, is called a breather. The same phenomena happen in the kink-anti-kink interaction.

Recently, Feng *et al.* [40] have studied the Akhmediev breathers for the nonlinear Schrödinger equation on the elliptic function background. The loop-like kink breather and its transition phenomena for the Vakhnenko equation have been investigated by Bang-Qing Li [41].

Breather solution can be obtained using the homoclinic test method, where the test function is considered as

$$\phi = exp[-p_1(\xi + m_1 y + c_1 t)] + \delta_1 \cos[p_2(\xi + m_2 y + c_2 t)] + \delta_2 exp[p_1(\xi + m_1 y + c_1 t)]$$

where $p_1, p_2, m_1, m_2, c_1, c_2, \delta_1, \delta_2$ are real undetermined constants. But this method is bound by one breather. To get multibreather, another method is useful, which is known as the complex conjugate method. To get multibreather using this method, first the multisoliton solutions of a considered model are needed. Now, the Hirota bilinear method (HBM) is appropriate to derive multisoliton solutions for NPDE. There is a key step in HBM where the bilinear formalism is obtained for a given evolution equation. To use complex conjugate method, we expand the function into a series of infinitesimal ϵ, as:

$$\phi = 1 + \phi^{(1)}\epsilon + \phi^{(2)}\epsilon^2 + \phi^{(3)}\epsilon^3 + \cdots$$

Then substituting this series expansion into the bilinear form, which is derived using the HBM, and comparing the coefficients of the same power with respect to ϵ, we get the expression of $\phi^{(1)}, \phi^{(2)}, \phi^{(3)}, \ldots$ Then, by selecting complex conjugate wave numbers, one can get multibreathers. From $2N$-soliton solutions, the formation of N-breather is possible. To get a minimum order breather, i.e., one breather of 2-soliton solutions is required,. For two breathers, the 4-soliton solution is needed. In this way, we can get multibreathers.

Breather Solutions for the Gardner Equation:

Suppose we consider the Gardner equation (GE) in the following form:

$$\phi_\tau + sc_1 B\phi\phi_\xi + BC\phi^2\phi_\xi + B\phi_{\xi\xi\xi} = 0 \tag{4.191}$$

Now, we take the transformation of dependent variable as follows:

$$\phi = 2[\tan^{-1}(g/f)]_\xi \tag{4.192}$$

Then the Hirota bilinear form of (4.191) is given by

$$(D_\tau + BD_\xi^3)(g.f) = 0 \tag{4.193}$$

and

$$\left(\frac{BC}{2}\right) D_\xi^2(f.f + g.g) - sc_1 BD_\xi(g.f) = 0 \tag{4.194}$$

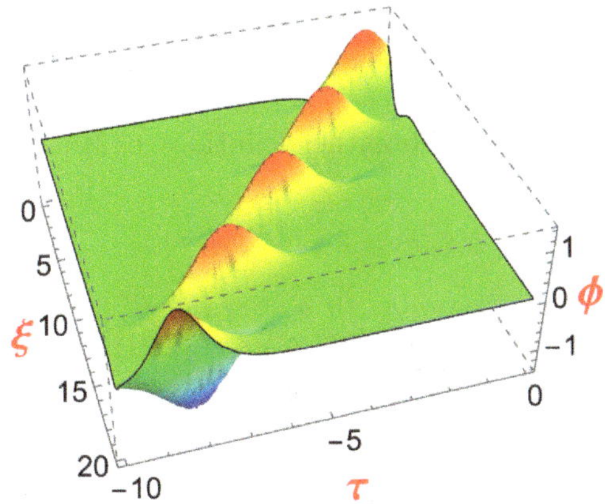

Figure 4.11. Two-soliton solution of Eq. (4.197).

To obtain 2-soliton solutions, we take

$$f = 1 + exp(\theta) + exp(\psi) + f_{12}exp(\theta + \psi), \tag{4.195}$$

$$g = 1 + A_1exp(\theta) + A_2exp(\psi) + A_{12}exp(\theta + \psi) \tag{4.196}$$

Using the values of f and g, from (4.193) and (4.194), we get the dispersion relation:

$$\Omega_i = Bp_i^3, \quad i = 1, 2$$

Hence, the two-soliton solutions of Eq. (4.191) is given by

$$\phi(\xi, \tau) = 2[\tan^{-1}(g/f)]_\xi \tag{4.197}$$

where

$$\theta = p_1\xi - Bp_1^3\tau, \quad \psi = p_2\xi - Bp_2^3\tau, \quad A_n = \frac{sc_1B + BCp_n}{sc_1B - BCp_n}, \quad n = 1, 2$$

and

$$f_{12} = \frac{(p_1 - p_2)^2[(sc_1B)^2 - (sc_1B)(BC)(p_1 + p_2) - (BC)^2p_1p_2]}{(p_1 + p_2)^2(sc_1B - BCp_1)(sc_1B - BCp_2)}$$

$$A_{12} = \frac{(p_1 - p_2)^2[(sc_1B)^2 + (sc_1B)(BC)(p_1 + p_2) - (BC)^2p_1p_2]}{(p_1 + p_2)^2(sc_1B - BCp_1)(sc_1B - BCp_2)}$$

where f, g are given by Eqs. (4.195) and (4.196). The breather solutions are derived by the direct substitution of complex conjugate wave numbers ($p_1 = m + in$, $p_2 = m - in$, where m, n are real) in two-soliton solutions.

Hence, the expression of breather solutions of Eq. (4.191) is given as follows:

$$\phi(\xi,\tau) = 2[\tan^{-1}(g/f)]_\xi \tag{4.198}$$

where

$$\begin{aligned} g &= 1 - \frac{n^2}{m^2}\left[\frac{(sc_1B)^2 + 2(sc_1B)(BC)m - (BC)^2(m^2+n^2)}{(sc_1B - BCm)^2 + (BC)^2n^2}\right] exp(2\Theta) \\ &\quad + 2\left[\zeta\cos(\Psi) - \eta\sin(\Psi)\right] exp(\Theta) \\ f &= 1 - \frac{n^2}{m^2}\left[\frac{(sc_1B)^2 - 2(sc_1B)(BC)m - (BC)^2(m^2+n^2)}{(sc_1B - BCm)^2 + (BC)^2n^2}\right] exp(2\Theta) \\ &\quad + 2\cos(\Psi) exp(\Theta) \\ \Theta &= m[\xi - B(m^2 - 3n^2)\tau], \quad \Psi = n[\xi - B(3m^2 - n^2)\tau] \\ \zeta &= \frac{(sc_1B)^2 - (BC)^2(m^2+n^2)}{(sc_1B - BCm)^2 + (BC)^2n^2}, \quad \eta = \frac{2(sc_1B)(BC)n}{(sc_1B - BCm)^2 + (BC)^2n^2}. \end{aligned}$$

4.18 Bell Polynomial

The Bell polynomial is a specific kind of generating function that plays a central role in combinatorics, particularly in expressing symmetric polynomials, partitions, and in many applications involving exponential generating functions. Eric Temple Bell introduced the Bell polynomial in his 1934 book *The Development of Mathematics.* However, the polynomial itself existed in various forms before this formalization, particularly in the context of symmetric functions, generating functions, and combinatorial structures.

One Dimensional:

C^∞ function is a function that is differentiable for all degrees of differentiation. C^∞ function is also called smooth. Let h be a C^∞ function of t, then the one-dimensional Bell polynomial is defined as

$$Y_{nt}(h) = Y_n(h_1, h_2, \ldots, h_{nt}) = e^{-h}\partial_t^n(e^h), \quad n = 1, 2, 3, \ldots \tag{4.199}$$

Several one-dimensional Bell polynomials can be obtained from the aforementioned statement as follows:

$$Y_t = h_t, \quad Y_{2t} = h_{2t} + h_t^2, \quad Y_{3t} = h_{3t} + 3h_th_{2t} + h_t^3, \ldots \tag{4.200}$$

The expressions (4.199) are produced by the formula given by

$$Y_{nt}(h) = \sum \frac{n!}{a_1!a_2!\cdots a_n!}\left(\frac{h_t}{1!}\right)^{a_1}\left(\frac{h_{2t}}{2!}\right)^{a_2}\cdots\left(\frac{h_{nt}}{n!}\right)^{a_n} \tag{4.201}$$

where the sum run over all partitions of $n = a_1 + 2a_2 + \cdots + na_n$.

Multidimensional:

Let us consider $h = h(t_1, t_2, \ldots, t_s)$ to be a C^∞ multivariable function. The multidimensional Bell polynomial can be written as

$$Y_{n_1t_1,\ldots,n_st_s}(h) = Y_{n_1,\ldots,n_s}(h_{m_1t_1,\ldots,m_st_s}) = e^{-h}\partial_{t_1}^{n_1}\ldots\partial_{t_s}^{n_s}(e^h) \tag{4.202}$$

where $h_{m_1t_1,\ldots,m_st_s} = \partial_{t_1}^{m_1}\ldots\partial_{t_s}^{m_s}h$, $m_i = 0, 1, \ldots, n_i$ and $i = 1, 2, \ldots, s$. $Y_{n_1t_1,\ldots,n_st_s}(h)$ represents the multivariable Bell polynomial with respect to $h_{m_1t_1,\ldots,m_st_s}$. When $h = h(t, z)$, the Bell polynomials can be calculated in the following way:

$$Y_{2t}(h) = h_{2t} + h_t^2, \quad Y_{3t}(h) = h_{3t} + 3h_th_{2t} + h_t^3$$
$$Y_{t,z}(h) = h_{t,z} + h_th_z, \quad Y_{2t,z}(h) = h_{2t,z} + h_zh_{2t} + 2h_{t,z}h_t + h_t^2h_z, \ldots$$

The multidimensional binary Bell polynomials as per Eq. (4.202) can be written as

$$\begin{aligned}&\mathscr{Y}_{n_1t_1,\ldots,n_st_s}(v, w)\\ &= y_{n_1t_1,\ldots,n_st_s}(h)\Bigg|_{h_{m_1t_1,\ldots,m_st_s}=\begin{cases}v_{m_1t_1,\ldots,m_st_s}, & m_1+m_2+\cdots+m_s \text{ is odd}\\ w_{m_1t_1,\cdots,m_st_s}, & m_1+m_2+\cdots+m_s \text{ is even}\end{cases}}\end{aligned} \tag{4.203}$$

With the help of (4.203), we can write the binary Bell polynomials as

$$\mathscr{Y}_t(v, w) = v_t, \quad \mathscr{Y}_{2t}(v, w) = w_{2t} + v_t^2, \quad \mathscr{Y}_{t,z}(v, w) = w_{t,z} + v_tv_z,$$
$$\mathscr{Y}_{2t,z}(v, w) = v_{2t,z} + v_zw_{2t} + 2w_{t,z}v_t + v_t^2v_z,$$
$$\mathscr{Y}_{3t}(v, w) = v_{3t} + 3v_tw_{2t} + v_t^3, \ldots$$

The relation between the standard Hirota bilinear expression $D_{t_1}^{n_1}\ldots D_{t_s}^{n_s}(h.g)$ and the binary Bell polynomial is

$$(hg)^{-1}D_{t_1}^{n_1}\ldots D_{t_s}^{n_s}(h.g) = \mathscr{Y}_{n_1t_1,\ldots,n_st_s}\left(v = \ln\frac{h}{g}, w = \ln hg\right), \tag{4.204}$$

Also

$$D_{t_1}^{n_1} \dots D_{t_s}^{n_s} h.g$$
$$= (\partial_{t_1} - \partial_{t_1'})^{n_1} \dots (\partial_{t_s} - \partial_{t_s'})^{n_s}$$
$$h(t_1, t_2, \dots, t_s).g(t_1', t_2', \dots, t_s')\big|_{t_1'=t_1, \dots t_s'=t_s} \quad (4.205)$$

In case when $h = g$, (4.205) is expressed as

$$(h)^{-2} D_{t_1}^{n_1} \dots D_{t_s}^{n_s} h.h = \mathscr{Y}_{n_1 t_1, \dots, n_s t_s}(v = 0, w = 2 \ln h)$$
$$= \begin{cases} 0, & n_1 + n_2 + \cdots + n_s \text{ is odd} \\ \mathscr{P}_{n_1 t_1, \cdots, n_s t_s}(p) & n_1 + n_2 + \cdots + n_s \text{ is even} \end{cases} \quad (4.206)$$

where $\mathscr{P}$-polynomials are the even ordered $\mathscr{Y}$-polynomials.

Here

$$\mathscr{P}_{2t}(p) = p_{2t}, \qquad \mathscr{P}_{t,z}(p) = p_{t,z}$$
$$\mathscr{P}_{3t,z}(p) = p_{3t,z} + 3p_{t,z}p_{2t}, \qquad \mathscr{P}_{4t}(p) = p_{4t} + 3p_{2t}^2. \quad (4.207)$$

The binary Bell polynomial $\mathscr{Y}_{n_1 t_1, \cdots, n_s t_s}(v, w)$ can be expressed as a linear combination of $\mathscr{P}$-polynomials and Bell polynomials $Y_{n_1 t_1, \cdots, n_s t_s}(v)$ as

$$(hg)^{-1} D_{t_1}^{n_1} \dots D_{t_s}^{n_s} h.g$$
$$= \mathscr{Y}_{n_1 t_1, \dots, n_s t_s}(v, w)\Big|_{v=\ln \frac{h}{g},\ w=\ln hg}$$
$$= \mathscr{Y}_{n_1 t_1, \dots, n_s t_s}(v, v + p)\Big|_{v=\ln \frac{h}{g},\ p=2\ln g}$$
$$= \sum_{m_1=0}^{n_1} \cdots \sum_{m_s=0}^{n_s} \prod_{i=1}^{s} \binom{n_i}{m_i} \mathscr{P}_{n_1 t_1, \dots, n_s t_s}(p) Y_{(n_1-m_1)t_1, \dots, (n_s-m_s)t_s}(v) \quad (4.208)$$

Using the Hope-Cole transformation $v = \ln \psi$, the Bell polynomial is

$$Y_{n_1 t_1, \dots, n_s t_s}(v = \ln \psi) = \frac{\psi_{n_1 t_1, \dots, n_s t_s}}{\psi} \quad (4.209)$$

Again with the help of (4.208), it can also be written as

$$(hg)^{-1} D_{t_1}^{n_1} \dots D_{t_s}^{n_s} h.g$$
$$= \psi^{-1} \sum_{m_1=0}^{n_1} \cdots \sum_{m_s=0}^{n_s} \prod_{i=1}^{s} \binom{n_i}{m_i} \mathscr{P}_{n_1 t_1, \dots, n_s t_s}(p) \psi_{(n_1-m_1)t_1, \dots, (n_s-m_s)t_s}. \quad (4.210)$$

4.18.1 *Damped-forced KdV equation*

Consider the damped-forced KdV equation to get the Hirota bilinear form using the Bell polynomial

$$\frac{\partial u}{\partial t}+\alpha u\frac{\partial u}{\partial x}+\beta\frac{\partial^3 u}{\partial x^3}+\gamma u=F(t) \tag{4.211}$$

α, γ are arbitrary constants, while β is t-dependent. We introduce a potential field p defined as

$$u=e^{-\gamma t}p_{xx}+e^{-\gamma t}\int e^{\gamma t}F(t)dt=a(t)p_{xx}+r(t) \tag{4.212}$$

where $a(t)=e^{-\gamma t}$ and $r(t)=e^{-\gamma t}\int e^{\gamma t}F(t)dt$.

By substituting (4.212) into Eq. (4.211) and integrating with respect to x, we obtain

$$E(p)=p_{x,t}+\frac{\alpha a(t)}{2}p_{2x}^2+\alpha r(t)p_{2x}+\beta p_{4x}=0 \tag{4.213}$$

Using Eq. (4.207) and the condition $a(t)\alpha = 6\beta$, Eq. (4.213) can be re-written as a combination of $\mathscr{P}$-polynomial expressions as follows

$$E(p)=\mathscr{P}_{x,t}(p)+\beta\mathscr{P}_{4x}(p)+\alpha r(t)\mathscr{P}_{2x}(p)=0 \tag{4.214}$$

Using the transformation $p = 2\ln f$ and Eq. (4.206), we can obtain the bilinear form of (4.211):

$$(D_tD_x+\alpha r(t)D_x^2+\beta D_x^4)f.f=0 \tag{4.215}$$

where $\beta=\alpha e^{-\gamma t}/6$ and $r(t)=e^{-\gamma t}\int e^{\gamma t}F(t)dt$.

4.18.2 *Damped-forced Burgers' equation*

To obtain the Hirota bilinear form of the damped-forced Burger equation with the help of the Bell polynomial. Let us consider the damped-forced Burger equation.

$$\frac{\partial u}{\partial t}+\alpha u\frac{\partial u}{\partial x}+\beta\frac{\partial^2 u}{\partial x^2}+\gamma u=F(t) \tag{4.216}$$

α, γ are arbitrary constants, and β is t dependent.

Now, we introduce potential fields v, w defined as

$$u = a(t)v_x + r(t) \tag{4.217}$$

where $a(t) = e^{-\gamma t}$ and $r(t) = e^{-\gamma t} \int e^{\gamma t} F(t)dt$.

$$w = \log(fg) = \log f + \log g, \quad v = \log(f/g) = \log f - \log g$$
$$w - v = p \implies w - p = v, \quad p = 2\log g$$

Now,

$$v_t + \frac{\alpha a(t)}{2} v_x^2 + \alpha r(t) v_x + \beta(w_{2x} - p_{2x}) = 0 \tag{4.218}$$

Using Eq. (4.18) and the condition $\alpha a(t) = 2\beta$ Eq. (4.218) may be re-written as a combination of binary Bell polynomial $\mathscr{Y}$ expressions as follows

$$\mathscr{Y}_t(v, w) + \beta \mathscr{Y}_{2x}(v, w) + \alpha r(t) \mathscr{Y}_x(v, w) = 0 \tag{4.219}$$

We obtain the bilinear form of (4.216) by transforming $v = \ln f/g, w = \ln fg$ and using Eq. (4.249).

$$(D_t + \beta D_x^2 + \alpha r(t) D_x) f.g = 0 \tag{4.220}$$
$$\beta D_x^2 (g.g) = 0 \tag{4.221}$$

where $\beta = \alpha e^{-\gamma t}/2$ and $r(t) = e^{-\gamma t} \int e^{\gamma t} F(t)dt$.

4.18.3 *Damped-forced ZK equation*

Let us consider the Damped-forced ZK equation is

$$\frac{\partial u}{\partial t} + \alpha u \frac{\partial u}{\partial x} + \beta \frac{\partial^3 u}{\partial x^3} + \gamma \frac{\partial^3 u}{\partial x \partial^2 y} + \delta u = F(t) \tag{4.222}$$

where α, β, γ, δ, are arbitrary constants. We introduce a potential field p defined as

$$u = e^{-\delta t} p_{x,y} + e^{-\delta t} \int e^{\delta t} F(t)dt = a(t) p_{x,y} + r(t) \tag{4.223}$$

where $a(t) = e^{-\delta t}$ and $r(t) = e^{-\delta t} \int e^{\delta t} F(t)dt$.

Substituting (4.223) in (4.222) and then integrating with respect to x, we get

$$E(p) = p_{y,t} + \frac{\alpha a(t)}{2} p_{x,y}^2 + \alpha r(t) p_{x,y} + \beta p_{3x,y} + \gamma p_{x,3y} = 0 \qquad (4.224)$$

Using Eq. (4.207) and considering $A = 2, B = 1, C = 1$, Eq. (4.224) may be re-written as a combination of $\mathscr{P}$-polynomial expressions as follows

$$E(p) = \mathscr{P}_{y,t}(p) + r(t)\mathscr{P}_{x,y}(p) + \mathscr{P}_{3x,y}(p) + \mathscr{P}_{x,3y}(p)$$
$$+ \mathscr{P}_{x,y}(p)\{a(t)\mathscr{P}_{x,y}(p) - 3\mathscr{P}_{2x}(p) - 3\mathscr{P}_{2y}(p)\} = 0$$

Therefore,

$$E(p) = \mathscr{P}_{y,t}(p) + r(t)\mathscr{P}_{x,y}(p) + \mathscr{P}_{3x,y}(p) + \mathscr{P}_{x,3y}(p) = 0 \qquad (4.225a)$$

$$E(p) = a(t)\mathscr{P}_{x,y}(p) - 3\mathscr{P}_{2x}(p) - 3\mathscr{P}_{2y}(p) = 0 \qquad (4.225b)$$

Using the transformation $p = 2\ln f$ and Eq. (4.206), we can obtain the bilinear form of (4.222):

$$(D_t D_y + r(t) D_x D_y + D_x^3 D_y + D_x D_{3y}) f.f = 0 \qquad (4.226a)$$

$$(a(t) D_x D_y - 3D_x^2 - 3D_y^2) f.f = 0 \qquad (4.226b)$$

where $a(t) = e^{-\delta t}$ and $r(t) = e^{-\delta t} \int e^{\delta t} F(t) dt$.

4.18.4 *Damped-forced KP equation*

The Damped-forced KP equation is

$$\frac{\partial}{\partial x}\left(\frac{\partial u}{\partial t} + \alpha u \frac{\partial u}{\partial x} + \beta \frac{\partial^3 u}{\partial x^3} + \gamma u\right) + \delta \frac{\partial^2 u}{\partial x^2} = \frac{\partial^2 h}{\partial x^2} \qquad (4.227)$$

α, γ and δ are arbitrary constants, while β is t-dependent. Integrating the above equation with respect to x, we get

$$u_t + \alpha u u_x + \beta u_{3x} + \gamma u + \delta \int u_{2y} dx = h_x$$

Putting $h = xF(t)$ in the above equation, we get

$$(u_t + \alpha u u_x + \beta u_{3x} + \gamma u)_x + \delta u_{2y} = F(t) \qquad (4.228)$$

Now, we introduce a potential field p defined as

$$u = e^{-\gamma t} p_{xx} + e^{-\gamma t} \int e^{\gamma t} F(t) dt = a(t) p_{xx} + r(t) \tag{4.229}$$

where $a(t) = e^{-\gamma t}$ and $r(t) = e^{-\gamma t} \int e^{\gamma t} F(t) dt$.

Substituting Eq. (4.229) into Eq. (4.228) and integrating twice with respect to x yields

$$E(p) = p_{x,t} + \frac{\alpha a(t)}{2} p_{2x}^2 + \alpha r(t) p_{2x} + \beta p_{4x} + \delta p_{2y} = 0 \tag{4.230}$$

Using Eq. (4.207) and the condition $\alpha a(t) = 6\beta$, Eq. (4.230) may be rewritten as a combination of $\mathscr{P}$-polynomial expressions as follows

$$E(p) = \mathscr{P}_{x,t}(p) + \alpha r(t) \mathscr{P}_{2x}(p) + \beta \mathscr{P}_{4x}(p) + \delta \mathscr{P}_{2y}(p) = 0 \tag{4.231}$$

Using the transformation $p = 2 \ln f$ and Eq. (4.206), we can obtain the bilinear form of (4.228):

$$(D_t D_x + \alpha r(t) D_x^2 + \beta D_x^4 + \delta D_y^2) f.f = 0 \tag{4.232}$$

where $a(t) = e^{-\gamma t}$ and $r(t) = e^{-\gamma t} \int e^{\gamma t} F(t) dt$.

4.18.5 *Damped-forced potential KdV equation*

Let us consider the Damped-forced potential KdV equation

$$\frac{\partial u}{\partial t} + \alpha \left(\frac{\partial u}{\partial x} \right)^2 + \beta \frac{\partial^3 u}{\partial x^3} + \gamma u = F(t) \tag{4.233}$$

A, D, are arbitrary constants, and B is t dependent.

We introduce a potential field p defined as

$$u = e^{-\gamma t} p_x + e^{-\gamma t} \int e^{\gamma t} F(t) dt = a(t) p_x + r(t) \tag{4.234}$$

where $a(t) = e^{-\gamma t}$ and $r(t) = e^{-\gamma t} \int e^{\gamma t} F(t) dt$.

Substituting Eq. (4.234) in Eq. (4.233), we get

$$E(p) = p_{x,t} + \alpha a(t) p_{2x}^2 + \beta p_{4x} = 0 \tag{4.235}$$

Using Eq. (4.207) and condition $\alpha a(t) = 3\beta$, Eq. (4.235) can be re-written as a combination of $\mathscr{P}$-polynomial expressions as follows

$$E(p) = \mathscr{P}_{x,t}(p) + \beta \mathscr{P}_{4x}(p) = 0 \tag{4.236}$$

Equation (4.206) and the transformation $p = 2\ln f$ provide the bilinear form of (4.233)

$$(D_t D_x + \beta D_x^4) f.f = 0 \tag{4.237}$$

where $\beta = \alpha e^{-\gamma t}/3$.

4.18.6 *Forced BBM equation*

Consider the Forced BBM equation

$$\frac{\partial u}{\partial t} + \frac{\partial u}{\partial x} + u\frac{\partial u}{\partial x} - \frac{\partial^3 u}{\partial^2 x \partial t} = F(t) \tag{4.238}$$

We introduce a potential field p defined as

$$u = p_{xx} + \int F(t)dt = p_{xx} + r(t) \tag{4.239}$$

where $r(t) = \int F(t)dt$.

Substituting (4.239) in Eq. (4.238) and integrating with respect to x, we get

$$E(p) = p_{x,t} + p_{2x} + \frac{1}{2}p_{2x}^2 + r(t)p_{2x} - p_{3x,t} = 0 \tag{4.240}$$

Using Eq. (4.207), then Eq. (4.240) can be re-written as a combination of $\mathscr{P}$-polynomial expressions as follows

$$E(p) = \mathscr{P}_{x,t}(p) + (1 + r(t))\mathscr{P}_{2x}(p) - \mathscr{P}_{3x,t}(p) + \mathscr{P}_{2x}(p)\Big[3\mathscr{P}_{x,t}(p) + \frac{1}{2}\mathscr{P}_{2x}(p)\Big] = 0$$

Therefore,

$$E(p) = \mathscr{P}_{x,t}(p) + (1 + r(t))\mathscr{P}_{2x}(p) - \mathscr{P}_{3x,t}(p) = 0 \tag{4.241a}$$

$$E(p) = 3\mathscr{P}_{x,t}(p) + \frac{1}{2}\mathscr{P}_{2x}(p) = 0 \tag{4.241b}$$

With the aid of Eq. (4.206) and the transformation $p = 2\ln f$, we obtain the bilinear form of 4.238

$$(D_x D_t + (1 + r(t))D_x^2 - D_x^3 D_t) f.f = 0 \tag{4.242a}$$

$$(D_x^2 + 6D_x D_t) f.f = 0 \tag{4.242b}$$

Note: The Hirota bilinear form is not deterministic if the damped term is included in the BBM equation due to the $\psi_{2x,t}$ term.

4.18.7 *Sawada–Kotera equation*

Let us consider the Sawada–Kotera equation

$$\frac{\partial u}{\partial t} + 15\frac{\partial}{\partial x}\left(u^3 + u\frac{\partial^2 u}{\partial x^2}\right) + \frac{\partial^5 u}{\partial x^5} = 0 \tag{4.243}$$

We introduce a potential field p defined as

$$\psi = p_{2x} \tag{4.244}$$

Substituting Eq. (4.244) in Eq. (4.243) and integrating with respect to x, we get

$$E(p) = p_{x,t} + 15a(t)p_{2x}^3 + 15p_{2x}p_{4x} + p_{6x} = 0 \tag{4.245}$$

Using Eq. (4.207), then Eq. (4.245) can be re-written as a combination of $\mathscr{P}$-polynomial expressions as follows

$$E(p) = \mathscr{P}_{x,t}(p) + \mathscr{P}_{6x}(p) = 0 \tag{4.246}$$

Using the transformation $p = 2\ln f$ with (4.206), we obtain the bilinear form of (4.243) as

$$(D_t D_x + D_x^6)f.f = 0 \tag{4.247}$$

Note: If we include the damped term and forced term in the SK equation, the Hirota bilinear form is not deterministic because of the ψ^3 term.

4.18.8 *Boussinesq equation*

Let us consider the Boussinesq equation

$$\psi_{tt} - \psi_{xx} - \psi_{4x} - 3(\psi^2)_{xx} = 0 \tag{4.248}$$

We introduce a potential field p defined as

$$u = p_{2x} \tag{4.249}$$

Substituting Eq. (4.249) in Eq. (4.248) and integrating two times with respect to x, we get

$$E(p) = p_{tt} - p_{2x} - p_{4x} - p_{2x}^2 = 0 \tag{4.250}$$

Using Eq. (4.207), then Eq. (4.250) can be re-written as a combination of $\mathscr{P}$-polynomial expressions as follows

$$E(p) = \mathscr{P}_{2t}(p) - \mathscr{P}_{4x}(p) - \mathscr{P}_{4x}(p) = 0 \tag{4.251}$$

Using the transformation $p = 2\ln f$ with (4.206), we obtain the bilinear form of (4.248) as

$$(D_t^2 - D_x^2 - D_x^4)f.f = 0 \tag{4.252}$$

Note: If we include a damped term and a forced term in the Boussinesq equation, the Hirota bilinear form is not deterministic because of the ψ_{2t} term.

4.19 Wronskian Technique

It is well known that the Wronskian is an important mathematical method for solving ordinary differential equations. It is named after Józef Hoene-Wroński (1776–1853), a Polish mathematician who introduced and thoroughly researched the concept in the early nineteenth century. Some nonlinear integrable systems can also be solved by using the Wronskian technique, which was initially developed for ODEs and then expanded to PDEs. The development of mathematical physics and integrable systems theory is strongly linked to the history of the approach to NLPDEs. For the purpose of testing linear independence and constructing fundamental sets of solutions, a Wronskian determinant is generated from the solutions of a linear homogeneous ODE and its derivatives. Wronskian theory was applied to linear PDEs, where it proved useful in the study of linear independence of solutions and the construction of basic sets of solutions. During the latter half of the 20th century, integrable systems — systems of NLPDEs with unique characteristics like soliton solutions, conservation laws, and infinite-dimensional symmetries — were largely understood by mathematicians and physicists through enormous advances. A tendency was noticed to link many integrable systems to auxiliary linear problems, which are frequently stated as Lax pairs or other linear systems. Between integrable NLPDEs and their linear equivalents, these linear issues offered a crucial link. Researchers in mathematics and physics discovered that the Wronskian approach could be modified to create solutions for specific integrable NLPDEs, building on the pioneering work of J. S. Lax and R. S. Phillips, among others. Creating determinants of solutions to related linear problems and using the resulting structures to produce precise solutions were the tasks involved in this adaptation. The KdV equation played a central role in the development of the Wronskian method for NLPDEs. Exact multisoliton solutions to the KdV equation were constructed using the Wronskian method, demonstrating its effectiveness in solving integrable systems.

The nonlinear Schrödinger equation, the Sine-Gordon equation, and several other soliton equations emerging in mathematical physics are examples of integrable NLPDEs that have been the subject of additional research and applications due to the effectiveness of the Wronskian technique for solving the KdV problem. Differential equations, linear algebra, analysis, and other branches of mathematics can all benefit from the use of the Wronskian concept. Because of its importance and adaptability, it is an essential tool for comprehending and resolving mathematical issues in a variety of fields, such as economics, engineering, and physics.

Definition and Basic Concept:

The Wronskian of a set of functions $\{\phi_1, \phi_2, \ldots, \phi_n\}$ is defined as the determinant of a matrix composed of these functions and their derivatives.

For ODEs, the Wronskian $W(y_1, y_2, \ldots, y_n)$ is given by:

$$\begin{vmatrix} y_1 & y_2 & \cdots & y_n \\ y_1' & y_2' & \cdots & y_n' \\ \vdots & \vdots & \ddots & \vdots \\ y_1^{(n-1)} & y_2^{(n-1)} & \cdots & y_n^{(n-1)} \end{vmatrix}$$

In PDEs, we analyze functions that are dependent on many variables, and the Wronskian incorporates partial derivatives. Assume $(u_1, u_2, \ldots, u_n)$ are PDE solutions; their Wronskian matrix may contain partial derivatives with respect to each independent variable, such as x and t. The Wronskian function is employed in PDEs to ensure that solutions to PDE systems are linearly independent. The general solution to a PDE is commonly depicted as a linear combination of linearly independent solutions, which explains this.

Let us consider a system of PDEs in variables x and t:

$$\frac{\partial u}{\partial t} = L(u)$$

where L is a differential operator. If u_1 and u_2 are solutions, their Wronskian $W(u_1, u_2)$ might be defined as:

$$W(u_1, u_2) = \begin{vmatrix} u_1 & u_2 \\ \frac{\partial u_1}{\partial x} & \frac{\partial u_2}{\partial x} \end{vmatrix} = u_1 \frac{\partial u_2}{\partial x} - u_2 \frac{\partial u_1}{\partial x}$$

If $W(u_1, u_2) \neq 0$, u_1 and u_2 are linearly independent.

For linear PDEs, linearly independent solutions form a basis for the solution space. The general solution is often a linear combination of these basis solutions. The Wronskian can be used to study the stability and behavior of solutions, particularly in perturbation analysis. In solving PDEs using methods like the separation of variables, the Wronskian ensures the orthogonality and completeness of eigenfunctions.

Importance: When solving NLPDEs, the Wronskian technique provides a number of significant benefits and insights. Certain kinds of NlPDEs can have exact, analytical solutions constructed using the Wronskian technique. For theoretical analysis and prediction, these solutions are invaluable as they offer accurate descriptions of the behavior of the system. The Wronskian method offers an organized and methodical way to solve NLPDEs. Many integrable NLPDEs can be solved using the Wronskian method, including the sine-Gordon equation, the nonlinear Schrödinger equation, and the KdV equation. Because of its adaptability, it is an effective tool in mathematical engineering and physics.

When creating multisoliton solutions—which depict the interactions of several solitary waves in nonlinear systems—the Wronskian technique works very well. These solutions are essential for soliton equations because they capture intricate nonlinear features.

The Wronskian technique's solutions frequently disclose the system's deep structural characteristics, such as hidden symmetries, conservation laws, and integrability features. These revelations offer a more thorough comprehension of the NLPDEs' underlying dynamics and behavior.

Although the computation of large determinants can be computationally demanding, the Wronskian technique's organized nature can effectively arrange complex calculations, especially for multisoliton solutions. With complex nonlinear systems, this computational efficiency is useful.

The Wronskian technique yields solutions that can be immediately interpreted in the context of physical processes, offering valuable insights into the practical applications of NLPDEs. For practical engineering and scientific reasons, it is imperative to comprehend the physical implications of the answers.

In summary, the Wronskian technique is an important tool for solving NLPDEs due to its ability to generate exact solutions, its systematic approach, versatility, ability to capture multisoliton phenomena, provision of structural insights, computational efficiency, and physical interpretability. It remains a valuable method for studying integrable systems and nonlinear wave phenomena in various fields of science and engineering.

4.20 Definition

Wronskian is a mathematical determinant whose first row consists of n functions of x and whose following rows consist of the successive derivatives of these same functions with respect to x. The Wronskian of two differentiable functions, f and g, is

$$W(f,g) = fg' - gf'$$

More generally, the Wronskian of n differentiable functions $\phi_1, \phi_2, \ldots, \phi_n$ is defined as the determinant

$$W(\phi_1, \phi_2, \ldots, \phi_n) = \begin{vmatrix} \phi_1 & \phi_1' & \cdots & \phi_1^{(n-1)} \\ \phi_2 & \phi_2' & \cdots & \phi_2^{(n-1)} \\ \vdots & \vdots & \ddots & \vdots \\ \phi_n & \phi_n' & \cdots & \phi_n^{(n-1)} \end{vmatrix} = (\widehat{n-1})$$

where $\phi_i = \phi_i(x)$. It can be written as $W(0,1,2,\ldots,(n-1)) = (\widehat{n-1})$, where the cap denotes the consecutive derivatives of the functions upto $(n-1)$ order.

4.20.1 *Derivatives of Wronskian (W)*

1. 1st order derivative of W is given by

$$\frac{dW}{dx} = \begin{vmatrix} \phi_1 & \phi_1' & \cdots & \phi_1^{(n)} \\ \phi_2 & \phi_2' & \cdots & \phi_2^{(n)} \\ \vdots & \vdots & \ddots & \vdots \\ \phi_n & \phi_n' & \cdots & \phi_n^{(n)} \end{vmatrix} + \cdots + \begin{vmatrix} \phi_1' & \phi_1' & \cdots & \phi_1^{(n-1)} \\ \phi_2' & \phi_2' & \cdots & \phi_2^{(n-1)} \\ \vdots & \vdots & \ddots & \vdots \\ \phi_n' & \phi_n' & \cdots & \phi_n^{(n-1)} \end{vmatrix}$$

$$= \begin{vmatrix} \phi_1 & \phi_1' & \cdots & \phi_1^{(n-2)} & \phi_1^{(n)} \\ \phi_2 & \phi_2' & \cdots & \phi_2^{(n-2)} & \phi_2^{(n)} \\ \vdots & \vdots & \ddots & \vdots & \vdots \\ \phi_n & \phi_n' & \cdots & \phi_n^{(n-2)} & \phi_n^{(n)} \end{vmatrix}$$

$$= (0,1,2,\ldots,n-2,n)$$

$$= (\widehat{n-2},n)$$

2. Similarly the second-order derivative of W is given by

$$\frac{d^2W}{dx^2} = (\widehat{n-3}, n-1, n) + (\widehat{n-2}, n+1)$$

3. Third-order derivative of W is given by

$$\frac{d^3W}{dx^3} = (\widehat{n-4}, n-2, n-1, n) + 2(\widehat{n-3}, n-1, n+1)$$

4. 4th order derivative of W is given by

$$\begin{aligned}\frac{d^4W}{dx^4} &= (\widehat{n-5}, n-3, n-2, n-1, n)\\ &\quad + 3(\widehat{n-4}, n-2, n-1, n+1) + 2(\widehat{n-3}, n, n+1)\\ &\quad + 3(\widehat{n-3}, n-1, n+2) + (\widehat{n-2}, n+3).\end{aligned}$$

4.20.2 *Properties*

A Wronskian W is an $N \times N$ determinant of a matrix with columns $S^{(0)}, \ldots, S^{(N-1)}$, where $S^{(0)} = (\phi_1(x,y,t), \ldots, \phi_N(x,y,t))^T$ and $S^{(j)} = \frac{\partial^j S^{(0)}}{\partial x^j}$, written $W = \left| S^{(0)}, \ldots, S^{(N-1)} \right|$ or in a more compact notation $W = (\widehat{N-1})$. Here $(\widehat{N-k})$, will represent the $(N-k+1)$ consecutive (ordered) columns $S^{(0)}, \ldots, S^{(N-k)}$, and in this notation $(N-k)$ will indicate the single column $S^{(N-k)}$. The most useful property of a Wronskian is the compactness with which its derivatives may be written. We now have:

$$\begin{aligned}\frac{\partial W}{\partial x} &= \sum_{j=0}^{N-1} \left| S^{(0)}, \ldots, S^{(j-1)}, \frac{\partial S^{(j)}}{\partial x}, S^{(j+1)}, \ldots, S^{(N-1)} \right|\\ &= \sum_{j=0}^{N-1} \left| S^{(0)}, \ldots, S^{(j-1)}, S^{(j+1)}, S^{(j+1)}, \ldots, S^{(N-1)} \right|\\ &= \left| S^{(0)}, \ldots, S^{(N-2)}, S^{(N)} \right| = (\widehat{N-2}, N)\end{aligned}$$

Since in all other terms there are two equal columns, in a similar way, higher-order x derivatives of W may be computed to give sums of terms that depend only on the order of the derivative and not upon N. Further, if $S^{(0)}$ is such that $\frac{\partial S^{(j)}}{\partial y} = \sum_{p=0}^{k} \alpha_p \frac{\partial^p S^{(j)}}{\partial x^p}$, and $\frac{\partial S^{(j)}}{\partial t} = \sum_{p=0}^{L} \beta_p \frac{\partial^p S^{(j)}}{\partial x^p}$, then the y and t derivatives of W may also be computed in a compact form.

For example, suppose

$$\frac{\partial S^{(j)}}{\partial y} = \alpha \frac{\partial^{2 S^{(j)}}}{\partial x^2}$$

then

$$\begin{aligned}
\frac{\partial W}{\partial y} &= \sum_{j=0}^{N-1} | \, S^{(0)}, \ldots, S^{(j-1)}, \alpha S^{(j+2)} S^{(j+1)}, \ldots, S^{(N-1)} | \\
&= \alpha \left[| \, S^{(0)}, \ldots, S^{(N-3)}, S(N), S^{(N-1)} | + | \, S^{(0)}, \ldots, S^{(N-2)}, S^{(N)} | \right] \\
&= \alpha \left[-(\widehat{N-3}, N-1, N) + (\widehat{N-2}, N+1) \right]
\end{aligned}$$

where we reorder the columns in the first term and hence introduce a minus sign. Now, $\frac{\partial W}{\partial y} = -\alpha P + \alpha Q$, where

$$P = \begin{vmatrix} \phi_1 & \phi_1' & \cdots & \phi_1^{(N-3)} & \phi_1^{(N-1)} & \phi_1^{(N)} \\ \phi_2 & \phi_2' & \cdots & \phi_2^{(N-3)} & \phi_2^{(N-1)} & \phi_2^{(N)} \\ \vdots & \vdots & \ddots & \vdots & \vdots & \vdots \\ \phi_N & \phi_N' & \cdots & \phi_N^{(N-3)} & \phi_N^{(N-1)} & \phi_N^{(N)} \end{vmatrix}$$

$$Q = \begin{vmatrix} \phi_1 & \phi_1' & \cdots & \phi_1^{(N-2)} & \phi_1^{(N+1)} \\ \phi_2 & \phi_2' & \cdots & \phi_2^{(N-2)} & \phi_2^{(N+1)} \\ \vdots & \vdots & \ddots & \vdots & \vdots \\ \phi_N & \phi_N' & \cdots & \phi_N^{(N-2)} & \phi_N^{(N+1)} \end{vmatrix}$$

$$= \alpha[-(\widehat{N-3}, N-1, N) + (\widehat{N-2}, N+1)]$$

The final property that we will describe makes use of the determinantal identity

$$\begin{aligned}
&\left(\sum_{i=0}^{N} \xi_i \right) | \, S^{(0)}, \ldots, S^{(N-1)} | \\
&\quad = \sum_{j=0}^{N-1} | \, S^{(0)}, \ldots, S^{(j-1)}, \xi S^{(j)}, S^{(j+1)}, \ldots, S^{(N-1)} |
\end{aligned}$$

where $\xi S^{(j)} = \left(\xi_1 \frac{\partial^j S_1}{\partial x^j}, \ldots, \xi_N \frac{\partial^j S_N}{\partial x^j} \right)^T$. Suppose $S^{(0)}$ is such that, for all $i = 1, 2, \ldots, N$; there is a k (the same for all i) such that $\frac{\partial^k S_i}{\partial x^k} = \xi_i S_i$, or more concisely, $\xi S^{(j)} = S^{(j+k)}$; then the above sum will reduce to the sum of only a few terms.

4.21 Plücker relation

The Plücker relation comes from algebraic geometry and linear algebra, specifically the study of determinants and Grassmannians. Given a set of vectors in a vector space, the Plücker relations are a set of polynomial equations that define the coordinates of the Grassmannian as a projective variety. The Grassmannian $Gr(k, n)$ is the space of all k-dimensional subspaces of an n-dimensional vector space.

Consider a $k \times n$ matrix A (with $k \leq n$). The $k \times k$ minors (determinants of $k \times k$ submatrices) of A are called the *Plücker coordinates.* These coordinates must satisfy certain quadratic relations, known as the *Plücker relations.*

For example, if $k = 2$, the Plücker relations for the minors p_{ij} (determinants of the 2×2 submatrices) are:

$$p_{ij}p_{kl} = p_{ik}p_{jl} - p_{il}p_{jk}$$

This equation expresses the fact that not all $k \times k$ minors can be independent; they satisfy these specific relations.

Also for the Grassmannian $Gr(2, 4)$, the Plücker coordinates are the minors of a 2×4 matrix. The Plücker relation in this case can be written as:

$$p_{12}p_{34} = p_{13}p_{24} - p_{14}p_{23}$$

Here, p_{ij} denotes the minor of the matrix formed by taking the ith and jth columns.

The Plücker relations are fundamental in the study of Grassmannians, as they describe the algebraic structure of the space of all k-dimensional subspaces in $\mathbb{C}^n$ (or $\mathbb{R}^n$) and have applications in various areas of mathematics, including geometry, combinatorics, and physics.

These relations ensure that the set of all $k \times k$ minors of a matrix is consistent with the geometric properties of the subspaces they describe.

4.22 Solution of the KdV Equation

The KdV equation is

$$u_t + 6uu_x + u_{3x} = 0 \tag{4.253}$$

Consider the transformation

$$u(x,t) = 2\frac{\partial^2}{\partial x^2}(\log f) = 2\frac{\partial}{\partial x}\left(\frac{f_x}{f}\right) = 2\left(\frac{f_{xx}}{f} - \frac{f_x^2}{f^2}\right)$$

Now,

$$\begin{aligned}
u_t &= 2\left[\frac{ff_{xxt} - f_{xx}f_t}{f^2} - \frac{f^2.2f_xf_{xt} - f_x^2.2ff_t}{f^4}\right] \\
&= 2\left[\frac{f_{xxt}}{f} - \frac{f_{xx}f_t}{f^2} - 2\frac{f_xf_{xt}}{f^2} + \frac{2f_x^2f_t}{f^3}\right]. \\
u_x &= 2\left[\frac{ff_{xxx} - f_{xx}f_x}{f^2} - \frac{f^2.2f_xf_{xx} - f_x^2.2ff_x}{f^4}\right] \\
&= 2\left[\frac{f_{xxx}}{f} - \frac{f_{xx}f_x}{f^2} - 2\frac{f_xf_{xx}}{f^2} + 2\frac{f_x^3}{f^3}\right] \\
&= 2\left[\frac{f_{xxx}}{f} - 3\frac{f_xf_{xx}}{f^2} + 2\frac{f_x^3}{f^3}\right]. \\
u_{2x} &= 2\left[\frac{ff_{4x} - f_{3x}f_x}{f^2} - 3\frac{f^2(f_xf_{3x} + f_{2x}^2) - f_x^2f_{2x}.2f}{f^4}\right. \\
&\quad \left. + 2\frac{f^3.3f_x^2f_{2x} - f_x^4 3f^2}{f^6}\right] \\
&= 2\left[\frac{f_{4x}}{f} - \frac{f_{3x}f_x}{f^2} - 3\frac{f_xf_{3x}}{f^2} - 3\frac{f_{2x}^2}{f^2} + 6\frac{f_x^2f_{2x}}{f^3} + 6\frac{f_x^2f_{2x}}{f^3} - 6\frac{f_x^4}{f^4}\right] \\
&= 2\left[\frac{f_{4x}}{f} - 4\frac{f_xf_{3x}}{f^2} + 12\frac{f_x^2f_{2x}}{f^3} - 3\frac{f_{2x}^2}{f^2} - 6\frac{f_x^4}{f^4}\right]
\end{aligned}$$

Now, we want to find, the solution of the KdV equation is in the Wronskian form. So initially we consider the solution of the quadratic function f in term of Wronskian form

$f = W(f_1, f_2, \ldots, f_N)$, then

$$f = \begin{vmatrix} f_1 & f_1' & \cdots & f_1^{(N-1)} \\ f_2 & f_2' & \cdots & f_2^{(N-1)} \\ \vdots & \vdots & \ddots & \vdots \\ f_N & f_N' & \cdots & f_N^{(N-1)} \end{vmatrix} = (\widehat{N-1})$$

where $f_i^{(n)} = \frac{\partial^n}{\partial x^n}(f_i)$, for all $i, n \in \mathbb{N}$.

Now, consider $N = 3$. Then

$$f = W(f_1, f_2, f_3) = \begin{vmatrix} f_1 & f_1' & f_1'' \\ f_2 & f_2' & f_2'' \\ f_3 & f_3' & f_3'' \end{vmatrix} = (\widehat{3-1}) = (\hat{2})$$

Therefore,

$$f_x = \begin{vmatrix} f_1' & f_1' & f_1'' \\ f_2' & f_2' & f_2'' \\ f_3' & f_3' & f_3^{v\prime} \end{vmatrix} + \begin{vmatrix} f_1 & f_1'' & f_1'' \\ f_2 & f_2'' & f_2'' \\ f_3 & f_3'' & f_3'' \end{vmatrix} + \begin{vmatrix} f_1 & f_1' & f_1^{(3)} \\ f_2 & f_2' & f_2^{(3)} \\ f_3 & f_3' & f_3^{(3)} \end{vmatrix}$$

$$= (\hat{1}, 3) = (\widehat{3-2}, 3)$$

Then we get the general form

$$f_x = (\widehat{N-2}, N)$$

Now,

$$f_{xx} = (\widehat{3-3}, 3-1, 3) + (\widehat{3-2}, 3+1)$$

Therefore,

$$f_{xx} = (\widehat{N-3}, N-1, N) + (\widehat{N-2}, N+1).$$

$$f_{xxx} = (\widehat{3-4}, 3-2, 3-1, 3) + 2(\widehat{3-3}, 3-1, 3+1) + (\widehat{3-2}, 3+2)$$

Therefore, in general

$$f_{3x} = (\widehat{N-4}, N-2, N-1, N) + 2(\widehat{N-3}, N-1, N+1) + (\widehat{N-2}, N+2)$$

Let

$$f = W(f_1, f_2, f_3, f_4) = \begin{vmatrix} f_1 & f_1' & f_1'' & f_1^{(3)} \\ f_2 & f_2' & f_2'' & f_2^{(3)} \\ f_3 & f_3' & f_3'' & f_3^{(3)} \\ f_4 & f_4' & f_4'' & f_4^{(3)} \end{vmatrix} = (\hat{3}) = (\widehat{4-1})$$

Therefore,

$$f_x = \begin{vmatrix} f_1 & f_1' & f_1'' & f_1^{(4)} \\ f_2 & f_2' & f_2'' & f_2^{(4)} \\ f_3 & f_3' & f_3'' & f_3^{(4)} \\ f_4 & f_4' & f_4'' & f_4^{(4)} \end{vmatrix} = (\widehat{2}, 4) = (\widehat{4-2}, 4)$$

Therefore,

$$f_{xx} = (\widehat{4-3}, 4-1, 4) + (\widehat{4-2}, 4+1)$$

Therefore,

$$f_{xxx} = (\widehat{4-4}, 4-2, 4-1, 4) + 2(\widehat{4-3}, 4-1, 4+1) + (\widehat{4-2}, 4+2).$$

$$\begin{aligned} f_{4x} &= (\widehat{-1}, 1, 2, 3, 4) + 3(\hat{0}, 2, 3, 5) + 2(\hat{1}, 4, 5) + 3(\hat{1}, 3, 6) + (\hat{2}, 7) \\ &= (\widehat{4-5}, 4-3, 4-2, 4-1, 4) + 3(\widehat{4-4}, 4-2, 4-1, 4+1) \\ &\quad + 2(\widehat{4-3}, 4, 4+1) + 3(\widehat{4-3}, 4-1, 4+2) + (\widehat{4-2}, 4+3) \end{aligned}$$

Therefore, in general, $f_{4x} = (\widehat{N-5}, N-3, N-2, N-1, N) + 3(\widehat{N-4}, N-2, N-1, N+1) + 2(\widehat{N-3}, N, N+1) + 3(\widehat{N-3}, N-1, N+2) + (\widehat{N-2}, N+3)$.

Now,

$$\frac{\partial f_i}{\partial t} = -4\frac{\partial^3 f_i}{\partial x^3}, \quad i = 1, 2, \ldots, N$$

Therefore,

$$\begin{aligned} f_t &= \begin{vmatrix} \frac{\partial f_1}{\partial t} & f_1' & f_1'' \\ \frac{\partial f_2}{\partial t} & f_2' & f_2'' \\ \frac{\partial f_3}{\partial t} & f_3' & f_3'' \end{vmatrix} + \begin{vmatrix} f_1 & \frac{\partial f_1'}{\partial t} & f_1'' \\ f_2 & \frac{\partial f_2'}{\partial t} & f_2'' \\ f_3 & \frac{\partial f_3'}{\partial t} & f_3'' \end{vmatrix} + \begin{vmatrix} f_1 & f_1' & \frac{\partial f_1''}{\partial t} \\ f_2 & f_2' & \frac{\partial f_2''}{\partial t} \\ f_3 & f_3' & \frac{\partial f_3''}{\partial t} \end{vmatrix}, \\ &= -4(\hat{-1}, 1, 2, 3) + 4(\hat{0}, 2, 4) - 4(\hat{1}, 5), \\ &= -4(\widehat{3-4}, 3-2, 3-1, 3) + 4(\widehat{3-3}, 3-1, 3+1) \\ &\quad - 4(\widehat{3-2}, 3+2) \end{aligned}$$

Therefore, in general,

$$\begin{aligned} f_t &= -4(\widehat{N-4}, N-2, N-1, N) + 4(\widehat{N-3}, N-1, N+1) \\ &\quad - 4(\widehat{N-2}, N+2) \end{aligned}$$

Therefore,

$$\begin{aligned} f_{xt} &= 4(\hat{0}, 3, 4) - 4(\hat{1}, 6), \\ &= 4(\widehat{3-3}, 3, 3+1) - 4(3-2, 3+3) \end{aligned}$$

Therefore, in general,

$$f_{xt} = 4(\widehat{N-3}, N, N+1) - 4(\widehat{N-2}, N+3)$$

Now,

$$f_{xxt} = 4\begin{vmatrix} f_1 & f_1^{(3)} & f_1^{(5)} \\ f_2 & f_2^{(3)} & f_2^{(5)} \\ f_3 & f_3^{(3)} & f_3^{(5)} \end{vmatrix} - 4\begin{vmatrix} f_1 & f_1'' & f_1^{(6)} \\ f_2 & f_2'' & f_2^{(6)} \\ f_3 & f_3'' & f_3^{(6)} \end{vmatrix}$$

$$+ 4\begin{vmatrix} f_1' & f_1^{(3)} & f_1^{(4)} \\ f_2' & f_2^{(3)} & f_2^{(4)} \\ f_3' & f_3^{(3)} & f_3^{(4)} \end{vmatrix} - 4\begin{vmatrix} f_1 & f_1' & f_1^{(7)} \\ f_2 & f_2' & f_2^{(7)} \\ f_3 & f_3' & f_3^{(7)} \end{vmatrix},$$

$$= 4(\hat{0}, 3, 5) - 4(\hat{0}, 2, 6) + 4(\widehat{-1}, 1, 3, 4) - 4(\hat{1}, 7),$$

$$= 4(\widehat{3-4}, 3-2, 3, 3+1) + 4(\widehat{3-3}, 3, 3+2)$$

$$- 4(\widehat{3-3}, 3-1, 3+3) - 4(\widehat{3-2}, 3+4)$$

In general,

$$f_{xxt} = 4(\widehat{N-4}, N-2, N, N+1) + 4(\widehat{N-3}, N, N+2)$$

$$- 4(\widehat{N-3}, N-1, N+3) - 4(\widehat{N-2}, N+4)$$

Now,

$$u_t = 2\frac{\partial}{\partial t}\left[\frac{\partial^2}{\partial x^2}(\log f)\right] = 2\frac{\partial}{\partial t}\left[\frac{\partial}{\partial x}\left(\frac{f_x}{f}\right)\right] = 2\frac{\partial}{\partial x}\left[\frac{\partial}{\partial t}\left(\frac{f_x}{f}\right)\right],$$

$$= 2\frac{\partial}{\partial x}\left[\frac{f f_{xt} - f_x f_t}{f^2}\right]$$

Putting in (4.253),

$$2\frac{\partial}{\partial x}\left[\frac{f f_{xt} - f_x f_t}{f^2}\right] + 6uu_x + u_{3x} = 0,$$

$$\Rightarrow 2\frac{\partial}{\partial x}\left[\frac{f f_{xt} - f_x f_t}{f^2}\right] + \frac{\partial}{\partial x}(3u^2) + \frac{\partial}{\partial x}(u_{2x}) = 0$$

Integrating both sides with respect to x, we get,

$$2\left[\frac{f f_{xt} - f_x f_t}{f^2}\right] + 3u^2 + u_{2x} = 0 \tag{4.254}$$

taking integrating constant as zero. Substituting the values of u and u_{2x} in (4.254), we get,

$$f f_{xt} - f_x f_t + f_{4x} f - 4 f_{3x} f_x + 3 f_{xx}^2 = 0 \tag{4.255}$$

Let us assume the solution of the quadratic form of the KdV equation is in the Wronskian form $f = W(f_1, f_2, \ldots, f_N)$, where $f_1, f_2, \ldots, f_N$ are differentiable function. They must be satisfy the following compatibility conditions as

$$-\frac{\partial^2 \phi_i}{\partial x^2} = \lambda_i \phi_i, \quad \frac{\partial \phi_i}{\partial t} = -4\frac{\partial^3 \phi_1}{\partial x^3}$$

This is the quadratic form of the KdV equation. Putting the values of the partial derivatives of f in the form of the Wronskian determinant into (4.255), we get,

$$\begin{aligned}
&6(\widehat{N-1}).(\widehat{N-3}, N, N+1) - 3(\widehat{N-1}).(\widehat{N-2}, N+3) \\
&\quad - 12(\widehat{N-2}, N).(\widehat{N-3}, N-1, N+1) + (\widehat{N-1})(\widehat{N-5}, N-3, \\
&N-2, N-1, N) + 3(\widehat{N-1}).(\widehat{N-4}, N-2, N-1, N+1) \\
&\quad + 3(\widehat{N-1}).(\widehat{N-3}, N-1, N+2) + 3[(\widehat{N-3}, N-1, N) \\
&\quad + (\widehat{N-2}, N+1)]^2 = 0
\end{aligned}$$

Consequently, the quadratic form (4.255) transferred to the Plücker relation for the Wronskian determinant. Thus, the Nth order Wronskian solutions for (4.253) can be expressed as

$$u = 2\frac{\partial^2}{\partial x^2}(\log f)$$

with

$$f = \begin{vmatrix} \phi_1 & \phi_1' & \cdots & \phi_1^{(N-1)} \\ \phi_2 & \phi_2' & \cdots & \phi_2^{(N-1)} \\ \vdots & \vdots & \ddots & \vdots \\ \phi_N & \phi_N' & \cdots & \phi_N^{(N-1)} \end{vmatrix} = (\widehat{N-1})$$

It is called the Wronskian solution. Wronskian solutions for the KdV equation require the following conditions:

$$-\frac{\partial^2 \phi_i}{\partial x^2} = \lambda_i \phi_i, \quad \frac{\partial \phi_i}{\partial t} = -4\frac{\partial^3 \phi_1}{\partial x^3}, \quad 1 \leq i \leq N.$$

Case 1: When $\lambda_i = 0$, we get

$$\frac{\partial^2 \phi_i}{\partial x^2} = 0,$$
$$\Rightarrow \frac{\partial \phi_i}{\partial x} = c_{1i},$$
$$\Rightarrow \phi_i = c_{1i}x + c_{2i},$$

where c_{1i} and c_{2i} are arbitrary functions of t.

Case 2: When $\lambda_i < 0$.
Let

$$\lambda_i = -\eta_i^2$$

Then

$$\frac{\partial^2 \phi_i}{\partial x^2} - \eta_i^2 \phi_i = 0$$

Therefore,

$$\phi_i = A(t)e^{\eta_i x} + B(t)e^{-\eta_i x}$$

where $A(t),\ B(t)$ are arbitrary functions of t.

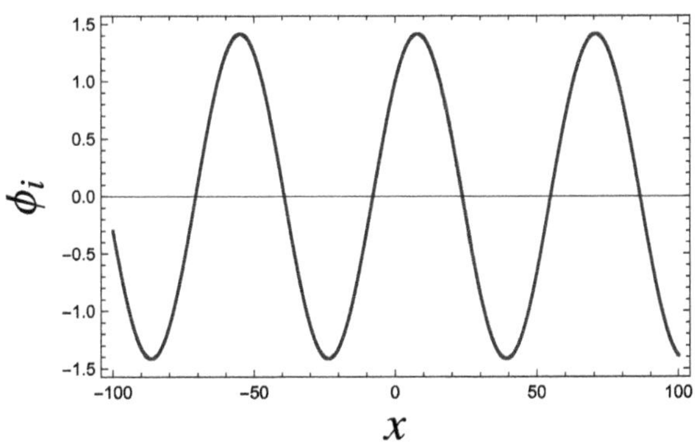

Figure 4.12. Solution of Eq. (4.253).

Therefore,

$$\frac{\partial \phi_i}{\partial t} = A'(t)e^{\eta_i x} + B'(t)e^{-\eta_i x}$$

and

$$\frac{\partial^3 \phi_i}{\partial x^3} = \eta_i^3 A(t)e^{\eta_i x} - \eta_i^3 B(t)e^{-\eta_i x}$$

Therefore,

$$\begin{aligned}
&\frac{\partial \phi_i}{\partial t} = -4\frac{\partial^3 \phi_i}{\partial x^3},\\
&\Rightarrow A'(t)e^{\eta_i x} + B'(t)e^{-\eta_i x} = -4\eta_i^3 A(t)e^{\eta_i x} + 4\eta_i^3 B(t)e^{-\eta_i x},\\
&\Rightarrow [A'(t) + 4\eta_i^3 A(t)]e^{\eta_i x} + [B'(t) - 4\eta_i^3 B(t)]e^{-\eta_i x} = 0
\end{aligned}$$

Therefore,

$$A(t) = c_i e^{-4\eta_i^3 t}, \quad B(t) = c_i' e^{4\eta_i^3 t}$$

Therefore,

$$\phi_i(x,t) = c_{1i}\cosh(\eta_i x - 4\eta_i^3 t) + c_{2i}\sinh(\eta_i x - 4\eta_i^3 t)$$

We are not much interested about the periodic solutions. However the solutions are given in brief. So,

$$\begin{aligned}
\phi_i(x,t) &= c_i e^{-4\eta_i^3 t} e^{\eta_i x} + c_i' e^{4\eta_i^3 t} e^{-\eta_i x}\\
&= c_i e^{(\eta_i x - 4\eta_i^3 t)} + c_i' e^{-(\eta_i x - 4\eta_i^3 t)}
\end{aligned}$$

For $c_i = c_i' = \frac{1}{2}$, solution $\phi_i(x,t)$ become in the following form

$$\begin{aligned}
\phi_i(x,t) &= \frac{1}{2}\left[e^{(\eta_i x - 4\eta_i^3 t)} + e^{-(\eta_i x - 4\eta_i^3 t)}\right]\\
&= \cosh(\eta_i x - 4\eta_i^3 t)
\end{aligned}$$

Again for $c_i = \frac{1}{2}$, $c_i' = -\frac{1}{2}$, solution $\phi_i(x,t)$ become in the following form

$$\begin{aligned}
\phi_i(x,t) &= \frac{1}{2}\left[e^{(\eta_i x - 4\eta_i^3 t)} - e^{-(\eta_i x - 4\eta_i^3 t)}\right]\\
&= \sinh(\eta_i x - 4\eta_i^3 t)
\end{aligned}$$

For **single-soliton** solution we take

$$\phi_1 = \cosh(\eta_1 x - 4\eta_1^3 t)$$

Substituting the value of $f = \phi_i$ independent variable transformation, we get

$$u = 2\frac{\partial^2}{\partial x^2}(\log \phi_1) = 2\frac{\partial^2}{\partial x^2}(\log \cosh(\eta_1 x - 4\eta_1^3 t)) = 2\eta_1^2 sech^2(\eta_1 x - 4\eta_1^3 t)$$

For **two-soliton** solution, we take $\phi_1 = \cosh(\eta_i x - 4\eta_i^3 t)$ and $\phi_2 = \sinh(\eta_2 x - 4\eta_2^3 t)$ corresponding Wronskian between two solutions ϕ_1 and ϕ_2 is

$$f = W(\phi_1, \phi_2)$$

Substituting the above relation into dependence variable transform we have two-soliton solution of the KdV equation

$$\begin{aligned} u &= 2\frac{\partial^2}{\partial x^2}(\log W(\phi_1, \phi_2)) \\ &= 2\frac{\partial^2}{\partial x^2}(\log W(\cosh(\eta_1 x - 4\eta_1^3 t), \sinh(\eta_2 x - 4\eta_2^3 t)) \\ &= 4\frac{(\eta_2^2 - \eta_1^2)[(\eta_2^2 - \eta_1^2) + \eta_1^2 \cosh(2\xi_2) + \eta_2^2 \cosh(2\xi_1)]}{[(\eta_2 - \eta_1)\cosh(\xi_2 + \xi_1) + (\eta_2 + \eta_1)\cosh(\xi_2 - \xi_1)]^2} \end{aligned}$$

where $\xi_1 = \eta_1 x - 4\eta_1^3 t$, $\xi_2 = \eta_2 x - 4\eta_2^3 t$.

For **singular-soliton** solution, we take

$$\phi_1 = \sinh(\eta_1 x - 4\eta_1^3 t)$$

substituting the value of $f = \phi_1$ in dependent variable transformation, we have

$$u = 2\frac{\partial^2}{\partial x^2}(\log \phi_1) = 2\frac{\partial^2}{\partial x^2}(\log \sinh(\eta_1 x - 4\eta_1^3 t)) = -2\eta_1^2 cosech^2(\eta_1 x - 4\eta_1^3 t)$$

For **singular two-soliton** solution, we take $\phi_1 = \sinh(\eta_i x - 4\eta_i^3 t)$ and $\phi_2 = \sinh(\eta_2 x - 4\eta_2^3 t)$ corresponding wronskian between two solutions ϕ_1 and ϕ_2 is

$$f = W(\phi_1, \phi_2)$$

Substituting the above relation into dependence variable transform we have two-soliton solution of the KdV equation

$$
\begin{aligned}
u &= 2\frac{\partial^2}{\partial x^2}(\log W(\phi_1, \phi_2)) \\
&= 2\frac{\partial^2}{\partial x^2}(\log W(\sinh(\eta_1 x - 4\eta_1^3 t), \sinh(\eta_2 x - 4\eta_2^3 t)) \\
&= 4\frac{(\eta_2^2 - \eta_1^2)[(\eta_1^2 - \eta_2^2) + \eta_1^2 \cosh(2\xi_2) - \eta_2^2 \cosh(2\xi_1)]}{[(\eta_2 - \eta_1)\sinh(\xi_2 + \xi_1) + (\eta_2 + \eta_1)\sinh(\xi_2 - \xi_1)]^2}
\end{aligned}
$$

where $\xi_1 = \eta_1 x - 4\eta_1^3 t$, $\xi_2 = \eta_2 x - 4\eta_2^3 t$.

Case 3: When $\lambda_i > 0$.
Let $\lambda_i = \eta_i^2$. Then,

$$\phi_i(x,t) = A(t)\cos(\eta_i x) + B(t)\sin(\eta_i x)$$

where $A(t)$, $B(t)$ are arbitrary functions of t.

Therefore,

$$\frac{\partial \phi_i}{\partial t} = A'(t)\cos(\eta_i x) + B'(t)\sin(\eta_i x)$$

and

$$
\begin{aligned}
\frac{\partial \phi_i}{\partial x} &= -\eta_i A(t)\sin(\eta_i x) + \eta_i B(t)\cos(\eta_i x), \\
\Rightarrow \frac{\partial^2 \phi_i}{\partial x^2} &= -\eta_i^2 A(t)\cos(\eta_i x) - \eta_i^2 B(t)\sin(\eta_i x), \\
\Rightarrow \frac{\partial^3 \phi_i}{\partial x^3} &= \eta_i^3 A(t)\sin(\eta_i x) - \eta_i^3 B(t)\cos(\eta_i x)
\end{aligned}
$$

Now,

$$
\begin{aligned}
&\frac{\partial \phi_i}{\partial t} = -4\frac{\partial^3 \phi_1}{\partial x^3}, \\
&\Rightarrow A'(t)\cos(\eta_i x) + B'(t)\sin(\eta_i x) = -4\eta_i^3 A(t)\sin(\eta_i x) + 4\eta_i^3 B(t)\cos(\eta_i x), \\
&\Rightarrow [A'(t) - 4\eta_i^3 B(t)]\cos(\eta_i x) + [B'(t) + 4\eta_i^3 A(t)]\sin(\eta_i x) = 0
\end{aligned}
$$

Therefore,

$$
\begin{aligned}
B(t) &= \frac{1}{4\eta_i^3}\Big[-4c_{1i}\eta_i^3 \sin(4\eta_i^3 t) + 4c_{2i}\eta_i^3 \cos(4\eta_i^3 t)\Big], \\
&= -c_{1i}\sin(4\eta_i^3 t) + c_{2i}\cos(4\eta_i^3 t)
\end{aligned}
$$

Therefore,

$$\begin{aligned}\phi_i(x,t) &= [c_{1i}\cos(4\eta_i^3 t) + c_{2i}\sin(4\eta_i^3 t)]\cos(\eta_i x)\\ &\quad + [-c_{1i}\sin(4\eta_i^3 t) + c_{2i}\cos(4\eta_i^3 t)]\sin(\eta_i x),\\ &= c_{1i}[\cos(4\eta_i^3 t)\cos(\eta_i x) - \sin(4\eta_i^3 t)\sin(\eta_i x)]\\ &\quad + c_{2i}[\sin(4\eta_i^3 t)\cos(\eta_i x) + \cos(4\eta_i^3 t)\sin(\eta_i x)],\\ &= c_{1i}\cos(\eta_i x + 4\eta_i^3 t) + c_{2i}\sin(\eta_i x + 4\eta_i^3 t)\end{aligned}$$

For $\lambda_i = 0$, the solution is called the rational solution.

For $\lambda_i < 0$, the solution is called a negaton solution and, for $\lambda_i > 0$, the solution is called the position solution.

4.23 Solutions of the Non-Planar KdV Equation

The cylindrical KdV equation is

$$\frac{\partial w}{\partial t} + \frac{\partial^3 w}{\partial x^3} - 6w\frac{\partial w}{\partial x} + \frac{1}{2t}w = 0 \tag{4.256}$$

Let the transformation be $w = \frac{-x}{12t} - \frac{1}{2t}u(z,\tau)$, where $z = -x\tau$, $t = \frac{1}{2\tau^2}$.

Now, $\tau^2 = \frac{1}{2t}$ then $\frac{\partial\tau}{\partial t} = -\tau^3$ and $z = -x\tau$ then $\frac{\partial z}{\partial t} = -x\frac{\partial\tau}{\partial t} = -z\tau^2$. So, $\frac{\partial z}{\partial x} = -\tau$.

Now,

$$w = \frac{z\tau}{6} - \tau^2 u(z,\tau)$$

Then

$$\begin{aligned}\frac{\partial w}{\partial t} &= \frac{1}{6}[-z\tau^3 - z\tau^3] + 2\tau^4 u(z,\tau) - \tau^2\left(\frac{\partial u}{\partial\tau}\frac{\partial\tau}{\partial t} + \frac{\partial u}{\partial z}\frac{\partial z}{\partial t}\right),\\ &= -\frac{z\tau^3}{3} + \tau^5 u_\tau + 2\tau^4 u(z,\tau) + z\tau^4 u_z.\\ \frac{\partial w}{\partial x} &= \frac{1}{6}\left(z\frac{\partial\tau}{\partial x} + \frac{\partial z}{\partial x}\tau\right) - \tau^2\left(\frac{\partial u}{\partial\tau}\frac{\partial\tau}{\partial x} + \frac{\partial u}{\partial z}\frac{\partial z}{\partial x}\right) - 2\tau\frac{\partial\tau}{\partial x}u(z,\tau),\\ &= -\frac{\tau^2}{6} + \tau^3 u_z.\\ \frac{\partial^2 w}{\partial x^2} &= -\frac{\tau}{3}\frac{\partial\tau}{\partial x} + \tau^3\frac{\partial u_z}{\partial x} + 3\tau^2\frac{\partial\tau}{\partial x}u_z = -\tau^4 u_{zz}.\\ \frac{\partial^3 w}{\partial x^3} &= -\tau^4\frac{\partial u_{zz}}{\partial x} - 4\tau^3 u_{zz}\frac{\partial\tau}{\partial x} = \tau^5 u_{zzz}\end{aligned}$$

Putting in (4.256), we get,

$$\begin{aligned}
&-\frac{z\tau^3}{3}+\tau^5u_\tau+2\tau^4u(z,\tau)+z\tau^4u_z+\tau^5u_{zzz}\\
&-6\left(\frac{z\tau}{6}-\tau^2u\right)\left(-\frac{\tau^2}{6}+\tau^3u_z\right)+\tau^2\left(\frac{z\tau}{6}-\tau^2u\right)=0, \qquad (4.257)\\
&\Rightarrow \tau^5u_{zzz}+6\tau^5uu_z+\tau^5u_\tau=0, \quad (\text{since } \tau^5\neq 0)\\
&\Rightarrow u_{zzz}+6uu_z+u_\tau=0
\end{aligned}$$

So, Eq. (4.256) reduces to the KdV equation with dependent variable u and independent variables z and τ. Since we have previously gotten the solution to the KdV equation using the Wronskian technique, by the above transformation, we can also get the solution to the cylindrical KdV equation. Hence, the solution of the cylindrical KdV equation (4.256) is given by

$$u=2\frac{\partial^2}{\partial z^2}(\log f)$$

with

$$f=\begin{vmatrix}\phi_1 & \phi_1' & \cdots & \phi_1^{(N-1)}\\ \phi_2 & \phi_2' & \cdots & \phi_2^{(N-1)}\\ \vdots & \vdots & \ddots & \vdots\\ \phi_N & \phi_N' & \cdots & \phi_N^{(N-1)}\end{vmatrix}=(\widehat{N-1})$$

where

$$\begin{aligned}
\phi_i(x,t)&=[c_{1i}\cos(4\eta_i^3\tau)+c_{2i}\sin(4\eta_i^3\tau)]\cos(\eta_iz)+[-c_{1i}\sin(4\eta_i^3\tau)\\
&\quad+c_{2i}\cos(4\eta_i^3\tau)]\sin(\eta_iz),\\
&=c_{1i}[\cos(4\eta_i^3\tau)\cos(\eta_iz)-\sin(4\eta_i^3\tau)\sin(\eta_iz)]\\
&\quad+c_{2i}[\sin(4\eta_i^3\tau)\cos(\eta_iz)+\cos(4\eta_i^3\tau)\sin(\eta_iz)],\\
&=c_{1i}\cos(\eta_iz+4\eta_i^3\tau)+c_{2i}\sin(\eta_iz+4\eta_i^3\tau)
\end{aligned}$$

and $\tau=\sqrt{\frac{1}{2t}}$, $z=-x\sqrt{\frac{1}{2t}}$.

4.24 Solution of KP Equation

The KP equation is

$$\frac{\partial}{\partial x}(u_t - 6uu_x + u_{3x}) + 3u_{2y} = 0 \tag{4.258}$$

Consider the transformation $u(x,t,y) = -2\frac{\partial^2}{\partial x^2}(\log f)$,

$$\text{where } f = \begin{vmatrix} f_1 & f_1' & \cdots & f_1^{(N-1)} \\ f_2 & f_2' & \cdots & f_2^{(N-1)} \\ \vdots & \vdots & \ddots & \vdots \\ f_N & f_N' & \cdots & f_N^{(N-1)} \end{vmatrix}, \quad f_i^{(n)} = \frac{\partial^n}{\partial x^n}(f_i), \quad \text{for all } i, n \in \mathbb{N}.$$

The quadratic form of the KP Eq. (4.258) is

$$(ff_{xt} - f_x f_t) + 3f_{xx}^2 + ff_{4x} - 4f_x f_{3x} + 3f_{yy}f - 3f_y^2 = 0 \tag{4.259}$$

Now,

$$\begin{aligned}
f_{xt} &= 4(\widehat{N-3}, N, N+1) - 4(\widehat{N-2}, N+3), \\
f_x &= (\widehat{N-2}, N), \\
f_t &= -4(\widehat{N-4}, N-2, N-1, N) + 4(\widehat{N-3}, N-1, N+1) \\
&\quad - 4(\widehat{N-2}, N+2), \\
f_{xx} &= (\widehat{N-3}, N-1, N) + (\widehat{N-2}, N+1), \\
f_{4x} &= (\widehat{N-5}, N-3, N-2, N-1, N) + 3(\widehat{N-4}, N-2, N-1, N+1) \\
&\quad + 2(\widehat{N-3}, N, N+1) + 3(\widehat{N-3}, N-1, N+2) + (\widehat{N-2}, N+3), \\
f_{3x} &= (\widehat{N-4}, N-2, N-1, N) + 2(\widehat{N-3}, N-1, N+1) + (\widehat{N-2}, N+2)
\end{aligned}$$

We know that $\frac{\partial f_i}{\partial y} = \frac{\partial^2 f_i}{\partial x^2}, i = 1, 2, 3, \ldots, N$. Let,

$$f = \begin{vmatrix} f_1 & f_1' & f_1'' \\ f_2 & f_2' & f_2'' \\ f_3 & f_3' & f_3'' \end{vmatrix} = (\hat{2})$$

Therefore,

$$f_y = -(\hat{0}, 2, 3) + (\hat{1}, 4) = -(\widehat{3-3}, 3-1, 3) + (\widehat{3-2}, 3+1)$$

In general, $f_y = -(\widehat{N-3}, N-1, N) + (\widehat{N-2}, N+1)$. Now,

$$\begin{aligned} f_{yy} &= (\hat{0},3,4) - (\hat{0},2,5) - (\widehat{-1},1,2,4) + (\hat{0},3,4) + (\hat{1},6), \\ &= 2(\widehat{3-3},3,3+1) - (\widehat{3-3},3-1,3+2) - (\widehat{3-4},3-2,3-1,3+1) \\ &\quad + (\widehat{3-2},3+3) \end{aligned}$$

In general,

$$\begin{aligned} f_{yy} = 2(\widehat{N-3}, N, N+1) - (\widehat{N-3}, N-1, N+2) \\ - (\widehat{N-4}, N-2, N-1, N+1) + (\widehat{N-2}, N+3). \end{aligned}$$

Putting these values in the L.H.S of (4.259), we get,

$$\begin{aligned} &= 12(\widehat{N-1})(\widehat{N-3}, N, N+1) - 12(\widehat{N-2}, N)(\widehat{N-3}, N-1, N+1) \\ &\quad + 12(\widehat{N-3}, N-1, N)(\widehat{N-2}, N+1) \\ &\quad + (\widehat{N-1})(\widehat{N-5}, N-3, N-2, N-1, N) \\ &= 0 \end{aligned}$$

This is nothing but the Plücker relation for Wronskian determinats. Therefore, we have shown that $f = f_N$ solves the KP equation (4.258). Hence, the Wronskian solution is given by $u = -2\frac{\partial^2}{\partial x^2}(\log f)$ with

$$f = \begin{vmatrix} \phi_1 & \phi_1' & \cdots & \phi_1^{(N-1)} \\ \phi_2 & \phi_2' & \cdots & \phi_2^{(N-1)} \\ \vdots & \vdots & \ddots & \vdots \\ \phi_N & \phi_N' & \cdots & \phi_N^{(N-1)} \end{vmatrix} = (\widehat{N-1})$$

under the conditions

$$\frac{\partial \phi_i}{\partial y} - \frac{\partial^2 \phi_i}{\partial x^2} = 0 \tag{4.260}$$

and

$$\frac{\partial \phi_i}{\partial t} + 4\frac{\partial^3 \phi_i}{\partial x^3} = 0 \tag{4.261}$$

Therefore, $(D^2 - D')\phi_i = 0$, where $D = \frac{\partial}{\partial x}$, $D' = \frac{\partial}{\partial y}$. Therefore,

$$\phi_i(x, y, t) = \sum_r A_{ir}(t)e^{a_{ir}x + b_{ir}y}$$

where, $a_{ir}^2 - b_{ir} = 0$ i.e., $b_{ir} = a_{ir}^2$. Then,

$$\begin{aligned} \phi_i(x, y, t) &= \sum_r A_{ir}(t)e^{a_{ir}x + a_{ir}^2 y}, \\ \Rightarrow \frac{\partial \phi_i}{\partial t} &= \sum_r A'_{ir}(t)e^{a_{ir}x + a_{ir}^2 y}, \\ \Rightarrow \frac{\partial^3 \phi_i}{\partial x^3} &= \sum_r A_{ir}(t)(a_{ir}^3)e^{a_{ir}x + a_{ir}^2 y} \end{aligned} \tag{4.262}$$

Putting in (4.261), we have

$$\begin{aligned} &\sum_r A'_{ir}(t)e^{a_{ir}x + a_{ir}^2 y} + 4\sum_r A_{ir}(t)(a_{ir}^3)e^{a_{ir}x + a_{ir}^2 y} = 0, \\ &\Rightarrow \sum_r A'_{ir}(t) + 4\sum_r A_{ir}(t)(a_{ir}^3) = 0 \end{aligned}$$

Therefore,

$$\begin{aligned} A'_{ir}(t) &= -4A_{ir}(t)(a_{ir}^3), \\ \Rightarrow \frac{dA_{ir}(t)}{A_{ir}(t)} &= -4a_{ir}^3 dt, \\ \Rightarrow A_{ir}(t) &= c_i e^{-4a_{ir}^3 t} \end{aligned}$$

Therefore,

$$\begin{aligned} \phi_i(x, y, t) &= \sum_r c_i e^{-4a_{ir}^3 t} e^{a_{ir}x + a_{ir}^2 y}, \\ &= \sum_r c_i e^{a_{ir}x + a_{ir}^2 y - 4a_{ir}^3 t} \end{aligned}$$

It is not necessary to stipulate the exact form of ϕ; it has only been necessary to specify that it satisfies the PDEs (4.260) and (4.261). To obtain the soliton solutions, we can write

$$\phi_j = e^{-l_j x + l_j^2 y + 4l_j^3 t} + a_{j0}(-1)^{j-1}e^{n_j x + n_j^2 y - 4l_j^3 t}$$

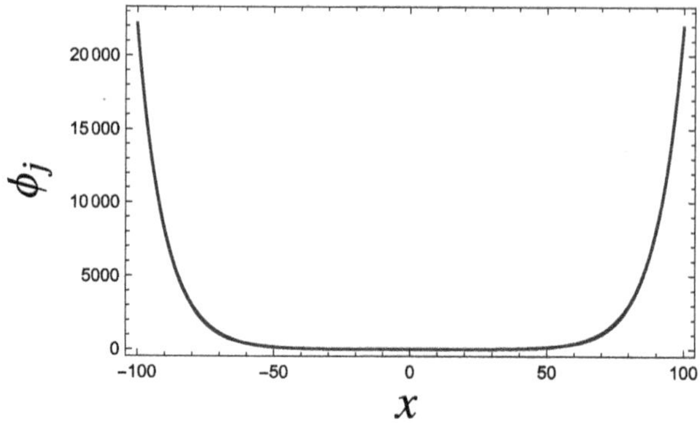

Figure 4.13. Soliton solution of Eq. (4.263).

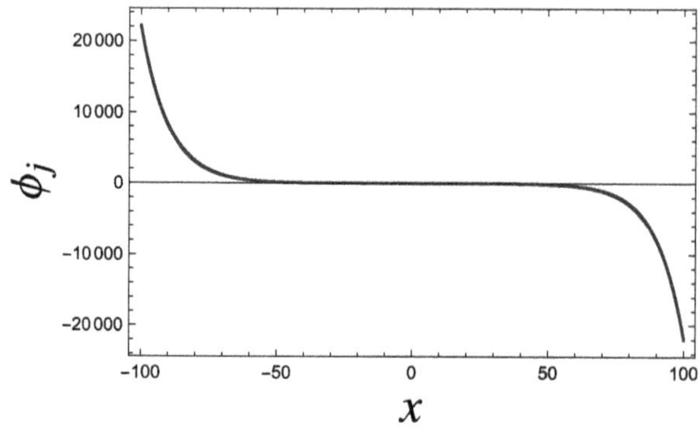

Figure 4.14. Soliton solution of Eq. (4.263).

where a_{j0} is an arbitrary phase factor. For the one-dimensional KP equation, this obviously becomes, on putting $l_j = n_j$,

$$\begin{aligned} \phi_j &= 2e^{n_j^2 y}(-1)^{j-1}\cosh(n_j x - 4n_j^3 t),\ j \text{ odd}, \\ \phi_j &= 2e^{n_j^2 y}(-1)^{j-1}\sinh(n_j x - 4n_j^3 t),\ j \text{ even}, \end{aligned} \tag{4.263}$$

when $a_{j0} = 1$. Therefore, the solution of the KP equation in Wronskian form is given by

$$u = -2\frac{\partial^2}{\partial x^2}(\log f)$$

Then for the KP equation,

$$\frac{\partial}{\partial x}(u_t - 6uu_x + u_{3x}) + 3u_{2y} = 0$$

consider the transformation $u = -2\frac{\partial^2}{\partial x^2}(\log f)$ with

$$f = \begin{vmatrix} \phi_1 & \phi_1' & \cdots & \phi_1^{(N-1)} \\ \phi_2 & \phi_2' & \cdots & \phi_2^{(N-1)} \\ \vdots & \vdots & \ddots & \vdots \\ \phi_N & \phi_N' & \cdots & \phi_N^{(N-1)} \end{vmatrix}$$

and ϕ_i's are given by the system (4.263).

4.25 Double–Wronskian Method

A powerful technique for analyzing and solving NLPDEs is the double-Wronskian method. This method extends the concept of the traditional Wronskian, incorporating a more complex determinant structure that allows for the construction of solutions to NLPDEs, particularly integrable systems such as soliton equations. For integrable NLPDEs, the double-Wronskian technique proved to be especially useful to generate multisoliton solutions. These solutions are essential for understanding the dynamics of nonlinear wave events, while they depict the interactions of numerous waves. This method's importance in mathematical physics was established by its ability to construct exact solutions to integrable PDEs. The efficacy and range of applications of the method have been expanded by using contemporary mathematical tools, geometric methods, computational techniques, and symbolic computation.

Two sets of functions, $\{f_j\}$ and $\{g_j\}$, are commonly involved in double-Wronskian matrices. The determinant of such a matrix is constructed using these functions and their derivatives. Let us take a matrix W, which may be in the form

$$\begin{vmatrix} f_1 & f_2 & \cdots & f_n & g_1 & g_2 & \cdots & g_n \\ f_1' & f_2' & \cdots & f_n' & g_1' & g_2' & \cdots & g_n' \\ \vdots & \vdots & \ddots & \vdots & \vdots & \vdots & \ddots & \vdots \\ f_1^{(n-1)} & f_2^{(n-1)} & \cdots & f_n^{(n-1)} & g_1^{(n-1)} & g_2^{(n-1)} & \cdots & g_n^{(n-1)} \end{vmatrix}$$

To construct exact soliton type solution of various NLPDEs like the Sine-Gordon equation, the nonlinear Schrödinger equation, and the KdV equation, the double-Wronskian method is employed. Breathers, other localized structures, and multisoliton solutions are common manifestations of these solutions.

Steps in the Double–Wronskian Method:

Step 1: Find linear equations whose solutions can be used to construct the double Wronskian determinant. These linear equations are often related to the Lax pair or other auxiliary linear problems associated with the NLPDE.

Step 2: Create a matrix with the solutions to these linear equations and their derivatives.

Step 3: Calculate the determinant of this matrix and ensure it satisfies the NLPDE.

Step 4: Substitute the determinant back into the original NLPDE to verify it indeed solves the equation.

Advantages and Applications:

The double Wronskian method provides a systematic methodology for finding exact solutions to NLPDEs that are sometimes difficult to get using other methods. It is useful to study multisoliton solutions and their interactions. This method is commonly used for integrable systems with an underlying mathematical structure that supports determinant-based solutions. It is a powerful tool for understanding nonlinear wave equations and integrable systems because it gives insight into their complex dynamics.

4.26 Double–Wronskian Solutions of the Boussinesq Systems

The Wronskian technique has been applied to a variety of nonlinear equations, including KdV and mKdV problems. In 1983, Nimmo and Freeman demonstrated that the nonlinear Schrödinger equation had a double Wronskian solution. Now we consider a (2+1)-dimensional system of NLPDEs

$$u_t + \alpha_1(t)u_{xy} + \alpha_2(t)(uw)_x + \alpha_3(t)v_x = 0, \tag{4.264}$$

$$\begin{aligned} v_t &+ \beta_1(t)(wv_x + 2vu_y + uv_y) + \beta_2(t)(u_x w_y - u_y^2) \\ &+ \beta_3(t)v_{xy} + \beta_4(t)u_{xyy} = 0, \end{aligned} \tag{4.265}$$

$$w_x = u_y \tag{4.266}$$

Consider transformations

$$u = -\frac{1}{2}(lnp)_x, \quad v = \frac{1}{2}b(lnp)_{xy} - pq$$

For the system (4.264–4.266), the integrability of the system needs the constraints $\beta_1(t) = \alpha_2(t) = 2\alpha_3(t) = -a(t)$, $\beta_2(t) = ba(t)$, $\alpha_1(t) = -\beta_3(t) = \frac{a(t)}{2}(1-b)$, $\beta_4(t) = b\frac{a(t)}{2}(b-2)$, where $a(t)$ is smooth arbitrary function and b is nonzero constant. Accordingly,

$$u = -\frac{1}{2}\frac{p_x}{p}, \quad v = \frac{1}{2}b\frac{\partial}{\partial x}\left(\frac{p_y}{p}\right) - pq = \frac{1}{2}b\left(\frac{p_{xy}}{p} - \frac{p_x p_y}{p^2}\right) - pq,$$

$$u_t = -\frac{1}{2}\left(\frac{p_{xt}}{p} - \frac{p_x p_t}{p^2}\right.,$$

$$u_x = -\frac{1}{2}\left(\frac{p_{2x}}{p} - \frac{p_x^2}{p^2}\right),$$

$$u_{xy} = -\frac{1}{2}\left(-\frac{p_{2x}p_y}{p^2} + \frac{p_{xxy}}{p} - \frac{2p_x p_{xy}}{p^2} + \frac{2p_x^2 p_y}{p^3}\right),$$

$$v_x = \frac{1}{2}b\left(\frac{p_{xxy}}{p} - \frac{2p_x p_{xy}}{p^2} - \frac{p_{xx}p_y}{p^2} + \frac{2p_x^2 p_y}{p^3}\right) - p_x q - pq_x$$

Since,

$$u_y = -\frac{1}{2}\left(\frac{p_{xy}}{p} - \frac{p_x p_y}{p^2}\right),$$

$$\Rightarrow \int u_y \partial x = -\frac{1}{2}\int\left(\frac{p_{xy}}{p} - \frac{p_x p_y}{p^2}\right)\partial x,$$

$$\Rightarrow w = -\frac{1}{2}\int\frac{\partial}{\partial x}\left(\frac{p_y}{p}\right)\partial x = -\frac{1}{2}\left(\frac{p_y}{p}\right)$$

Putting the constraint in the system (4.264–4.265), we get

$$u_t + \frac{a(t)}{2}(1-b)u_{xy} - a(t)(uw)_x - \frac{a(t)}{2}v_x = 0, \tag{4.267}$$

$$v_t - a(t)(wv_x + 2vu_y + uv_y) + ba(t)(u_x w_y - u_y^2) - \frac{a(t)}{2}(1-b)v_{xy}$$
$$+\frac{ba(t)}{2}(b-2)u_{xyy} = 0 \tag{4.268}$$

Putting the values of u_t, u_{xy}, $(uw)_x$, v_x in (4.267), and after some algebraic manipulations, we get

$$p_t + a(t)\left(\frac{1}{2}p_{xy} - qp^2\right) = 0$$

Similarly Eq. (4.268) can be transformed into

$$q_t - a(t)\left(\frac{1}{2}q_{xy} - pq^2\right) = 0$$

Therefore, the transformed system is

$$\begin{aligned} p_t + a(t)\left(\frac{1}{2}p_{xy} - qp^2\right) &= 0 \\ q_t - a(t)\left(\frac{1}{2}q_{xy} - pq^2\right) &= 0 \end{aligned} \tag{4.269}$$

Consider the rational transformations $p = \frac{g}{f}$, $q = \frac{h}{f}$. So $p_t = \frac{g_t}{f} - \frac{gf_t}{f^2}$, $p_x = \frac{g_x}{f} - \frac{gf_x}{f^2}$, $p_{xy} = \frac{g_{xy}}{f} - \frac{g_x f_y}{f^2} - \frac{gf_{xy}+g_y f_x}{f^2} + \frac{2gf_x f_y}{f^3}$, $q_t = \frac{h_t}{f} - \frac{hf_t}{f^2}$, $q_x = \frac{h_x}{f} - \frac{hf_x}{f^2}$, $q_{xy} = \frac{h_{xy}}{f} - \frac{h_x f_y}{f^2} - \frac{hf_{xy}+h_y f_x}{f^2} + \frac{2hf_x f_y}{f^3}$.

Then from first equation of the system (4.269), we get

$$\frac{fg_t - gf_t}{f^2} + \frac{1}{2}a(t)\left[\frac{fg_{xy} - gf_{xy}}{f^2} - \frac{g_x f_y + g_y f_x}{f^2} + 2\frac{gf_x f_y}{f^3} - 2\frac{hg^2}{f^3}\right] = 0 \tag{4.270}$$

Now,

$$D_t(g.f) = \left(\frac{\partial}{\partial t} - \frac{\partial}{\partial t'}\right)(g(x,y).f(x',y'))|_{t=t'} = g_t f - gf_t$$

and

$$\begin{aligned} D_x D_y(g.f) &= \left(\frac{\partial}{\partial x} - \frac{\partial}{\partial x'}\right)\left(\frac{\partial}{\partial y} - \frac{\partial}{\partial y'}\right)(g(x,y,t).f(x',y'),t')|_{x=x',y=y'}, \\ &= g_{xy}f - g_y f_x - g_x f_y + gf_{xy} \end{aligned}$$

So,

$$\begin{aligned} D_x D_y(f.f) &= \left(\frac{\partial}{\partial x} - \frac{\partial}{\partial x'}\right)\left(\frac{\partial}{\partial y} - \frac{\partial}{\partial y'}\right)(f(x,y,t).f(x',y'),t')|_{x=x',y=y'}, \\ &= 2(ff_{xy} - f_x f_y) \end{aligned}$$

Therefore,

$$\begin{aligned} &D_x D_y(f.f) + 2gh = 0, \\ &\Rightarrow 2(ff_{xy} - f_x f_y) + 2gh = 0, \\ &\Rightarrow 2gh = 2(f_x f_y - ff_{xy}) \end{aligned}$$

Now, from (4.270)

$$\frac{fg_t - gf_t}{f^2} + \frac{1}{2}a(t)\left[\frac{fg_{xy} + gf_{xy}}{f^2} - \frac{g_x f_y + g_y f_y}{f^2}\right] = 0,$$

$$\Rightarrow \left(D_t + \frac{1}{2}a(t)D_x D_y\right)(g.f) = 0$$

and from the 2nd equation of the system (4.269), we have,

$$\frac{fh_t - hf_t}{f^2} - \frac{a(t)}{2}\left[\frac{fh_{xy} - fh_{xy}}{f^2} - \frac{h_x f_y + h_y f_x}{f^2} + \frac{2hf_x f_y}{f^3} - \frac{2gh^2}{f^2}\right] = 0 \tag{4.271}$$

Now,

$$D_t(h.f) = h_t f - hf_t$$

and

$$D_x D_y(h.f) = h_{xy} f - h_y f_x - h_x f_y + hf_{xy}$$

From (4.271),

$$\frac{fh_t - hf_t}{f^2} - \frac{a(t)}{2}\left[\frac{fh_{xy} - fh_{xy}}{f^2} - \frac{h_x f_y + h_y f_x}{f^2}\right] = 0$$

$$\Rightarrow \left(D_t - \frac{1}{2}a(t)D_x D_y\right)(h.f) = 0$$

Therefore, the Hirota bilinear form is

$$\begin{aligned} &D_x D_y(f.f) + 2gh = 0, \\ &\left(D_t + \frac{1}{2}a(t)D_x D_y\right)(g.f) = 0, \\ &\left(D_t - \frac{1}{2}a(t)D_x D_y\right)(h.f) = 0 \end{aligned} \tag{4.272}$$

For convenience, we adopt the Freeman and Nimmo notation for the double Wronskian determinant, as follows:

$$W^{N+1,M+1}(\phi;\psi) = det(\phi, \partial\phi, \ldots, \partial^N\phi; \psi, \partial\psi, \ldots, \partial^M\psi) = |\widehat{N}; \widehat{M}|$$

where

$$\phi = (\phi_1(x), \phi_2(x), \ldots, \phi_{(N+M+2)}(x))^T,$$

$$\psi = (\psi_1(x), \psi_2(x), \ldots, \psi_{(N+M+2)}(x))^T$$

Now, we shall present a double Wronskian solution for the bilinear system (4.272) under a sufficient condition given as a system of linear PDEs.

Theorem: Let ϕ and ψ satisfy the following linear system

$$\phi_x = -K\phi,$$
$$\phi_y = \phi_x,$$
$$\phi_t = -A(t)\phi_{xx},$$
$$\psi_x = K\psi,$$
$$\psi_y = \psi_x,$$
$$\psi_t = B(t)\psi_{xx}$$

where $K = (k_{ij})_{(N+M+2)X(N+M+2)}$ is a matrix with arbitrary constants entries k_{ij}, and $A(t) = (a_{ij}(t))_{(N+M+2)X(N+M+2)}$, is a matrix with arbitrary smooth functions.

Here, $\sum_{i-1}^{N+M+2} a_{ii}(t) = a(t)$, but $B(t) = (b_{ij}(t))_{(N+M+2)X(N+M+2)}$ is with entries defined to be

$$b_{ij}(t) = a_{ij}(t) \quad \text{if} \quad i = j,$$

and

$$b_{ij}(t) = -a_{ij}(t) \quad \text{if} \quad i \neq j$$

Then the following double Wronskian determinants,

$$f = W^{N+1,M+1}(\phi;\psi), \quad g = 2W^{N+2,M}(\phi;\psi), \quad h = -2W^{N,M+2}(\phi;\psi)$$

In order to prove this theorem, we need the following lemmas:

Lemma 1: The Plücker relation for the Wronskian determinant is

$$|B,a,b||B,c,d| - |B,a,c||B,b,d| + |B,a,d||B,c,b| = 0 \tag{4.273}$$

where B is an $N \times (N-2)$ matrix; a, b, c and d represent the N dimensional column vectors.

Lemma 2: Suppose Θ is an $N \times N$ matrix with column vector set Θ_j. Ω is an $N \times N$ operator matrix with column vector set Ω_j, and each entry Ω_{js} is an operator. Then we have

$$\sum_{j=1}^{N} |\Omega_j * \Theta| = \sum_{j=1}^{N} |(\Omega^T)_j * \Theta^T| \tag{4.274}$$

where, for any N-order column vectors A_j and B_j, we define

$$A_j@B_j = (A_{1j}B_{1j}, A_{2j}B_{2j}, \ldots, A_{Nj}B_{Nj})^T \tag{4.275}$$

and

$$|A_j * \Theta| = |\Theta_1, \Theta_2, \ldots, \Theta_{j-1}, A_j@\Theta_j, \Theta_{j+1}, \ldots, \Theta N| \tag{4.276}$$

Let ϕ and ψ satisfy the above systems and set

$$\Theta = W^{N+1,M+1}(\phi;\psi),$$
$$\text{and } \ \Omega = (\Omega_{ij})_{(N+M+2)X(N+M+2)}$$

where $\Omega = -\partial x$, if $1 \le i \le N+M+2; 1 \le j \le N+1$, and $\Omega = \partial x$, if $1 \le i \le N+M+2; N+2 \le j \le N+M+2$. In order to simplify the notations, we are going to use the following notations for Wronskian determinants:

$$\begin{aligned}
d_1 &= |\hat{N};\hat{M}|,\\
d_2 &= |\widehat{(N-1)}, N+1; \hat{M}|,\\
d_3 &= |\hat{N}; \widehat{(M-1)}, M+1|,\\
d_4 &= |\widehat{(N-2)}, N, N+1; \hat{M}|,\\
d_5 &= |\widehat{(N-1)}, N+2; \hat{M}|,\\
d_6 &= |\widehat{(N-1)}, N+1; \widehat{(M-1)}, M+1|,\\
d_7 &= |\hat{N}; \widehat{(M-2)}, M, M+1|,\\
d_8 &= |\hat{N}; \widehat{(M-1)}, M+2|,\\
d_9 &= |\widehat{(N+1)}; \widehat{(M-1)}|,\\
d_{10} &= |\hat{N}, N+2; \widehat{(M-1)}|,\\
d_{11} &= |\widehat{(N+1)}; \widehat{(M-2)}, M|,\\
d_{12} &= |\widehat{(N-1)}, N+1, N+2; \widehat{(M-1)}|,\\
d_{13} &= |\hat{N}, N+3; \widehat{(M-1)}|,
\end{aligned}$$

$$d_{14} = |\hat{N}, N+2; \widehat{(M-2)}, M|,$$

$$d_{15} = |\widehat{(N+1)}; \widehat{(M-3)}, M-1, M|,$$

$$d_{16} = |\widehat{(N+1)}; \widehat{(M-2)}, M+1|,$$

$$d_{17} = |\widehat{(N-1)}; \widehat{(M+1)}|$$

Then, by using the above lemmas, we get the following identities:

$$(d_4 + d_5 - 2d_6 + d_7 + d_8)d_1 = (d_2 - d_3)^2 \tag{4.277}$$

$$(d_{12} + d_{13} - 2d_{14} + d_{15} + d_{16})d_1 = (d_{10} - d11)(d_2 - d_3) \tag{4.278}$$

$$\text{and} \quad (d_4 + d_5 - 2d_6 + d_7 + d_8)d_9 = (d_2 - d_3)(d_{10} - d_{11}) \tag{4.279}$$

Now, we will prove the theorem.

Proof: Under the linear conditions in the theorem, we can compute the following derivatives:

$$f = W^{N+1,M+1}(\phi;\psi) = |\hat{N};\hat{M}| = |\phi, \partial\phi, \ldots, \partial^N\phi; \psi, \partial\psi, \ldots, \partial^M\psi|,$$

$$\Rightarrow f_x = |\widehat{(N-1)}, N+1; \hat{M}| = d_2 + d_3.$$

$$\Rightarrow f_{xy} = |\widehat{(N-2)}, N, N+1; \hat{M}| + |\widehat{(N-1)}, N+2; \hat{M}| + 2|\widehat{(N-1)}, N+1;$$
$$\widehat{(M-1)}, M+1| + |\hat{N}; \widehat{(M-2)}, M, M+1| + |\hat{N}; \widehat{(M-1)}, M+2|,$$
$$= d_4 + d_5 + 2d_6 + d_7 + d_8$$

Now,

$$g = 2W^{N+2,M}(\phi;\psi) = 2|\phi, \partial\phi, \ldots, \partial^{(N+1)}\phi; \psi, \partial\psi, \ldots, \partial^{(M-1)}\psi|$$
$$= 2|\widehat{(N+1)}, \widehat{(M-1)}|,$$

$$\Rightarrow g_x = 2|\hat{N}, N+2; \widehat{M-1}| + 2|\widehat{N+1}; \widehat{M-2}, M| = 2d_{10} + 2d_{11}.$$

$$\Rightarrow g_{xy} = 2|\widehat{(N-1)}, N+1, N+2; \widehat{(M-1)}| + 2|\hat{N}, N+3; \widehat{(M-1)}|$$
$$+ 4|\hat{N}, N+2; \widehat{(M-2)}, M| + 2|\widehat{(N+1)}; \widehat{(M-3)}, M-1, M|$$
$$+ 2|\widehat{(N+1)}; \widehat{(M-2)}, M+1|,$$
$$= 2d_{12} + 2d_{13} + 4d_{14} + 2d_{15} + 2d_{16}$$

and

$$\begin{aligned}
f_t &= a(t)|\phi, \partial\phi, \dots, \partial^{N-2}\phi, \partial^N\phi, \partial^{N-2}\phi; \psi, \partial\psi, \dots, \partial^M\psi| \\
&\quad - a(t)|\phi, \partial\phi, \dots, \partial^{N-1}\phi, \partial^{N+2}\phi; \psi, \partial\psi, \dots, \partial^M\psi| \\
&\quad - a(t)|\phi, \partial\phi, \dots, \partial^N\phi; \psi, \partial\psi, \dots, \partial^{M-2}\psi, \partial^M\psi, \partial^{M-2}\psi| \\
&\quad + a(t)|\phi, \partial\phi, \dots, \partial^N\phi; \psi, \partial\psi, \dots, \partial^{M-1}\psi, \partial^{M+2}\psi|, \\
&= -a(t)[-|\widehat{(N-2)}, N, N+1; \hat{M}| + |\widehat{(N-1)}, N+2; \hat{M}| + |\hat{N}; \widehat{(M-2)}, \\
&\quad M, M+1| - |\hat{N}; \widehat{(M-1)}, M+2|], \\
&= -a(t)[-d_4 + d_5 + d_7 - d_8]
\end{aligned}$$

and

$$\begin{aligned}
g_t &= 2a(t)[|\widehat{(N-1)}, N+1, N+2; \widehat{(M-1)}| - |\hat{N}, N+3; \widehat{(M-1)}| \\
&\quad - |\widehat{(N+1)}; \widehat{(M-3)}, M-1, M| + |\widehat{(N+1)}; \widehat{(M-2)}, M+1|], \\
&= 2a(t)(d_{12} - d_{13} - d[15] + d_{16})
\end{aligned}$$

Using the identity (4.277) and **lemma 1**, we can verify the first equation of the system (4.272).

$$\begin{aligned}
&D_xD_y(f.f) + 2gh \\
&\quad = 2[ff_{xy} - f_xf_y + gh] \\
&\quad = 2[d_1(d_4 + d_5 + 2d_6 + d_7 + d_8) - (d_2 + d_3)^2] + 2.2d_9(-2d_{17}) \\
&\quad = 2(d_2 - d_3)^2 - 2(d_2 + d_3)^2 + 8d_1d_6 - 8d_9d_{17} \\
&\quad = -8(d_2d_3 - d_1d_6 + d_9d_{17}) \\
&\quad = 0
\end{aligned}$$

Using the identities (4.278) and (4.279), we compute

$$\begin{aligned}
\left(g_t + \frac{1}{2}a(t)g_{xx}\right) f = \Bigg[&2a(t)(d_{12} - d_{13} - d_{15} + d_{16}) \\
&+ \frac{1}{2}a(t)(2d_{12} + 2d_{13} + 4d_{14} + 2d_{15} + 2d_{16})\Bigg] d_1
\end{aligned}$$

$$= a(t)d_1[2d_{12} - 2d_{13} - 2d_{15} + 2d_{16} + d_{12} + d_{13} + 2d_{14} + d_{15} + d_{16}]$$

$$= 4a(t)d_1(d_{12} + d_{16} - a(t)(d_{10} - d_{11})(d_2 - d_3), \quad \text{and}$$

$$\left(\frac{1}{2}a(t)f_{xx} - f_t\right)g$$

$$= \left[\frac{1}{2}a(t)(d_4 + d_5 + 2d_6 + d_7 + d_8 + a(t)(-d_4 + d_5 + d_7 - d_8)\right]d_9$$

$$= \frac{a(t)}{2}d_9(-d_4 + 3d_5 + 2d_6 + 3d_7 - d_8)$$

$$= \frac{1}{2}[4a(t)d_9(d_5 + d_7) - a(t)(d_4 + d_5 - 2d_6 + d_7 + d_8)d_9]$$

$$= \frac{1}{2}[4a(t)d_9(d_5 + d_7) - a(t)(d_2 - d_3)(d_{10} - d_{11})]$$

Hence, the second equation of (4.272) reduces to

$$\left(D_t + \frac{1}{2}a(t)D_xD_y\right)(g.f)$$

$$= 4a(t)\left([d_1d_{12} - d_2d_{12} + d_9d_5] + [d_1d_{15} - d_3d_{11} + d_9d_7]\right)$$

Again, by **lemma 1**, the right-hand side of the above equality is zero. Similarly, the third equation of the system, (4.272), can be verified. Therefore, it is shown that the double Wronskian determinants f, g, and h solve the bilinear system (4.272) under the linear condition given in the theorem.

4.27 Soliton Solution in Double–Wronskian Form

Now, we will present the soliton solutions derived by the double Wronskian technique. The solution to the linear system examined in the preceding theorem is as follows:

$$\phi = e^{-(Kx+Ky+K^2\int A(t)dt)}C, \quad \psi = e^{Kx+Ky+K^2\int B(t)dt}D \tag{4.280}$$

where $C^T = (C_1, C_2, \ldots, C_{(N+M+2)})$ and $D^T = (D_1, D_2, \ldots, D_{(N+M+2)})$ are arbitrary real constant vectors. If $K = \text{diag}\,(k_1, k_2, \ldots, k_{(N+M+2)})$ and $A(t) = \text{diag}\,(a_1, a_2, \ldots, a_{(N+M+2)})$, then we have

$$\phi_j = C_je^{-\eta_j}, \quad \psi_j = D_je^{\eta_j}, \quad \text{for } j = 1, 2, \ldots, (N+M+2) \tag{4.281}$$

where $\eta_j = k_jx + k_jy + k_j^2\int a_j(t)dt$.

One-soliton solution:

Let us consider $N = M = 0$, $C_1 = D_1 = D_2 = 1$, and $C_2 = -1$ then we get $K = \text{diag } (k_1, k_2)$, $A(t) = \text{diag } (a_1(t), a_2(t))$, $\phi_1 = e^{-\eta_1}$, $\phi_2 = e^{-\eta_2}$, $\psi_1 = e^{\eta_1}$, and $\psi_1 = e^{\eta_2}$.

Therefore, $\phi = (e^{-\eta_1}, -e^{-\eta_2})^T$ and $\psi = (e^{\eta_1}, e^{\eta_2})^T$

Hence,

$$f = W^{1,1}(\phi; \psi) = |\widehat{0}; \widehat{0}| = |\phi; \psi| = \begin{vmatrix} e^{-\eta_1} & e^{\eta_1} \\ -e^{-\eta_2} & e^{\eta_2} \end{vmatrix} = e^{\eta_2 - \eta_1} + e^{-(\eta_2 - \eta_1)},$$

$$= 2\cosh(\eta_2 - \eta_1)$$

and

$$g = 2W^{2,0}(\phi; \psi) = 2|\widehat{1}| = 2|\phi, \partial\phi| = \begin{vmatrix} e^{-\eta_1} & -k_1 e^{-\eta_1} \\ -e^{\eta_2} & k_2 e^{-\eta_2} \end{vmatrix},$$

$$= 2[k_2 e^{-(\eta_1+\eta_2)} - k_1 e^{-(\eta_1+\eta_2)}]$$

Thus,

$$p = \frac{g}{f} = \frac{2(k_2 - k_1)e^{-(\eta_1+\eta_2)}}{2\cosh(\eta_2 - \eta_1)}$$

and

$$u = -\frac{1}{2}(lnp)_x = -\frac{1}{2}\left(ln\frac{(k_2 - k_1)e^{-(\eta_1+\eta_2)}}{\cosh(\eta_2 - \eta_1)}\right) \tag{4.282}$$

which is a one-soliton solution for u.

Now,

$$h = -2W^{0,2}(\phi; \psi) = -2|\widehat{1}| = -2|\psi, \partial\psi| = -2\begin{vmatrix} e^{\eta_1} & k_1 e^{\eta_1} \\ e^{\eta_2} & k_2 e^{\eta_2} \end{vmatrix}$$

$$= -2(k_2 - k_1)e^{(\eta_1+\eta_2)}$$

Therefore,

$$q = \frac{h}{f} = \frac{-(k_2 - k_1)e^{(\eta_1+\eta_2)}}{\cosh(\eta_2 - \eta_1)}$$

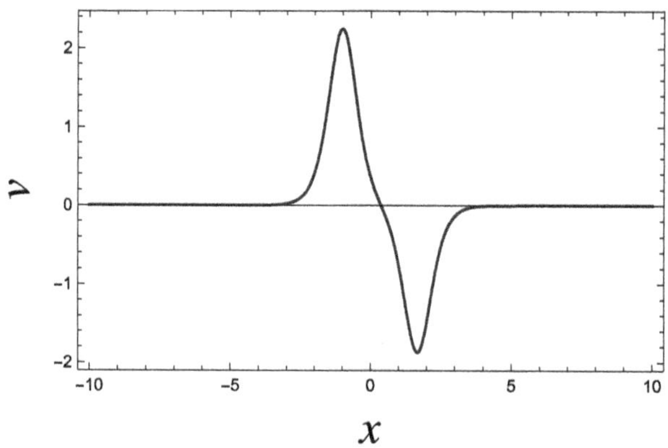

Figure 4.15. One-soliton solution of Eq. (4.283) for v.

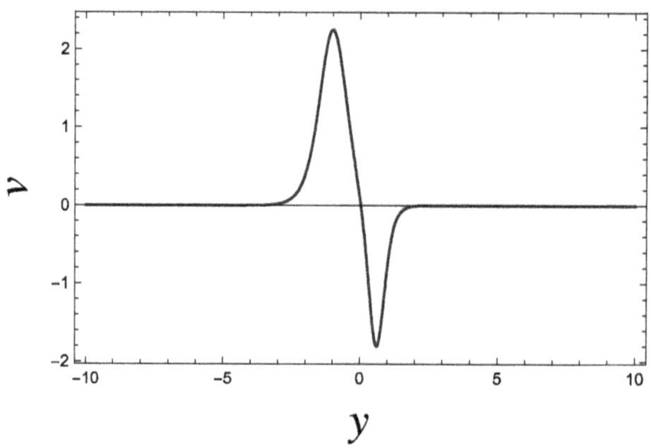

Figure 4.16. One-soliton solution of Eq. (4.283) for v.

Hence,

$$\begin{aligned} v &= \frac{1}{2}b(lnp)_{xy} - pq, \\ &= \frac{1}{2}b\left(ln\frac{(k_2-k_1)e^{-(\eta_1+\eta_2)}}{\cosh(\eta_2-\eta_1)}\right)_{xy} + (k_2-k_1)^2 sech^2(\eta_2-\eta_1) \end{aligned} \tag{4.283}$$

This is a one-soliton solution for v. Therefore, the one-soliton solution is given by (4.282) and (4.283).

Two-soliton solution:

Let us consider $M = 1$, $N = 0$, $C_i = D_i = 1$, $i = 1, 2, 3$. So, we get $\phi_1 = e^{-\eta_1}$, $\phi_2 = e^{-\eta_2}$, $\phi_3 = e^{-\eta_3}$, $\psi_1 = e^{\eta_1}$, $\psi_2 = e^{\eta_2}$, $\psi_3 = e^{\eta_3}$.

Therefore,

$$f = W^{1,2}(\phi;\psi) = |\widehat{0};\widehat{1}| = |\phi;\psi| = \begin{vmatrix} e^{-\eta_1} & e^{\eta_1} & k_1 e^{\eta_1} \\ e^{-\eta_2} & e^{\eta_1} & k_2 e^{\eta_2} \\ e^{-\eta_3} & e^{\eta_3} & k_1 e^{\eta_3} \end{vmatrix},$$

$$= (k_3 - k_2)e^{-\eta_1+\eta_2+\eta_3} - (k_3 - k_1)e^{\eta_1-\eta_2+\eta_3} + (k_2 - k_1)e^{\eta_1+\eta_2-\eta_3},$$

and

$$g = 2W^{2,1}(\phi;\psi) = 2|\widehat{1};\widehat{0}| = 2\begin{vmatrix} e^{-\eta_1} & -k_1 e^{-\eta_1} & e^{\eta_1} \\ e^{-\eta_2} & -k_2 e^{-\eta_2} & e^{\eta_2} \\ e^{-\eta_3} & -k_3 e^{-\eta_3} & e^{\eta_3} \end{vmatrix}$$

$$= 2[(-k_3 + k_2)e^{\eta_1-\eta_2-\eta_3} + (k_3 - k_1)e^{-\eta_1+\eta_2-\eta_3} + (k_1 - k_2)e^{-\eta_1-\eta_2+\eta_3}]$$

Therefore,

$$u = -\frac{1}{2}\left(ln\left(\frac{\lambda_1}{\lambda_2}\right)\right)_x \tag{4.284}$$

where $\lambda_1 = [(-k_3 + k_2)e^{\eta_1-\eta_2-\eta_3} - (k_3 - k_1)e^{-\eta_1+\eta_2-\eta_3} + (k_1 - k_2)e^{-\eta_1-\eta_2+\eta_3}]$, and $\lambda_2 = [(k_3 - k_2)e^{-\eta_1+\eta_2+\eta_3} - (k_3 - k_1)e^{\eta_1-\eta_2+\eta_3} + (k_2 - k_1)e^{\eta_1+\eta_2-\eta_3}]$

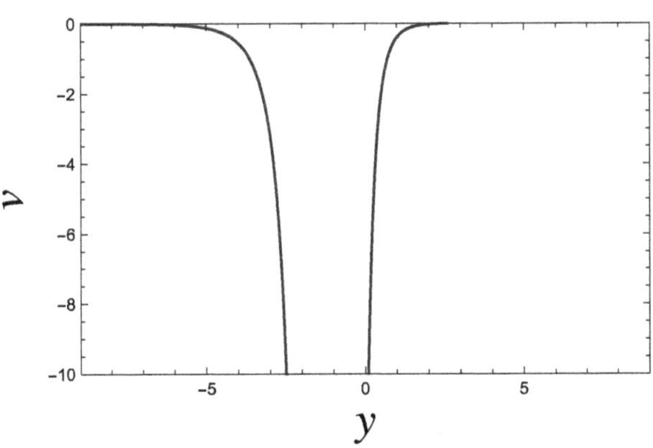

Figure 4.17. Singular-soliton solution of Eq. (4.285) for v.

Now,

$$h = -2W^{0,3}(\phi;\psi) = -2|\widehat{2}| = -2|\psi, \partial\psi, \partial^2\psi| = -2\begin{vmatrix} e^{\eta_1} & k_1e^{\eta_1} & k_1^2e^{\eta_1} \\ e^{\eta_2} & k_2e^{\eta_2} & k_2^2e^{\eta_2} \\ e^{\eta_3} & k_3e^{\eta_3} & k_3^2e^{\eta_3} \end{vmatrix},$$

$$= 2[k_2k_3(-k_3+k_2)e^{\eta_1+\eta_2+\eta_3} - k_1k_3(-k_3+k_1)e^{\eta_1+\eta_2+\eta_3}$$
$$+ k_1k_2(k_1-k_2)e^{\eta_1+\eta_2+\eta_3}]$$

Thus,

$$pq = \frac{gh}{f^2}$$
$$= 4K\frac{(k_2-k_3)e^{2\eta_1} - (-k_3+k_1)e^{2\eta_2} + (k_1-k_2)e^{2\eta_3}}{[(k_3-k_2)e^{-\eta_1+\eta_2+\eta_3} - (k_3-k_1)e^{\eta_1-\eta_2+\eta_3} + (k_2-k_1)e^{\eta_1+\eta_2-\eta_3}]^2}$$

where $K = \Pi_{1\le i<j\le 3}(k_j - k_i)$.

Hence, we can write,

$$v = \frac{1}{2}b(lnp)_{xy} - pq,$$
$$= \frac{1}{2}b\left(ln\frac{\lambda_3}{\lambda_4}\right)_{xy}$$
$$+ 4K\frac{(k_2-k_3)e^{2\eta_1} - (-k_3+k_1)e^{2\eta_2} + (k_1-k_2)e^{2\eta_3}}{[(k_3-k_2)e^{-\eta_1+\eta_2+\eta_3} - (k_3-k_1)e^{\eta_1-\eta_2+\eta_3} + (k_2-k_1)e^{\eta_1+\eta_2-\eta_3}]^2} \tag{4.285}$$

where $\lambda_3 = [(-k_3+k_2)e^{\eta_1-\eta_2-\eta_3} + (k_3-k_1)e^{-\eta_1+\eta_2-\eta_3} + (k_1 - k_2)e^{-\eta_1-\eta_2+\eta_3}]$, and $\lambda_4 = (k_3-k_2)e^{-\eta_1+\eta_2+\eta_3} - (k_3-k_1)e^{\eta_1-\eta_2+\eta_3} + (k_2 - k_1)e^{\eta_1+\eta_2-\eta_3}$ Therefore, the two-soliton solutions are given by (4.284) and (4.285).

In a similar way, one can obtain the N-soliton solution easily.

4.28 Darboux Transformation (DT)

Darboux transformations (DT) have their roots in the differential geometry of surfaces and were first proposed by Gaston Darboux in 1882. They represent a significant and well-researched link to contemporary soliton theory and integrable system theory. These transformations are utilized in the current theory of integrable systems to create PDE solutions starting from known solutions, even from trivial ones. The Darboux theorem says that *a Sturm–Liouville problem is covariant with respect to a linear transformation.* It is a well-known analytical technique for solving a nonlinear evolution equation. In this method, a pair of linear systems (Lax pair) $L\psi = \lambda\psi$ and $\psi_t = M\psi$, corresponding to Lax operators L and M of said evolution equation, is shown to be covariant under the DT and generates new exact solutions from old exact solutions (seed solutions) of the nonlinear evolution equation. Therefore, if we perform repeated DT, such as one, two, or three-folded, or, more generally, N-folded in a pair of linear systems (Lax pair), we can obtain a new exact solution from an earlier exact solution at each stage. Therefore, we can use the DT to create a new exact solution from an existing old one. Obtaining a new ψ from an old ψ is the main idea of DT. Hence, DT may be used to obtain several (infinite) kinds of nonlinear structures, including periodic and multisolitons, from an NLPDE.

4.29 Darboux Theorem

Let us consider the Sturm–Liouville problem

$$\psi''(x) + (\lambda - u(x))\psi(x) = 0 \tag{4.286}$$

Let for the spectral parameter $\lambda = \lambda_1$, $\psi = \psi_1(x)$ be a particular solution of (4.286). Then, according to Darboux, a new improved solution $\psi[1]$ can be obtained, defined as

$$\psi[1] = \left(\frac{d}{dx} - \frac{\psi_1'}{\psi_1}\right)\psi \tag{4.287}$$

$$= \left(\frac{d}{dx} - \frac{d}{dx}\log(\psi_1)\right)\psi \tag{4.288}$$

i.e., $\psi[1]$ should satisfy (4.286). So,

$$\psi''[1] + (\lambda - u[1])\psi[1] = 0 \tag{4.289}$$

where $u[1]$ is given by $u - 2\frac{d}{dx}\left(\frac{\psi_1'}{\psi_1}\right)$.

Note: (i) The Sturm–Liouville problem is covariant with respect to the Darboux transformation.

(ii) On each stage, a new $u[1]$ is to be determined, and accordingly, we get a new solution after every step.

4.30 Darboux Covariant Theorem

Statement: Let us assume that $\psi = \psi_1$, a particular solution to the Sturm–Liouville Eq. (4.286), exists for the spectral parameter $\lambda = \lambda_1$. Consider the DT on ψ defined by

$$\psi[1] = \left(\frac{d}{dx} - l\right)\psi \tag{4.290}$$

where $l = \frac{\psi_1'}{\psi_1}$. Then $\psi[1]$ obeys the following equation:

$$\psi''[1] + (\lambda - u[1])\psi[1] = 0 \tag{4.291}$$

where $u[1] = u - 2\frac{d}{dx}(l)$.

Proof: Let ψ_1 be a particular solution of Eq. (4.286) for the spectral parameter $\lambda = \lambda_1$. So,

$$\psi_1'' + (\lambda_1 - u)\psi_1 = 0 \tag{4.292}$$

Now,

$$\begin{aligned}
\psi[1] &= \left(\frac{d}{dx} - l\right)\psi = \psi' - l\psi, \\
\psi'[1] &= \psi'' - l'\psi - l\psi' = (u - \lambda - l')\psi - l\psi', \\
\psi''[1] &= (u' - l'')\psi + (u - \lambda - l')\psi' - l\psi'' - l'\psi' \\
&= (u' - l'' - lu + l\lambda)\psi + (u - \lambda - 2l')\psi'
\end{aligned}$$

Therefore,

$$\begin{aligned}
&\psi''[1] + (\lambda - u[1])\psi[1] \\
&\quad = (u' - l'' - lu + l\lambda)\psi + (u - \lambda - 2l')\psi' + (\lambda - u[1])(\psi' - l\psi) \\
&\quad = (u' - l'' - lu + lu[1])\psi + (u - 2l' - u[1])\psi'
\end{aligned}$$

If the coefficients of ψ and ψ' are zero, then ψ is arbitrary. We assume that the coefficient of ψ' vanishes when looking for new potential $u[1]$ through old potential u. Consequently, we have

$$u[1] = u - 2l' \tag{4.293}$$

so $u - 2l' - u[1] = 0$. Additionally $(u' - l'' - lu + lu[1]) = 0$.

Using (4.293), the coefficient of ψ can be written as

$$\begin{aligned} u' - l'' - lu + lu[1] &= u' - l'' - lu + l(u - 2l') \\ &= \frac{d}{dx}(u - l' - l^2) \end{aligned} \tag{4.294}$$

Now,

$$l = \frac{d}{dx}(\log \psi_1) = \frac{\psi_1'}{\psi_1}, \quad l' = \frac{\psi_1 \psi_1'' - (\psi_1')^2}{\psi_1^2}$$

$$u - l' - l^2 = u - \frac{\psi_1 \psi_1'' - (\psi_1')^2}{\psi_1^2} - \frac{(\psi_1')^2}{\psi_1^2} = \frac{u\psi_1 - \psi_1''}{\psi_1}$$

Since $\psi_1 \neq 0$ and from Eq. (4.292), we have $u - l' - l^2 = \lambda_1$. So

$$u - l'' - lu + lu[1] = \frac{d}{dx}(\lambda_1) = 0$$

Hence,

$$\psi''[1] + (\lambda - u[1])\psi[1] = 0 \tag{4.295}$$

where $u[1] = u - 2l' = u - 2\frac{d^2}{dx^2}(\log \psi_1)$.

Therefore, $\psi[1]$ is a solution of the Sturm–Liouville equation (4.286), corresponding to the new potential function $u[1] = u - 2\frac{d^2}{dx^2}(\log \psi_1)$. In other words, we may state that equation (4.286) is covariant with respect to the DT.

4.31 DT for ODE

Schrödinger equation with a harmonic potential:

Let us consider the equation

$$-\psi''(x) + x^2\psi(x) = \lambda\psi(x) \tag{4.296}$$

where λ is the spectral parameter.

Let $\psi(x) = \exp(\frac{x^2}{m})$ is a trial solution of Eq. (4.296) and m is a non-zero real number. Now, $\psi'(x) = \frac{2x}{m}\exp(\frac{x^2}{m})$, $\psi''(x) = \frac{2}{m}\exp(\frac{x^2}{m}) + \frac{4x^2}{m^2}\exp(\frac{x^2}{m})$.

Putting the value of $\psi(x)$ and $\psi''(x)$ in Eq. (4.296), we get

$$x^2\left(1 - \frac{4}{m^2}\right) - \left(\lambda + \frac{2}{m}\right) = 0 \tag{4.297}$$

If $(1 - \frac{4}{m^2}) = 0$, and $(\lambda + \frac{2}{m}) = 0$, then $\psi(x) = \exp(\frac{x^2}{m})$, satisfies Eq. (4.296). So, $m = 2,\ \lambda = -1$ or $m = -2,\ \lambda = 1$.

Therefore, $\psi = \exp(\frac{x^2}{2})$ and $\psi = \exp(-\frac{x^2}{2})$ is the solution of Eq. (4.296) with $\lambda = -1$ and $\lambda = 1$, respectively.

Consider the potential $u = x^2$ and the initial solution $\psi_1 = \exp(\frac{x^2}{2})$ for $\lambda = -1$. Using DT on ψ in (4.296), we get

$$\psi[1] = \left(\frac{d}{dx} - x\right)\psi, \quad u[1] = u - 2\frac{dh}{dx} = x^2 - 2, \quad h = \frac{\psi_1'}{\psi_1} = \frac{xe^{x^2/2}}{e^{x^2/2}} = x$$

which satisfy the equation

$$\begin{aligned} &\psi''[1] + (\lambda - u[1])\psi[1] = 0 \\ &\Longrightarrow \psi''[1] + (\lambda - x^2 + 2)\psi[1] = 0 \\ &\Longrightarrow -\psi''[1] + x^2\psi[1] = (\lambda + 2)\psi[1] \end{aligned}$$

For example, we consider

$$\psi = \psi[0] = \exp\left(-\frac{x^2}{2}\right) = (-1)^0 \exp\left(-\frac{x^2}{2}\right) = (-1)^0 H_0 \exp\left(-\frac{x^2}{2}\right)$$

with the eigen value $\lambda = 1$ gives

$$\begin{aligned} \psi[1] &= \left(\frac{d}{dx} - x\right)\psi = \left(\frac{d}{dx} - x\right)\psi[0] = \left(\frac{d}{dx} - x\right)\exp\left(-\frac{x^2}{2}\right) \\ &= (-1)^1 2x \exp\left(-\frac{x^2}{2}\right) \\ &= (-1)^1 H_1 \exp\left(-\frac{x^2}{2}\right) \end{aligned}$$

with the eigen value $\lambda + 2 = 3$. Where H_1 is the hermite polynomial of order 1. Again applying the DT on $\psi[1]$, we get

$$\begin{aligned} \psi[2] &= \left(\frac{d}{dx} - x\right)\psi[1] = \left(\frac{d}{dx} - x\right)\left(\frac{d}{dx} - x\right)\psi[0] \\ &= \left(\frac{d}{dx} - x\right)^2 \exp\left(-\frac{x^2}{2}\right) \\ &= (-1)^2(4x^2 - 2)\exp\left(-\frac{x^2}{2}\right) \\ &= (-1)^2 H_2 \exp\left(-\frac{x^2}{2}\right) \end{aligned}$$

with eigen value $\lambda + 2 = 5$.

Again applying DT on $\psi[2]$, we get

$$\psi[3] = \left(\frac{d}{dx} - x\right)\psi[2] = \left(\frac{d}{dx} - x\right)^3 \psi[0]$$
$$= \left(\frac{d}{dx} - x\right)^3 \exp\left(-\frac{x^2}{2}\right)$$
$$= (-1)^3(8x^3 - 12x)\exp\left(-\frac{x^2}{2}\right)$$
$$= (-1)^3 H_3 \exp\left(-\frac{x^2}{2}\right)$$

with the eigen value $\lambda + 2 = 7$.

Applying the DT on ψ, n-times, we get

$$\psi[n] = \left(\frac{d}{dx} - x\right)\left(\frac{d}{dx} - x\right)\left(\frac{d}{dx} - x\right)\cdots\left(\frac{d}{dx} - x\right)\psi[0]$$
$$= \left(\frac{d}{dx} - x\right)^n \psi[0]$$
$$= \left(\frac{d}{dx} - x\right)^n \exp\left(-\frac{x^2}{2}\right)$$
$$= (-1)^n H_n \exp\left(-\frac{x^2}{2}\right)$$

Thus, we obtained

$$H_n = (-1)^n \exp\left(\frac{x^2}{2}\right)\left(\frac{d}{dx} - x\right)^n \exp\left(\frac{-x^2}{2}\right)$$

where H_n is the Hermite Polynomial of order n.

4.32 DT for Generalized Lax Equation

Let us construct the DT for the generalized Lax equation (4.66) through its Lax pair. The Lax operators [29] for Eq. (4.66) are defined by

$$L = \partial_x^2 + \frac{1}{10}\gamma u I \tag{4.298}$$

$$M = -16\partial_x^5 - 4\gamma u\partial_x^3 - 6\gamma u_x\partial_x^2 - \left(\frac{3}{10}\gamma^2 u^2\right. \tag{4.299}$$

$$\left.+ 5\gamma u_{xx}\right)\partial_x - \left(\frac{3}{10}\gamma^2 u u_x + \frac{3}{2}\gamma u_{xxx}\right)I \tag{4.300}$$

where I is an identity operator. The Lax pair system of equations corresponding to the Lax operators L and M is given by

$$\psi_{xx} = \lambda\psi - \frac{1}{10}\gamma u\psi,$$
$$\psi_t = -16\psi_{xxxxx} - 4\gamma u\psi_{xxx} - 6\gamma u_x\psi_{xx} \tag{4.301}$$
$$- \left(\frac{3}{10}\gamma^2u^2 + 5\gamma u_{xx}\right)\psi_x - \left(\frac{3}{10}\gamma^2 uu_x + \frac{3}{2}\gamma u_{xxx}\right)\psi$$

where $\psi = \psi(x,t)$ is the non-zero eigenfunction and λ is a non-zero spectral parameter. The compatibility condition between two Eqs. (4.301), yields the Lax compatibility equation $L_t + (LM - ML) = 0$, which is equivalent to Eq. (4.66).

The one-fold DT on ψ in Eqs. (4.301) is defined as

$$\psi[1] = \psi_x - l\psi, \quad l = \partial_x(\ln\psi_1) \tag{4.302}$$

where ψ_1 is a particular non-trivial solution of Eq. (4.301) at $\lambda = \lambda_1$. The above Lax pair system of Eq. (4.301) can be re-expressed using the DT (4.302) in the following form

$$\bar{L}\psi[1] = \lambda\psi[1], \tag{4.303}$$
$$\psi[1]_t = \bar{M}\psi[1], \tag{4.304}$$

where

$$\bar{L} = \partial_x^2 + \frac{1}{10}\gamma u[1]I$$
$$\bar{M} = -16\partial_x^5 - 4\gamma u[1]\partial_x^3 - 6\gamma u_x[1]\partial_x^2 - \left(\frac{3}{10}\gamma^2u^2[1] + 5\gamma u_{xx}[1]\right)\partial_x$$
$$- \left(\frac{3}{10}\gamma^2 u[1]u_x[1] + \frac{3}{2}\gamma u_{xxx}[1]\right)I$$

Using (4.301) and (4.302) and after some direct calculations, we have the following results:

$$\psi_x[1] = \left(\lambda - \frac{1}{10}\gamma u - l_x\right)\psi - l\psi_x,$$
$$\psi_{xx}[1] = \left(\lambda - 2l_x - \frac{1}{10}\gamma u\right)\psi_x + \left(-\frac{1}{10}\gamma u_x - l_{xx} - l\lambda + \frac{1}{10}l\gamma u\right)\psi$$

To find the value of $u[1]$, we substitute the values of $\psi_x[1]$ and $\psi_{xx}[1]$ into Eqs. (4.303) and (4.304) and simplifying, then we have the following result:

$$u[1] = u + \frac{20}{\gamma}\partial_x^2(\ln\psi_1) \tag{4.305}$$

Hence, $\psi[1]$ satisfies Eq. (4.301) corresponding to $u[1]$; on the other hand, we can say that Eq. (4.301) are covariant under the Darboux transformation. The compatibility condition between two Eqs. (4.303) and (4.304) yields the Lax compatibility equation $\bar{L}_t + (\bar{L}\bar{M} - \bar{M}\bar{L}) = 0$, which is equivalent to Eq. (4.66) with respect to the potential $u[1]$. So, the DT generated a new solution $u[1]$ from the old solution u of Eq. (4.66).

4.32.1 *One-soliton solutions and periodic solutions*

Let us consider the initial trivial solution $u = u_0$ of Eq. (4.66), where u_0 being a real constant. Therefore, Lax pair linear system of Eq. (4.301) reduces to

$$\psi_{xx} + \left(\frac{1}{10}\gamma u_0 - \lambda\right)\psi = 0, \tag{4.306}$$

$$\psi_t + 16\psi_{xxxxx} + 4\gamma u_0\psi_{xxx} + \frac{3}{10}\gamma^2u_0^2\psi_x = 0 \tag{4.307}$$

Solving Eqs. (4.306) and (4.307) together for the spectral parameter $\lambda = \lambda_1$, we obtain the results as follows:

for $(\frac{1}{10}\gamma u_0 - \lambda_1) = -k_1^2$,

$$\begin{aligned}\psi_1 &= \cosh(k_1\eta_1),\\ \psi_1 &= \sinh(k_1\eta_1),\\ \psi_1 &= c_1\cosh(k_1\eta_1) + c_2\sinh(k_1\eta_1)\end{aligned} \tag{4.308}$$

for $(\frac{1}{10}\gamma u_0 - \lambda_1) = k_1^2$,

$$\begin{aligned}\psi_1 &= \cos(k_1\delta_1),\\ \psi_1 &= \sin(k_1\delta_1),\\ \psi_1 &= c_1\cos(k_1\delta_1) + c_2\sin(k_1\delta_1)\end{aligned} \tag{4.309}$$

where $\eta_1 = x - (16k_1^4 + 4\gamma u_0k_1^2 + \frac{3}{10}\gamma^2u_0^2)t - x_1$, $\delta_1 = x - (16k_1^4 - 4\gamma u_0k_1^2 + \frac{3}{10}\gamma^2u_0^2)t - y_1$, x_1, y_1, c_1, and c_2 are arbitrary constants.

Substituting Eqs. (4.308) and (4.309) into Eq. (4.305), we obtain

$$\begin{aligned} u_1[1] &= u_0 + \frac{20}{\gamma}k_1^2\text{sech}^2(k_1\eta_1), \\ u_2[1] &= u_0 - \frac{20}{\gamma}k_1^2\text{cosech}^2(k_1\eta_1), \\ u_3[1] &= u_0 + \frac{20k_1^2(c_1^2 - c_2^2)}{\gamma[c_1\cosh(k_1\eta_1) + c_2\sinh(k_1\eta_1)]^2}, \end{aligned} \tag{4.310}$$

$$\begin{aligned} u_4[1] &= u_0 - \frac{20}{\gamma}k_1^2\sec^2(k_1\delta_1), \\ u_5[1] &= u_0 - \frac{20}{\gamma}k_1^2\text{cosec}^2(k_1\delta_1), \\ u_6[1] &= u_0 - \frac{20k_1^2(c_1^2 + c_2^2)}{\gamma[c_1\cos(k_1\delta_1) + c_2\sin(k_1\delta_1)]^2} \end{aligned} \tag{4.311}$$

The set of solutions (4.311) are new exact one-soliton solutions and periodic solutions of Eq. (4.66). Here, $u_1[1]$, $u_2[1]$, $u_3[1]$ are new exact one-soliton solutions and $u_4[1]$, $u_5[1]$, $u_6[1]$ are periodic solutions.

4.32.2 *Two-soliton solutions and periodic solutions*

Again applying Darboux transformation on $\psi[1]$ in Eqs. (4.303) and (4.304) defined by

$$\psi[2] = \psi_x[1] - l_1\psi[1],\ l_1 = \partial_x(\ln\psi_2[1]) \tag{4.312}$$

where $\psi_2[1]$ is a non-trivial solution of the Lax pair linear system of Eqs. (4.303) and (4.304), for the spectral parameter $\lambda = \lambda_2$. Then the Lax pair linear system of Eqs. (4.303) and (4.304) is covariant under the Darboux transformation. Thus, $\psi[2]$ obeys the following equations

$$\psi_{xx}[2] = \lambda\psi[2] - \frac{1}{10}\gamma u[2]\psi[2], \tag{4.313}$$

$$\begin{aligned} \psi_t[2] = &-16\psi_{xxxxx}[2] - 4\gamma u[2]\psi_{xxx}[2] - 6\gamma u_x[2]\psi_{xx}[2] \\ &- \left(\frac{3}{10}\gamma^2u^2[2] + 5\gamma u_{xx}[2]\right)\psi_x[2] \\ &- \left(\frac{3}{10}\gamma^2u[2]u_x[2] + \frac{3}{2}\gamma u_{xxx}[2]\right)\psi[2] \end{aligned} \tag{4.314}$$

where the new solution of Eq. (4.66) corresponding to old exact solution $u[1]$ is

$$u[2] = u[1] + \frac{20}{\gamma}\partial_x^2(\ln \psi_2[1]) \tag{4.315}$$

Now, for $\psi_2[1]$, with spectral parameter λ_2, we solve Eq. (4.303) and (4.304) corresponding to known exact solutions of $u_1[1]$, $u_2[1]$, $u_4[1]$ and $u_5[1]$ of Eq. (4.66).

Case-1:

Solving Eqs. (4.303) and (4.304) when known exact solution $u[1] = u_1[1]$, we obtain the results as

for $\lambda = \lambda_2$, $(\frac{1}{10}\gamma u_0 - \lambda_2) = -k_2^2$,

$$\begin{aligned} \psi_2[1] &= k_2 \cosh(k_2\eta_2) - k_1 \tanh(k_1\eta_1)\sinh(k_2\eta_2), \\ \psi_2[1] &= k_2 \sinh(k_2\eta_2) - k_1 \tanh(k_1\eta_1)\cosh(k_2\eta_2), \end{aligned} \tag{4.316}$$

for $\lambda = \lambda_2$, $(\frac{1}{10}\gamma u_0 - \lambda_2) = k_2^2$,

$$\begin{aligned} \psi_2[1] &= k_2 \cos(k_2\delta_2) - k_1 \tanh(k_1\eta_1)\sin(k_2\delta_2), \\ \psi_2[1] &= -k_2 \sin(k_2\delta_2) - k_1 \tanh(k_1\eta_1)\cos(k_2\delta_2) \end{aligned} \tag{4.317}$$

where $\eta_2 = x - (16k_2^4 + 4\gamma u_0 k_2^2 + \frac{3}{10}\gamma^2 u_0^2)t - x_2$, $\delta_2 = x - (16k_2^4 - 4\gamma u_0 k_2^2 + \frac{3}{10}\gamma^2 u_0^2)t - y_2$ and x_2, y_2 are arbitrary constants.

Substituting (4.316) and (4.317) into Eq. (4.315) and simplifying, we obtain

$$\begin{cases} u_7[2] = u_0 + \frac{40(k_2^2-k_1^2)[(k_2^2-k_1^2)+k_1^2\cosh(2k_2\eta_2)+k_2^2\cosh(2k_1\eta_1)]}{\gamma[(k_2-k_1)\cosh(k_2\eta_2+k_1\eta_1)+(k_2+k_1)\cosh(k_2\eta_2-k_1\eta_1)]^2}, \\ u_8[2] = u_0 + \frac{40(k_2^2-k_1^2)[-(k_2^2-k_1^2)+k_1^2\cosh(2k_2\eta_2)-k_2^2\cosh(2k_1\eta_1)]}{\gamma[(k_2-k_1)\sinh(k_2\eta_2+k_1\eta_1)+(k_2+k_1)\sinh(k_2\eta_2-k_1\eta_1)]^2}, \\ u_9[2] = u_0 + \frac{10(k_2^2+k_1^2)[-(k_2^2+k_1^2)+k_1^2\cos(2k_2\delta_2)-k_2^2\cosh(2k_1\eta_1)]}{\gamma[(k_2\cos(k_2\delta_2)\cosh(k_1\eta_1)-k_1\sin(k_2\delta_2)\sinh(k_1\eta_1)]^2}, \\ u_{10}[2] = u_0 + \frac{10(k_2^2+k_1^2)[-(k_2^2+k_1^2)-k_1^2\cos(2k_2\delta_2)-k_2^2\cosh(2k_1\eta_1)]}{\gamma[(k_2\sin(k_2\delta_2)\cosh(k_1\eta_1)+k_1\cos(k_2\delta_2)\sinh(k_1\eta_1)]^2} \end{cases} \tag{4.318}$$

The set of solutions (4.318) are new exact two-soliton solutions and periodic two-soliton solutions of Eq. (4.66). Here, $u_7[2]$ and $u_8[2]$, are new exact two-soliton solutions and $u_9[2]$ and $u_{10}[2]$ are new exact periodic two-soliton solutions.

Case-2:

Solving Eqs. (4.303) and (4.304) together when known exact solution $u[1] = u_2[1]$, we obtain the results as,
for $\lambda = \lambda_2$, $(\frac{1}{10}\gamma u_0 - \lambda_2) = -k_2^2$,

$$\begin{aligned} \psi_2[1] &= k_2 \cosh(k_2\eta_2) - k_1 \coth(k_1\eta_1)\sinh(k_2\eta_2), \\ \psi_2[1] &= k_2 \sinh(k_2\eta_2) - k_1 \coth(k_1\eta_1)\cosh(k_2\eta_2). \end{aligned} \tag{4.319}$$

For $\lambda = \lambda_2$, $(\frac{1}{10}\gamma u_0 - \lambda_2) = k_2^2$,

$$\begin{aligned} \psi_2[1] &= k_2 \cos(k_2\delta_2) - k_1 \coth(k_1\eta_1)\sin(k_2\delta_2), \\ \psi_2[1] &= -k_2 \sin(k_2\delta_2) - k_1 \coth(k_1\eta_1)\cos(k_2\delta_2) \end{aligned} \tag{4.320}$$

Substituting Eqs. (4.319) and (4.320) in Eq. (4.315) and simplifying, we obtain

$$\begin{cases} u_{11}[2] = u_0 + \frac{40(k_2^2-k_1^2)[(k_2^2-k_1^2)-k_1^2\cosh(2k_2\eta_2)-k_2^2\cosh(2k_1\eta_1)]}{\gamma[(k_2-k_1)\cosh(k_2\eta_2+k_1\eta_1)-(k_2+k_1)\cosh(k_2\eta_2-k_1\eta_1)]^2}, \\ u_{12}[2] = u_0 + \frac{40(k_2^2-k_1^2)[-(k_2^2-k_1^2)-k_1^2\cosh(2k_2\eta_2)+k_2^2\cosh(2k_1\eta_1)]}{\gamma[(k_2-k_1)\sinh(k_2\eta_2+k_1\eta_1)-(k_2+k_1)\sinh(k_2\eta_2-k_1\eta_1)]^2}, \\ u_{13}[2] = u_0 + \frac{10(k_2^2+k_1^2)[(k_2^2+k_1^2)-k_1^2\cos(2k_2\delta_2)-k_2^2\cosh(2k_1\eta_1)]}{\gamma[(k_2\cos(k_2\delta_2)\sinh(k_1\eta_1)-k_1\sin(k_2\delta_2)\cosh(k_1\eta_1)]^2}, \\ u_{14}[2] = u_0 + \frac{10(k_2^2+k_1^2)[(k_2^2+k_1^2)+k_1^2\cos(2k_2\delta_2)-k_2^2\cosh(2k_1\eta_1)]}{\gamma[(k_2\sin(k_2\delta_2)\sinh(k_1\eta_1)+k_1\cos(k_2\delta_2)\cosh(k_1\eta_1)]^2} \end{cases} \tag{4.321}$$

The set of solutions (4.321) are new exact two-soliton solutions and periodic two-soliton solutions of Eq. (4.66). Here, $u_{11}[2]$, $u_{12}[2]$ are new exact two-soliton solutions and $u_{13}[2]$, $u_{14}[2]$ are new exact periodic two-soliton solutions.

Case-3:

Solving Eqs. (4.303) and (4.304) together when known exact solution $u[1] = u_4[1]$, we obtain the results as
for$\lambda = \lambda_2$, $(\frac{1}{10}\gamma u_0 - \lambda_2) = -k_2^2$,

$$\begin{aligned} \psi_2[1] &= k_2 \cosh(k_2\eta_2) + k_1 \tan(k_1\delta_1)\sinh(k_2\eta_2), \\ \psi_2[1] &= k_2 \sinh(k_2\eta_2) + k_1 \tan(k_1\delta_1)\cosh(k_2\eta_2) \end{aligned} \tag{4.322}$$

For $\lambda = \lambda_2$, $(\frac{1}{10}\gamma u_0 - \lambda_2) = k_2^2$,

$$\begin{aligned} \psi_2[1] &= k_2 \cos(k_2\delta_2) + k_1 \tan(k_1\delta_1)\sin(k_2\delta_2), \\ \psi_2[1] &= -k_2 \sin(k_2\delta_2) + k_1 \tan(k_1\delta_1)\cos(k_2\delta_2) \end{aligned} \tag{4.323}$$

Substituting Eqs. (4.322) and (4.323) into Eq. (4.315) and simplifying, we obtain

$$\begin{cases} u_{15}[2] = u_0 + \frac{10(k_2^2+k_1^2)[(k_2^2+k_1^2)-k_1^2\cosh(2k_2\eta_2)+k_2^2\cos(2k_1\delta_1)]}{\gamma[(k_2\cosh(k_2\eta_2)\cos(k_1\delta_1)+k_1\sinh(k_2\eta_2)\sin(k_1\delta_1)]^2}, \\ u_{16}[2] = u_0 + \frac{10(k_2^2+k_1^2)[-(k_2^2+k_1^2)-k_1^2\cosh(2k_2\eta_2)-k_2^2\cos(2k_1\delta_1)]}{\gamma[(k_2\sinh(k_2\eta_2)\cos(k_1\delta_1)+k_1\cosh(k_2\eta_2)\sin(k_1\delta_1)]^2}, \\ u_{17}[2] = u_0 + \frac{40(k_2^2-k_1^2)[-(k_2^2-k_1^2)-k_1^2\cos(2k_2\delta_2)-k_2^2\cos(2k_1\delta_1)]}{\gamma[(k_2+k_1)\cos(k_2\delta_2+k_1\delta_1)+(k_2-k_1)\cos(k_2\delta_2-k_1\delta_1)]^2}, \\ u_{18}[2] = u_0 + \frac{40(k_2^2-k_1^2)[-(k_2^2-k_1^2)+k_1^2\cos(2k_2\delta_2)-k_2^2\cos(2k_1\delta_1)]}{\gamma[(k_2-k_1)\sin(k_2\delta_2+k_1\delta_1)+(k_2+k_1)\sin(k_2\delta_2-k_1\delta_1)]^2} \end{cases} \tag{4.324}$$

The set of solutions (4.324) are new exact periodic two-soliton solutions and periodic solutions of Eq. (4.66). Here, $u_{15}[2]$, $u_{16}[2]$ are new exact periodic two-soliton solutions and $u_{17}[2]$, $u_{18}[2]$ are new exact periodic solutions.

Case-4:

Solving Eqs. (4.303) and (4.304) together when known exact solution $u[1] = u_5[1]$, we obtain the results as,

for $\lambda = \lambda_2$, $(\frac{1}{10}\gamma u_0 - \lambda_2) = -k_2^2$,

$$\begin{aligned} \psi_2[1] &= k_2\cosh(k_2\eta_2) - k_1\cot(k_1\delta_1)\sinh(k_2\eta_2), \\ \psi_2[1] &= k_2\sinh(k_2\eta_2) - k_1\cot(k_1\delta_1)\cosh(k_2\eta_2) \end{aligned} \tag{4.325}$$

For $\lambda = \lambda_2$, $(\frac{1}{10}\gamma u_0 - \lambda_2) = k_2^2$,

$$\begin{aligned} \psi_2[1] &= k_2\cos(k_2\delta_2) - k_1\cot(k_1\delta_1)\sin(k_2\delta_2), \\ \psi_2[1] &= -k_2\sin(k_2\delta_2) - k_1\cot(k_1\delta_1)\cos(k_2\delta_2) \end{aligned} \tag{4.326}$$

Substituting Eqs. (4.325) and (4.326) in Eq. (4.315) and simplifying, we obtain

$$\begin{cases} u_{19}[2] = u_0 + \frac{10(k_2^2+k_1^2)[(k_2^2+k_1^2)-k_1^2\cosh(2k_2\eta_2)-k_2^2\cos(2k_1\delta_1)]}{\gamma[(k_2\cosh(k_2\eta_2)\sin(k_1\delta_1)-k_1\sinh(k_2\eta_2)\cos(k_1\delta_1)]^2}, \\ u_{20}[2] = u_0 + \frac{10(k_2^2+k_1^2)[-(k_2^2+k_1^2)-k_1^2\cosh(2k_2\eta_2)+k_2^2\cos(2k_1\delta_1)]}{\gamma[(k_2\sinh(k_2\eta_2)\sin(k_1\delta_1)-k_1\cosh(k_2\eta_2)\cos(k_1\delta_1)]^2}, \\ u_{21}[2] = u_0 + \frac{40(k_2^2-k_1^2)[-(k_2^2-k_1^2)-k_1^2\cos(2k_2\delta_2)+k_2^2\cos(2k_1\delta_1)]}{\gamma[(k_2-k_1)\sin(k_2\delta_2+k_1\delta_1)-(k_2+k_1)\sin(k_2\delta_2-k_1\delta_1)]^2}, \\ u_{22}[2] = u_0 + \frac{40(k_2^2-k_1^2)[-(k_2^2-k_1^2)+k_1^2\cos(2k_2\delta_2)+k_2^2\cos(2k_1\delta_1)]}{\gamma[(k_2+k_1)\cos(k_2\delta_2-k_1\delta_1)-(k_2-k_1)\cos(k_2\delta_2+k_1\delta_1)]^2} \end{cases} \tag{4.327}$$

The set of solutions (4.327) are new exact periodic two-soliton solutions and periodic solutions of Eq. (4.66). Here, $u_{19}[2]$, $u_{20}[2]$ are new exact periodic two-soliton solutions and $u_{21}[2]$, $u_{22}[2]$ are new exact periodic solutions.

4.32.3 *Three-soliton solutions and periodic solutions*

Three folded DT on $\psi[2]$ in Eqs. (4.313) and (4.314) defined by

$$\psi[3] = \psi_x[2] - l_2\psi[2], \quad l_2 = \partial_x(\ln \psi_3[2]) \tag{4.328}$$

where $\psi_3[2]$ is a non-trivial solution of the Lax pair linear system of Eqs. (4.313) and (4.314), for the spectral parameter $\lambda = \lambda_3$. Then the Lax pair linear system of Eqs. (4.313) and (4.314) is covariant under the Darboux transformation. Thus, $\psi[3]$ obeys the following equations

$$\psi_{xx}[3] = \lambda\psi[3] - \frac{1}{10}\gamma u[3]\psi[3], \tag{4.329}$$

$$\begin{aligned}\psi_t[3] &= -16\psi_{xxxxx}[3] - 4\gamma u[3]\psi_{xxx}[3] - 6\gamma u_x[3]\psi_{xx}[3] \\ &\quad - \left(\frac{3}{10}\gamma^2 u^2[3] + 5\gamma u_{xx}[3]\right)\psi_x[3] \\ &\quad - \left(\frac{3}{10}\gamma^2 u[3]u_x[3] + \frac{3}{2}\gamma u_{xxx}[3]\right)\psi[3]\end{aligned} \tag{4.330}$$

where the new solution of Eq. (4.66) corresponding to old solution $u[2]$ of Eq. (4.66) is

$$u[3] = u[2] + \frac{20}{\gamma}\partial_x^2(\ln \psi_3[2]) \tag{4.331}$$

Using the three folded DT to get the new solution of Eq. (4.66) from old solution $u[2]$, we can solve Eqs. (4.313) and (4.314) for $\psi_3[2]$ at $\lambda = \lambda_3$ with different values of u[2], which was previously obtained. Then we get the new solution $u[3]$ of Eq. (4.66) by Eq. (4.331). So by the repeated DT on ψ, we are able to compute three-soliton, periodic solution and more generally n-soliton solution.

4.33 MKdV Equation Using Matrix Form

Step 1: Consider the mKdV equation

$$u_t + 6u^2u_x + u_{xxx} = 0 \tag{4.332}$$

and

$$\phi_x = U\phi, \quad \phi_t = V\phi \tag{4.333}$$

where U and V are 2×2 matrix and entries of U and V are $u(x,t)$ and λ (a parameter).

Step 2: Find the fundamental solution of the system of Eq. (4.333) taking $u = 0$. Now,

$$\phi(x,t,\lambda) = \begin{pmatrix} \phi_{11}(x,t,\lambda) & \phi_{12}(x,t,\lambda) \\ \phi_{21}(x,t,\lambda) & \phi_{22}(x,t,\lambda) \end{pmatrix}$$

is a fundamental solution of the system of Eq. (4.333).

Step 3: Chose a σ for μ_1 and $\lambda = \lambda_1$, like

$$\sigma = \frac{\phi_{22}(x,t,\lambda_1) + \mu_1\phi_{21}(x,t,\lambda_1)}{\phi_{12}(x,t,\lambda_1) + \mu_1\phi_{11}(x,t,\lambda_1)}$$

To determine D-matrix, $D = \lambda I - S$, we find the matrix $S = H\Lambda H^{-1}$ only.

Step 4: For finding H and Λ matrix:

$\begin{pmatrix} \alpha \\ \beta \end{pmatrix} = \begin{pmatrix} \mu_1\phi_{11}(x,t,\lambda_1) + \phi_{12}(x,t,\lambda_1) \\ \mu_1\phi_{21}(x,t,\lambda_1) + \phi_{22}(x,t,\lambda_1) \end{pmatrix}$ is a soltion of Eq. (4.333) for $\lambda = \lambda_1$.

$\begin{pmatrix} -\beta \\ \alpha \end{pmatrix} = \begin{pmatrix} -\mu_1\phi_{21}(x,t,\lambda_1) - \phi_{22}(x,t,\lambda_1) \\ \mu_1\phi_{11}(x,t,\lambda_1) + \phi_{12}(x,t,\lambda_1) \end{pmatrix}$ is a soltion of Eq. (4.333) for $\lambda = -\lambda_1$.

Now, H and Λ is chosen as

$$H = \begin{pmatrix} \alpha & -\beta \\ \beta & \alpha \end{pmatrix} \quad \text{and} \quad \Lambda = \begin{pmatrix} \lambda_1 & 0 \\ 0 & -\lambda_1 \end{pmatrix}$$

Step 5: Now, consider $\phi'(x,t,\lambda) = D\phi$ and ϕ' satisfies the system of equation

$$\phi'_x = U'\phi', \quad \phi'_t = V'\phi' \tag{4.334}$$

where $U' = DUD^{-1} + D_xD^{-1}$, $V' = DVD^{-1} + D_tD^{-1}$, with $u' = u + \frac{4\lambda_1\sigma}{1+\sigma^2}$.

Step 6: This DT (for the mKdV equation) in matrix form can be done successively and purely algebraiacally as

$$(u,\phi) \to (u',\phi') \to (u'',\phi'') \to \cdots$$

4.34 Solution of mKdV Equation

The well-known mKdV equation is

$$u_t + 6u^2 u_x + u_{xxx} = 0 \tag{4.335}$$

Lax pair of Eq. (4.335) is given as follows:

$$\begin{aligned} \phi_x = U\phi &= \begin{pmatrix} \lambda & u \\ -u & -\lambda \end{pmatrix} \phi \\ \phi_t = V\phi &= \begin{pmatrix} A & B \\ C & -A \end{pmatrix} \phi \end{aligned} \tag{4.336}$$

where $A = -4\lambda^3 - 2u^2\lambda$, $B = -4u\lambda^2 - 2u_x\lambda - 2u^3 - u_{xx}$, $C = 4u\lambda^2 - 2u_x\lambda + 2u^3 + u_{xx}$ and $\phi_{xt} = \phi_{tx}$.

Here, U and V satisfies the relation

$$U_t - V_x + [U, V] = 0$$

Now,

$$\phi(x, t, \lambda) = \begin{pmatrix} e^{\lambda x - 4\lambda^3 t} & 0 \\ 0 & e^{-\lambda x + 4\lambda^3 t} \end{pmatrix} \tag{4.337}$$

is the fundamental solution of the system of Eq. (4.336) after taking $u = 0$.

For $\lambda = \lambda_1 \neq 0$ and $\mu_1 = \exp(2\alpha_1 > 0)$, we choose

$$\sigma = \sigma_1 = \exp(-2\lambda_1 x + 8\lambda_1^3 t - 2\alpha_1) = \exp(-\nu_1),$$

where it is clear that $\nu_1 = 2\lambda_1 x - 8\lambda_1^3 t + 2\alpha_1$.

Further, $\begin{pmatrix} \alpha \\ \beta \end{pmatrix} = \begin{pmatrix} \exp(\lambda_1 x - 4\lambda_1^3 t + \alpha_1) \\ \exp(-\lambda_1 x + 4\lambda_1^3 t) \end{pmatrix}$ is a soltion of Eq. (4.335) for $\lambda = \lambda_1$.

$\begin{pmatrix} -\beta \\ \alpha \end{pmatrix} = \begin{pmatrix} -\exp(-\lambda_1 x + 4\lambda_1^3 t) \\ \exp(\lambda_1 x - 4\lambda_1^3 t + \alpha_1) \end{pmatrix}$ is a soltion of Eq. (4.335) for $\lambda = -\lambda_1$

$$H = \begin{pmatrix} \alpha & -\beta \\ \beta & \alpha \end{pmatrix} \quad \text{and} \quad \Lambda = \begin{pmatrix} \lambda_1 & 0 \\ 0 & -\lambda_1 \end{pmatrix}$$

$$S = H\Lambda H^{-1} = \frac{\lambda_1}{\cosh \nu_1} \begin{pmatrix} \sinh \nu_1 & 1 \\ 1 & -\sinh \nu_1 \end{pmatrix}$$

Hence, the required Darboux matrix is

$$D = \lambda I - \frac{\lambda_1}{\cosh \nu_1}\begin{pmatrix} \sinh \nu_1 & 1 \\ 1 & -\sinh \nu_1 \end{pmatrix} \tag{4.338}$$

where I is a 2×2 identity matrix and $\nu_1 = 2\lambda_1 x - 8\lambda_1^3 t + 2\alpha_1$.

Now, the new solution $\phi'(x,t,\lambda) = D\phi$ satisfies the system of equation

$$\begin{aligned} \phi'_x &= U'\phi', \\ \phi'_t &= V'\phi'. \end{aligned} \tag{4.339}$$

where

$$U' = \begin{pmatrix} \lambda & u' \\ -u' & -\lambda \end{pmatrix},$$

$$V' = \begin{pmatrix} -4\lambda^3 - 2u'^2\lambda & -4u'\lambda^2 - 2u'_x\lambda - 2u'^3 - u'_{xx} \\ 4u'\lambda^2 - 2u'_x\lambda + 2u'^3 + u'_{xx} & 4\lambda^3 + 2u'^2\lambda \end{pmatrix}$$

with $u' = \frac{4\lambda_1\sigma_1}{1+\sigma_1^2} = 2\lambda_1 \text{sech}(2\lambda_1 x - 8\lambda_1^3 t + 2\alpha_1)$.

4.34.1 *Gardner equation*

Consider the Gardner equation

$$u_t + auu_x + bu^2u_x + cu_{xxx} = 0 \tag{4.340}$$

where x denotes the scaled space variable along the direction of the wave mode, t denotes the scaled time, and $u(x,t)$ represents the amplitude of the wave model. The coefficients of the quadratic and cubic nonlinear components are a and b, respectively, while c reflects a dispersive effect. a, b, c are constant and $a = k_1c$, $b = k_2c$ (k_1, k_2 are non-zero constants).

Let's examine the Lax pair of (4.340) first to verify its integrability. Using Ablowitz–Kaup–Newell–Segur (AKNS) approach, Lax pair of (4.340) is given by

$$\psi_x = U\psi = \begin{pmatrix} \lambda & u \\ -\frac{1}{6}(k_2u + k_1) & -\lambda \end{pmatrix}\psi, \tag{4.341}$$

$$\psi_t = V\psi = \begin{pmatrix} A & B \\ C & -A \end{pmatrix}\psi \tag{4.342}$$

where $A = -4c\lambda^3 - \frac{1}{3}c(k_2u^2 + k_1u)\lambda - \frac{1}{6}k_1cu_x$, $B = -(4cu\lambda^2 + 2cu_x\lambda + \frac{1}{3}k_2cu^3 + \frac{1}{3}k_1cu^2 + cu_{xx})$, $C = \frac{2}{3}c(k_1 + k_2u)\lambda^2 - \frac{1}{3}k_2cu_x\lambda + \frac{1}{18}k_2^2cu^3 + \frac{1}{9}k_1k_2cu^2 + \frac{1}{18}k_1^2cu + \frac{1}{6}k_2cu_{xx}$.

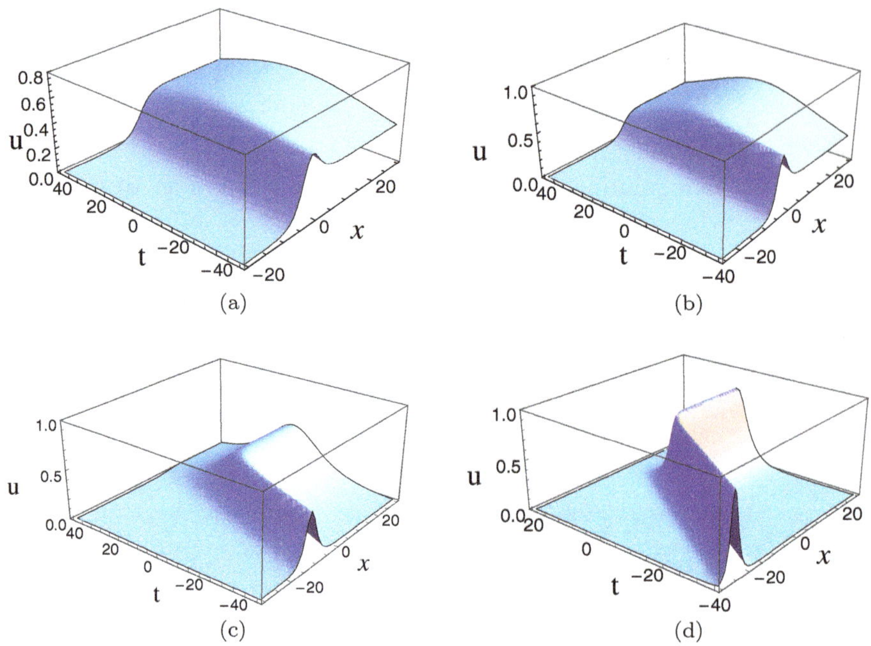

Figure 4.18. 3D profiles of the solution (4.345), (a) when $\lambda_1 = 0.25; c = 0.25; \mu_1 = 0.5; k_1 = 0.5$; (b) when $\lambda_1 = 0.3; c = 0.25; \mu_1 = 0.5; k_1 = 0.5$; (c) when $\lambda_1 = 0.25; c = 1; \mu_1 = 0.5; k_1 = 0.5$; (d) when $\lambda_1 = 0.4; c = 1; \mu_1 = 0.5; k_1 = 0.5$.

λ is a parameter independent of x and t and $\psi = \psi(x, t, \lambda)$. The Lax equation is

$$U_t - V_x + [U, V] = 0$$

which is equivalent to Eq. (4.340).

The gauge transformation is

$$\bar{\psi}(x, t, \lambda) = D(x, t, \lambda)\psi(x, t, \lambda) \tag{4.343}$$

where

$$D(x, t, \lambda) = \lambda I_2 - \frac{\lambda_1}{1 + \gamma^2}\begin{pmatrix} 1 - \gamma^2 & 2\gamma \\ 2\gamma & \gamma^2 - 1 \end{pmatrix} \tag{4.344}$$

D is the Darboux matrix of the transformation.

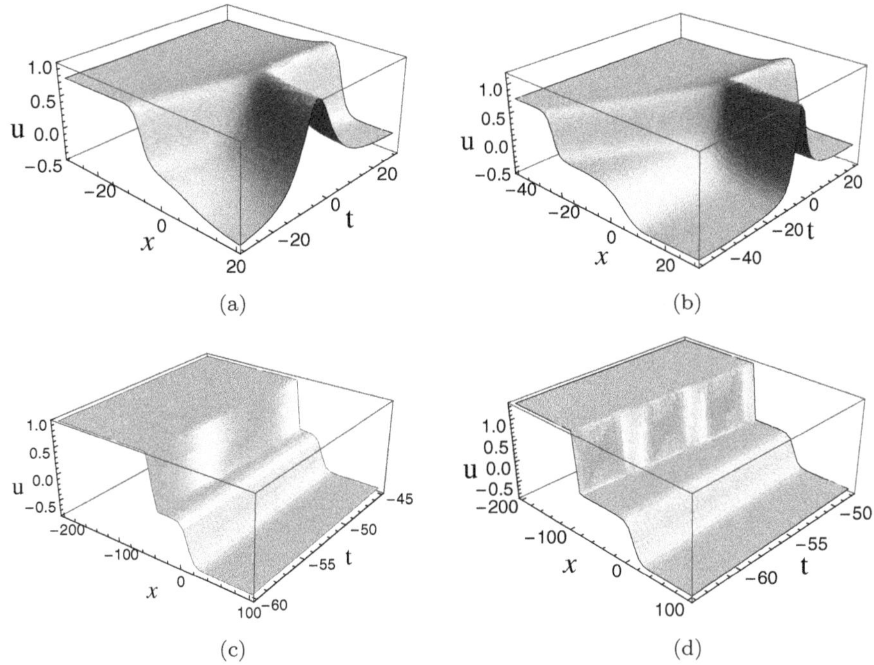

Figure 4.19. 3D Interaction of the two-soliton via solution (4.346) for (a) when $\lambda_1 = 0.5; c = 0.43; \mu_1 = 0.5; k_1 = 0.5; \mu_2 = 0.5; \lambda_2 = 0.4$; (b) when $\lambda_1 = 0.12; c = 0.8; \mu_1 = 0.5; k_1 = 0.25; \mu_2 = 0.5; \lambda_2 = 0.4$; (c) when $\lambda_1 = 0.1; c = 0.8; \mu_1 = 0.5; k_1 = 0.5; \mu_2 = 0.5; \lambda_2 = 0.5$; (d) when $\lambda_1 = 0.1; c = 0.7; \mu_1 = 0.5; k_1 = 0.5; \mu_2 = 0.5; \lambda_2 = 0.7$.

Using the above transformation one-soliton solution u_1 of Eq. (4.340) is given by

$$u_1 = \frac{4\lambda_1(e^{-v_1} + e^{-v_2})}{1 + e^{-2v_1} + e^{-2v_2} + 2e^{-v_1-v_2}} \tag{4.345}$$

where $v_1 = 2\lambda_1 x - 8c\lambda_1^3 t - 2\mu_1$ and $v_2 = -\frac{2}{3}ck_1\lambda_1^2 t - 4c\lambda_1^3 t$.

In the same way, two-soliton solution of Eq. (4.340) is given by

$$u_2 = \frac{4\lambda_1(e^{-v_1} + e^{-v_2})}{1 + e^{-2v_1} + e^{-2v_2} + 2e^{-v_1-v_2}} + \frac{4\lambda_2\bar{\gamma}_2}{1 + \bar{\gamma}_2^2} \tag{4.346}$$

where $\bar{\gamma}_2 = \frac{\bar{X}}{\bar{Y}}$, $\bar{X} = (\lambda_2 + \lambda_1) + (\lambda_2 - \lambda_1)(e^{-2v_1} + e^{-2v_2} + 2e^{-v_1-v_2}) - 2\lambda_1 e^{-2v_3}(e^{-v_1} + e^{-v_2}) + e^{-2v_4}(\lambda_2 + \lambda_1) + e^{-2v_4}(\lambda_2 - \lambda_1)(e^{-2v_1} + e^{-2v_2} + 2e^{-v_1-v_2})$, $\bar{Y} = -2\lambda_1(e^{-v_1} + e^{-v_2}) + e^{-2v_3}(\lambda_2 - \lambda_1) + e^{-2v_3}(\lambda_2 + \lambda_1)(e^{-2v_1} + e^{-2v_2} + 2e^{-v_1-v_2}) - 2\lambda_1 e^{-2v_4}(e^{-v_1} + e^{-v_2})$, $v_3 = \mu_2 + 4c\lambda_2^3 t - \lambda_2 x$, $v_4 = \mu_2 + 2c\lambda_2^3 t - \frac{1}{3}ck_1\lambda_2^2 t - \lambda_2 x$.

4.35 Bäcklund Transformation (BT)

A. V. Bäcklund introduced the Bäcklund transformation (BT) in 1880, and its origin also lies in differential geometry. It is a transformation between surfaces with three dimensions. From a known trivial solution, the BT can be found to generate a series of solutions to a PDE. BT further transforms a NLPDE into another PDE. Hence, the compatibility of the first and second PDEs becomes compulsory. The Miura transformation and the Gardner transformation are two significant examples of BT. In the theory of integrable systems, they are seen as relations between solutions of the same PDEs (auto BT) or as relations between solutions of two different PDEs (hetro BT). Therefore, BT does not depend on the definition of the Lax pair, yet both DTs and BTs serve the same purpose; they are used to construct non-trivial solutions starting from trivial ones.

4.35.1 *Definition*

Consider two functions u and v, which satisfy two differential equations

$$P[u] = 0 \tag{4.347}$$

$$Q[v] = 0 \tag{4.348}$$

where P and Q are differential operators.

If there exists a pair of relations

$$R_1[u, v] = 0, \quad R_2[u, v] = 0 \tag{4.349}$$

such that if $P[u] = 0$, then (4.349) can be solved for v to give a solution of (4.348). On the other hand, if $Q[v] = 0$, then (4.349) can be solved for u to give a solution of (4.347). Then (4.349) is called a Bäcklund transform (BT). If $P = Q$, (4.349) is called an Auto Bäcklund transform (a-BT). If $P \neq Q$ then (4.349) is called a hetero Bäcklund transform. (4.347) and (4.348) are called the integrability conditions for the BT relations given by (4.349). (4.349) can be integrated for v if the integrability condition $P[u] = 0$ is satisfied. (4.349) can be integrated for u if the integrability condition $Q[v] = 0$ is satisfied.

4.35.2 *Auto-Bäcklund transformation*

An auto-Bäcklund transformation is a special type of BT where the transformation relates different solutions of the same equation. The classic example of an auto-Bäcklund transformation for the Sine-Gordon equation. Also, the simplest examples of BT are the well-known Cauchy–Riemann equations in complex analysis for the analyticity of a complex function, $f = u(x,t) + iv(x,t)$. Functions $u = u(x,t)$ and $v(x,t)$ satisfy the Laplace equation and are harmonic, namely,

$$\frac{\partial^2 u}{\partial x^2} + \frac{\partial^2 u}{\partial t^2} = 0, \tag{4.350}$$

$$\frac{\partial^2 v}{\partial x^2} + \frac{\partial^2 v}{\partial t^2} = 0 \tag{4.351}$$

if the following Cauchy–Riemann (C–R) equation holds.

$$\frac{\partial u}{\partial x} = \frac{\partial v}{\partial t}, \tag{4.352}$$

$$\frac{\partial u}{\partial t} = -\frac{\partial v}{\partial x} \tag{4.353}$$

It can be easily seen that $u_x = v_t$, $u_{xx} = v_{tx} = (v_x)_t = -u_{tt}$. Therefore, $u_{xx} + u_{tt} = 0$. Similarly, it can be seen that (4.353) produces (4.351). As (4.350) and (4.351) are the same Laplace equation. So, (4.352) and (4.353) are called auto-Bäcklund transformation. One can solve the Laplace equation by solution (4.352) and (4.353) without solving the Laplace equation. If u is known, then v can be obtained easily. Similarly, if v is known, then for a similar complex function $f_1 = v(x,t) + iw(x,t)$, $w(x,t)$ can be obtained, and w also satisfies the Laplace equation. In the same way, one can obtain infinite solutions of the Laplace equation.

Equations (4.352) and (4.353) constitute an auto-BT for Eqs. (4.350) and (4.351) and can be used to construct solutions of the same equations, starting with known ones. Hence, C–R equation is an example of auto-BT.

For instance, consider the simple solution $v(x,t) = xt$. Then, according to a second solution of Eqs. (4.352) and (4.353), u has to satisfy $\frac{\partial u}{\partial x} = x$ and $\frac{\partial u}{\partial t} = -t$, given by $u = \frac{1}{2}(x^2 - t^2)$.

4.35.3 *Hetero-Bäcklund transformation*

A hetero-Bäcklund transformation links solutions of two different PDEs. This contrasts with an auto-Bäcklund transformation, where the solutions pertain to the same equation. A hetero-Bäcklund transformation is particularly useful in demonstrating connections between different equations or mapping one integrable system into another. Let us consider the KdV equation

$$\frac{\partial u}{\partial t} - 6u\frac{\partial u}{\partial x} + \frac{\partial^3 u}{\partial x^3} = 0 \tag{4.354}$$

and the potential KdV equation is given by

$$\frac{\partial v}{\partial t} - a\left(\frac{\partial v}{\partial x}\right)^2 + \frac{\partial^3 v}{\partial x^3} = 0 \tag{4.355}$$

Let us consider the transformation

$$\frac{\partial v}{\partial x} = \frac{3u}{a}, \quad \frac{\partial v}{\partial t} = -\frac{3}{a}\frac{\partial^2 v}{\partial x^2} + \frac{9u^2}{a} \tag{4.356}$$

From (4.356), we can obtain $u = \frac{a}{3}\frac{\partial v}{\partial x}$ so $\frac{\partial u}{\partial x} = \frac{a}{3}\frac{\partial^2 v}{\partial x^2}$, $\frac{\partial^3 u}{\partial x^3} = \frac{a}{3}\frac{\partial^4 v}{\partial x^4}$, $\frac{\partial u}{\partial t} = \frac{a}{3}\frac{\partial^2 v}{\partial t \partial x}$. Therefore, Eq. (4.354) becomes

$$\begin{aligned}
&\frac{a}{3}\frac{\partial^2 v}{\partial t \partial x} - 6\left(\frac{a}{3}\frac{\partial v}{\partial x}\right)\left(\frac{a}{3}\frac{\partial^2 v}{\partial x^2}\right) + \frac{a}{3}\frac{\partial^4 v}{\partial x^4} \\
\implies &\frac{\partial v}{\partial t} - a\left(\frac{\partial v}{\partial x}\right)^2 + \frac{\partial^3 v}{\partial x^3} = 0
\end{aligned}$$

Also one can obtain obtain KdV equation from potential KdV equation. Now, $\frac{\partial v}{\partial x} = \frac{3u}{a}$, $\frac{\partial v}{\partial t} = -\frac{3}{a}\frac{\partial^2 u}{\partial x^2} + \frac{9u^2}{a}$.

$$\begin{aligned}
&\frac{\partial^2 v}{\partial t \partial x} = \frac{a}{3}\frac{\partial u}{\partial t} \\
\implies &\frac{\partial}{\partial x}\left(\frac{\partial v}{\partial t}\right) = \frac{a}{3}\frac{\partial u}{\partial t} \\
\implies &\frac{\partial}{\partial x}\left(-\frac{3}{a}\frac{\partial^2 u}{\partial x^2} + \frac{9u^2}{a}\right) = \frac{a}{3}\frac{\partial u}{\partial t} \\
\implies &-\frac{3}{a}\frac{\partial^3 u}{\partial x^3} + \frac{9}{a}2a\frac{\partial u}{\partial x} = \frac{a}{3}\frac{\partial u}{\partial t} \\
\implies &\frac{\partial u}{\partial t} - 6u\frac{\partial u}{\partial x} + \frac{\partial^3 u}{\partial x^3} = 0
\end{aligned}$$

which is a KdV equation. As the KdV equation (4.354) can be solved for u then the potential KdV equation (4.355) can be solved by using (4.356). This is the beauty of BT. This BT is called *Hetero BT* which gives the

KdV equation for the variable u and a potential KdV equation for the variable v. So starting from a trivial solution of the KdV equation, one can obtain the solution of the potential KdV equation, and repeatedly this process is infinite solutions are obtained.

4.36 Soliton Solution of the Sine-Gordon Equation

Let us consider the Sine-Gordon equation

$$\frac{\partial^2 u}{\partial x \partial t} = \sin(u), \quad u = u(x,t) \tag{4.357}$$

Let us consider the relationships between functions $u = u(x,t)$ and $v = v(x,t)$ as

$$\frac{1}{2}\left(\frac{\partial u}{\partial t} + \frac{\partial v}{\partial t}\right) = \alpha \sin\left(\frac{u-v}{2}\right) \tag{4.358}$$

$$\frac{1}{2}\left(\frac{\partial u}{\partial x} - \frac{\partial v}{\partial x}\right) = \frac{1}{\alpha} \sin\left(\frac{u+v}{2}\right) \tag{4.359}$$

where α is a real non-zero constant. Differentiating Eq. (4.358) with respect to x and Eq. (4.359) with respect to t, we obtain

$$\frac{1}{2}\left(\frac{\partial^2 u}{\partial x \partial t} + \frac{\partial^2 v}{\partial x \partial t}\right) = \cos\left(\frac{u-v}{2}\right) \sin\left(\frac{u+v}{2}\right), \tag{4.360}$$

$$\frac{1}{2}\left(\frac{\partial^2 u}{\partial t \partial x} - \frac{\partial^2 v}{\partial t \partial x}\right) = \cos\left(\frac{u+v}{2}\right) \sin\left(\frac{u-v}{2}\right) \tag{4.361}$$

Considering the compatibilities of the above equations, namely $\frac{\partial^2 u}{\partial x \partial t} = \frac{\partial^2 u}{\partial t \partial x}$ and $\frac{\partial^2 v}{\partial x \partial t} = \frac{\partial^2 v}{\partial t \partial x}$, and adding Eqs. (4.360) and (4.361) one can show that u obeys Eq. (4.357). The same holds true for v as well after Eqs. (4.360) and (4.361) are subtracted. As a result, the auto Bäcklund transformation for the Sine-Gordon equation is represented by relations (4.358) and (4.359).

Clearly, $v = 0$ is a solution to Eq. (4.357). Now, putting the value $v = 0$ in (4.358) and (4.359), we obtain

$$\frac{\partial u}{\partial x} = \frac{2}{\alpha} \sin\left(\frac{u}{2}\right) \tag{4.362}$$

$$\frac{\partial u}{\partial t} = 2\alpha \sin\left(\frac{u}{2}\right) \tag{4.363}$$

Integrating both Eqs. (4.362) and (4.363), we obtain

$$2\log\left(\tan\frac{u}{4}\right)+\psi(t)=\frac{2x}{\alpha}, \tag{4.364}$$

$$2\log\left(\tan\frac{u}{4}\right)+\phi(x)=2\alpha t \tag{4.365}$$

subtracting between Eqs. (4.364), (4.365), we obtain

$$\psi(t)-\frac{2x}{\alpha}=\phi(x)-2\alpha t \tag{4.366}$$

From which is follows that for compatibility, we must have

$$\psi(t)=K-2\alpha t, \quad \phi(x)=K-\frac{2x}{\alpha} \tag{4.367}$$

where K is arbitrary constant. Thus, on adding Eqs. (4.364) and (4.365) and using the result (4.367), we get

$$4\log\left(\tan\frac{u}{4}\right)+\phi(x)+\psi(t)=2\alpha t+\frac{2x}{\alpha}$$

$$\Longrightarrow 4\log\left(\tan\frac{u}{4}\right)+2K-2\alpha t-\frac{2x}{\alpha}=2\alpha t+\frac{2x}{\alpha}$$

$$\Longrightarrow \tan\frac{u}{4}=C\exp\left(\alpha t+\frac{x}{\alpha}\right)$$

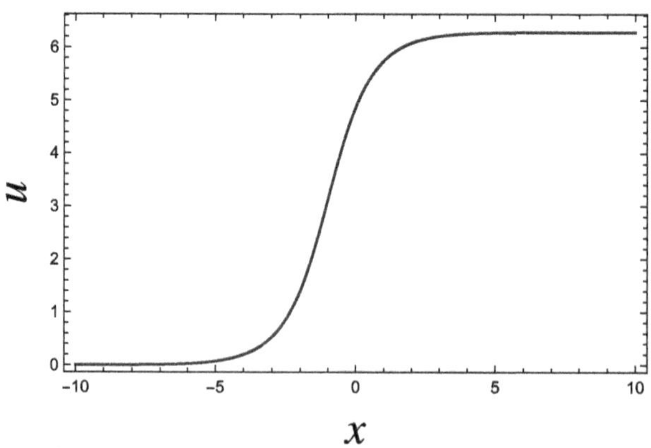

Figure 4.20. Solution of Eq. (4.368).

Hence, the new solution to the Sine-Gordon equation (4.357) can be written as

$$u = 4\arctan\left[C\exp\left(\alpha t + \frac{x}{\alpha}\right)\right] \tag{4.368}$$

where $C = \exp(-\frac{K}{2})$. Note that we can improve the solution further.

4.37 Liouville Equation

Let us consider the Liouville equation

$$\frac{\partial^2 u}{\partial x \partial t} = \exp(u), \quad u = u(x,t) \tag{4.369}$$

Now, considering the following relations between functions $u = u(x,t)$, and $v = v(x,t)$

$$\frac{\partial u}{\partial x} + \frac{\partial v}{\partial x} = \sqrt{2}\exp\left(\frac{u-v}{2}\right), \tag{4.370}$$

$$\frac{\partial u}{\partial t} - \frac{\partial v}{\partial t} = \sqrt{2}\exp\left(\frac{u+v}{2}\right) \tag{4.371}$$

Differentiating Eq. (4.370) with respect to t, and Eq. (4.371) with respect to x, we obtain

$$\frac{\partial^2 u}{\partial t \partial x} + \frac{\partial^2 v}{\partial t \partial x} = \exp(u) \tag{4.372}$$

$$\frac{\partial^2 u}{\partial x \partial t} - \frac{\partial^2 v}{\partial x \partial t} = \exp(u) \tag{4.373}$$

The above equations are compatible, namely $\frac{\partial^2 u}{\partial x \partial t} = \frac{\partial^2 u}{\partial t \partial x}$ and $\frac{\partial^2 v}{\partial x \partial t} = \frac{\partial^2 v}{\partial t \partial x}$. We conclude that u obeys the Liouville equation (4.369) by adding Eqs. (4.372) and (4.373). Additionally, v follows the following different equation after subtraction of Eqs. (4.372) and (4.373), thus

$$\frac{\partial^2 v}{\partial x \partial t} = 0 \tag{4.374}$$

Hence, relations (4.370) and (4.371) are hetero-Bäcklund transformation for Liouville equation.

In this case, the BT is connected to different PDEs. The solution of Eq. (4.374) is easily obtained, and one can exploit it to find the solutions of the Liouville equation. Let the solution of Eqs. (4.369) and (4.374) is

$$v = \phi(x) + \psi(t) \tag{4.375}$$

where ϕ and ψ are two arbitrary functions of x and t, respectively. Substituting $v = \phi(x) + \psi(t)$ in (4.370), we get

$$\begin{aligned}
&\frac{\partial u}{\partial x} + \frac{\partial \phi}{\partial x} = \sqrt{2}\exp\left(\frac{u - \phi - \psi}{2}\right) \\
&\Longrightarrow \frac{\partial u}{\partial x} + \frac{\partial \phi}{\partial x} = \sqrt{2}\exp\left(\frac{u + \phi - \psi - 2\phi}{2}\right) \\
&\Longrightarrow \exp\left[-\frac{(u+\phi)}{2}\right]\left(\frac{\partial u}{\partial x} + \frac{\partial \phi}{\partial x}\right) = \sqrt{2}\exp\left(-\phi - \frac{\psi}{2}\right).
\end{aligned} \tag{4.376}$$

By integrating Eq. (4.376) with respect to x, we obtained

$$\begin{aligned}
&-2\exp\left[-\frac{(u+\phi)}{2}\right] = \sqrt{2}\exp\left(-\frac{\psi}{2}\right)\left[\int \exp(-\phi)dx + f(t)\right] \\
&\Longrightarrow \exp\left[-\frac{(u+\phi-\psi)}{2}\right] = -\frac{1}{\sqrt{2}}\left[\int \exp(-\phi)dx + f(t)\right] \\
&\Longrightarrow \exp\left(-\frac{\phi}{2}\right)\exp\left[-\frac{(u-\psi)}{2}\right] = -\frac{1}{\sqrt{2}}\left[\int \exp(-\phi)dx + f(t)\right]
\end{aligned} \tag{4.377}$$

Now, differentiating Eq. (4.377) with respect to t, we get

$$\exp\left(-\frac{\phi}{2}\right)\exp\left[-\frac{(u-\psi)}{2}\right]\left(\frac{\partial u}{\partial t} - \frac{\partial \psi}{\partial t}\right) = \sqrt{2}f'(t) \tag{4.378}$$

Again substituting $v = \phi(x) + \psi(t)$ in (4.371), we get,

$$\begin{aligned}
&\frac{\partial u}{\partial t} - \frac{\partial \psi}{\partial t} = \sqrt{2}\exp\left(\frac{u + \phi + \psi}{2}\right) \\
&\Longrightarrow \frac{\partial u}{\partial t} - \frac{\partial \psi}{\partial t} = \sqrt{2}\exp\left(\frac{u + \phi - \psi + 2\psi}{2}\right) \\
&\Longrightarrow \exp\left(-\frac{\phi}{2}\right)\exp\left[-\frac{(u-\psi)}{2}\right]\left(\frac{\partial u}{\partial t} - \frac{\partial \psi}{\partial t}\right) = \sqrt{2}\exp(\psi)
\end{aligned} \tag{4.379}$$

From Eqs. (4.378) and (4.379), we get

$$f'(t) = \exp(\psi) \tag{4.380}$$

By integrating the above equation with respect to t, we obtained,

$$f(t) = \int \exp(\psi)dt + C \tag{4.381}$$

where C is the integrating constant. Now, substituting the value of $f(t)$ in Eq. (4.377), we get

$$\exp\left(-\frac{\phi}{2}\right)\exp\left[-\frac{(u-\psi)}{2}\right] = -\frac{1}{\sqrt{2}}\left[\int \exp(-\phi)dx + \int \exp(\psi)dt + C\right]$$

$$\implies \exp\left(-\frac{u}{2}\right) = -\frac{1}{\sqrt{2}}\exp\left[\frac{(\phi-\psi)}{2}\right]\left[\int \exp(-\phi)dx + \int \exp(\psi)dt + C\right]$$

$$\implies \exp(u) = \frac{2\exp(\psi-\phi)}{[\int \exp(-\phi)dx + \int \exp(\psi)dt + C]^2}$$

$$\implies u = \log 2 + \psi - \phi - 2\log\left[\int \exp(-\phi)dx + \int \exp(\psi)dt + C\right]$$

Hence, the new exact solution of the Liouville equation is

$$u(x,t) = \psi(t) - \phi(x) - 2\log\left[\int \exp(-\phi(x))dx + \int \exp(\psi(t))dt + C\right] + \log 2 \tag{4.382}$$

References

[1] I. N. Sneddon, *Elements of Partial Differential Equations*, Dover Publications, INC, Mineola, New York, 2006.

[2] B. B. Kadomtsev and V. I. Petviashvili, *Sov. Phys. Dokl.* 15, 539–541 (1970).

[3] P. Chatterjee, K. Roy, and U. N. Ghosh, *Waves and Wave Interaction in Plasmas,* World Scientific Publishing Co. Ptc. Ltd., Singapore, 2023.

[4] G. B. Whitham, *Linear and Nonlinear Waves*, John wiley & sons, New York, 1973.

[5] Abdul- Majid Wazwaz, *Partial Differential Equation and Solitary waves Theory,* Springer, Higher Education Press, Beijing, 2009.

[6] I. Shingareva and C. Lizarraga-celaya, *Solving Nonlinear Partial Differential Equations with Maple and Mathematica,* Springerwien New York, 2011.

[7] L. Debnath, *Nonlinear Partial Differential Equations for Scientists and Engineers,* Birkhauser, Boston, 2011.

[8] G. Baumann, *Mathematica for Theoretical Physics*, Springer, New York, 2005.

[9] J. Weiss, M. Tabor, and G. Carnevale, The Painleve property for partial differential equation, *J. Math. Phys.*, 24 (3), 522, 1983.

[10] X. U. Gui, L. Zhi-Bin, A maple package for the Painleve test of nonlinear partial differential equations, *Chin. Phys. Lett.* 20, 7, 975 (2003).

[11] W. Hereman and S. Angenent, *The Painleve Test for NonLinear Ordinary and Partial Differential Equations,* University of wisconsin-madison center for the mathematical science, Technical summary report, 1989.

[12] X. Lu, Y.F. Hua, S.J. Chen, and X.F. Tang. *Integrability Characteristics of a Novel (2+1)-Dimensional Nonlinear Model: Painlevé Analysis, Soliton Solutions, Backlund Transformation, Lax Pair and Infinitely Many Conservation Laws. Communications in Nonlinear Science and Numerical Simulation,* 95, 105612 (2021).

[13] H. Brezis and F. Browder. Partial differential equations in the 20th century, *Advances in Mathematics*, 135(1), 76 (1998).

[14] R. Hirota, *Phys. Rev, Lett.* 27, 1192 (1971).

[15] M. Ablowitz and H. Segur, Solitons and the Inverse Scattering Transform, *SIAM,* Philadelphia, (1981).

[16] M. Ablowitz and P. Clarkson, Solitons, *Nonlinear Evolution Equations and Inverse Scattering*, Cambridge University Press, Cambridge, (1991).

[17] P. Lax, Integrals of nonlinear equations of evolution and solitary waves, *Comm. Pure Applied Math.,* 21 (5), 467 (1968).

[18] A. R. Its, *The Riemann-Hilbert Problem and Integrable Systems,* Notices of the AMS, 50(11), 1389 (2003).

[19] R. M. Miura, C. S. Gardner, M. D. Kruskal, and Korteweg-de Vries equation and generalizations. II. Existence of conservation laws and constants of motion, *J.Math. Phys.,* 9(8), 1204 (1967).

[20] S. Clifford, J. M. Gardner, M. D. Kruskal, R. M. Miura, *Method for solving the Korteweg-deVries equation.Physical Review Letters,* 19 (19), 1095 (1967).

[21] R. Hirota, Exact N-soliton solutions of the wave equation of long waves in shallow-water and in nonlinear lattices. *Journal of Mathematical Physics*, 14 (7), 810 (1973).

[22] R. Hirota, M. Ito, Resonance of solitons in one dimension. *Journal of the Physical Society of Japan,* 52 (3), 744 (1983).

[23] A. M. Wazwaz, Solitons and singular solitons for the Gardner–KP equation. *Applied Mathematics and Computation,* 204 (1), 162 (2008).

[24] P. G. Drazin and R.S. Johnson. *Solitons: An Introduction.* Cambridge University Press, 2 (1989).

[25] R. C. Davidson, *Methods in Nonlinear Plasma Theory.* Academic (1972).

[26] U. N. Ghosh, S. Nasipuri, and P. Chatterjee, Study of lump soliton structures in quantum electron-positron-ion magnetoplasma. *Contribution to Plasma Physics,* 62 (8), (2022).

[27] N. Freeman and J. J. C. Nimmo, Soliton solutions to Korteweg- de Vries and Kadomtsev-Petviashvili equations: the wronskian technique, *Phys. Lett. A*, 95, 1 (1983).

[28] J. J. C. Nimmo, The use of Backlund transformations in obtaing N-soliton solutions in Wronskian form, *Phys. Lett. A,* 99, 279 (1983).

[29] W. X. Ma, Wronskians, generalized Wronskians and solutions to the Kprteweg-de Varies equation, *Chaos Solitons Fractals*, 19, 163, 2004.

[30] W. X. Ma and Y. You, Solving the Korteweg-de Varies equation by its bilinear form: Wronskian solutions, *Trans. Amer. Math. Soc,* 3577, 1753 (2005).

[31] Y. Yao, D. Zhang, and D. Chen, The double Wronskian solutions to the Kadomtset-Petviashvili equation, *Modern Phys. Lett. B* . 22, 621 (2008).

[32] A. M. S. Jabr, *Nonlinear Evolution Equations and Inverse Scattering,* Thesis, Zarqa University, 2016.

[33] C. Gu, H. Hu, and Z. Zhou, *Darboux transformations in integrable systems theory and their applications to geometry,* Springer, 2005.

[34] M. J. Ablowitz and H. Segur, Solitons and Inverse Scattering Transform-Society for Industrial Mathematics, *SIAM Studies in Applied Mathematics,* 4, (2000).

[35] S. Hovikov, S. V. Manakov, L. P. Pitaevskii, and V. E. Zakharov, *Theory of solitons: The Inverse Scattering Method, Contemporary Soviet Mathematics,* (1984).

[36] P. L. Vu, *Inverse Scattering Problems and Their Application to Nonlinear Integrable Equations*-CRC Press, Chapman & Hall (2023).

[37] A. Abdeljabbar, *New Double Wronskian Exact Solutions for a Generalized (2+1)-Dimensional Nonlinear System with Variable, Partial Differential Equations in Applied Mathematics Coefficients,* 3, 100022 (2021).

[38] R. Hirota, A new form of Backlund transformations and its relation to the inverse scattering problem, *Progr Theoret Phys,* 5, 1498 (1974).

[39] V. B. Matveev and M. A. Salle, *Darboux Transformations and Solitons,* Springer-Verlag, Berlin (1991).

[40] B. Feng, L. Ling, D. A. Takahashi, Multi-breather and higher-order rogue waves for the nonlinear Schrodinger equation on the elliptic function back ground, *Studies in Applied Mathematics*, **144**(1), 46–101, 201).

[41] B. Q Li, Loop-like Kink breather and its transition phenomena for the Vakhnenko equation arising from high-frequency wave propagation in electromagnetic physics, *Applied Mathematics Letters*, **112**, 106822 (2021).

Index